ACCRETION DISKS AND MAGNETIC FIELDS IN ASTROPHYSICS

ASTROPHYSICS AND SPACE SCIENCE LIBRARY

A SERIES OF BOOKS ON THE RECENT DEVELOPMENTS
OF SPACE SCIENCE AND OF GENERAL GEOPHYSICS AND ASTROPHYSICS
PUBLISHED IN CONNECTION WITH THE JOURNAL
SPACE SCIENCE REVIEWS

VOLUME 156
PROCEEDINGS

ACCRETION DISKS AND MAGNETIC FIELDS IN ASTROPHYSICS

PROCEEDINGS OF THE
EUROPEAN PHYSICAL SOCIETY STUDY CONFERENCE,
HELD IN NOTO (SICILY), ITALY, JUNE 16-21, 1988

edited by

G. BELVEDERE
Institute of Astronomy, University of Catania, Italy

KLUWER ACADEMIC PUBLISHERS
DORDRECHT / BOSTON / LONDON

Library of Congress Cataloging in Publication Data

Accretion disks and magnetic fields in astrophysics : proceedings of
the European Physical Society study conference held in Noto, Sicily,
Italy, from 16 to 21 June 1988 / edited by Gaetano Belvedere.
p. cm. -- (Astrophysics and space science library)
Includes index.
ISBN 0-7923-0295-8
1. Accretion (Astrophysics)--Congresses. 2. Magnetic fields
(Cosmic physics)--Congresses. 3. Astrophysics--Congresses.
I. Belvedere, G. II. European Physical Society. III. Series.
QB466.A25A22 1989
523.01'88--dc20 89-33105

ISBN 0-7923-0295-8

Published by Kluwer Academic Publishers,
P.O. Box 17, 3300 AA Dordrecht, The Netherlands.

Kluwer Academic Publishers incorporates
the publishing programmes of
D. Reidel, Martinus Nijhoff, Dr W. Junk and MTP Press.

Sold and distributed in the U.S.A. and Canada
by Kluwer Academic Publishers,
101 Philip Drive, Norwell, MA 02061, U.S.A.

In all other countries, sold and distributed
by Kluwer Academic Publishers Group,
P.O. Box 322, 3300 AH Dordrecht, The Netherlands.

printed on acid free paper

Printed in The Netherlands

UNDER

THE HIGH PATRONAGE

OF THE PRESIDENT

OF THE ITALIAN REPUBLIC

TABLE OF CONTENTS

(*) Review paper

INTRODUCTORY TALK[(*)]

Authorities, Colleagues, Ladies and Gentlemen,

On behalf of the Scientific and Local Organizing Committees, it is an honour for me to welcome the participants in the European Physical Society Study Conference "Accretion Disks and Magnetic Fields in Astrophysics".

We are very pleased that this Meeting be held in Noto, a City that boasts a glorious history, wealthy cultural traditions and beautiful works of baroque art.

The main purpose of the Conference is to try to set up some unified picture of the dynamical interaction, under many different conditions, of rotation, turbulence and magnetic fields. We attempt to bring together colleagues with different expertises, who have worked at different aspects of the problem, with relation to different objects and topics, ranging from star formation, stellar activity and close binaries, to neutron stars, X-ray galactic sources and active galaxies.

For instance, dynamo processes similar to those invoked to explain solar and stellar activity have been implicated recently in studying galactic magnetic fields, dynamical processes occuring in the atmospheres of neutron stars, and the origin of jets in protostellar objects and Active Galactic Nuclei.

Relativistic and non - thermal processes give ample evidence of the role played by magnetic fields in solar and stellar flares, in cataclysmic variables, neutron stars and nuclei of galaxies.

Instabilities studied in solar activity energetic phenomena and in magnetic accretion disks around collapsed objects may shed light on some properties of more massive disks around galactic central engines.

The choice of topics and the list of invited and contributed talks appears a natural one for comparison of outstanding different phenomena.

It emphasizes the universality of the underlying physical processes and mechanisms in spite of different

----

(*) Given at the Opening Ceremony.

masses, sizes, energy contents and timescales.

Further it emphasizes the opportunity of getting together scientists who, even working apparently in different research fields, have a common scientific background, which may favour the exchange of ideas and help to attain the main aim of the Conference, namely to find a common denominator beyond the phenomenological differences.

During the Meeting, we will discuss the physics of accretion disks and the phenomenology of various astrophysical systems, where accretion disks are supposed to exist, thanks also to the increasing observational evidence.

We will also discuss the important role of magnetic fields in determining the structure, dynamics and energetics of disks, and in explaining the presence of high velocity flows and jets.

We will also attempt to define a boundary between physical properties and phenomenologies which depend on mass, size and the nature of the central object, and those which are common to different astrophysical systems independently of their peculiar nature.

To get these objectives, an essential role will be played by ample discussion.

We have to consider this Meeting as a real Workshop, where the room devoted to public and private exchange of ideas and suggestions is perhaps the most relevant element to produce remarkable results.

We hope that the present Conference may contribute to set up an unified view of the related topics, stimulating collaboration among colleagues and other future meetings.

We want to express our sincere gratitude to the President of the Italian Republic, Hon. Prof. Francesco Cossiga, for benevolently according His High Patronage to the Conference.

Warm thanks are due to the Rector of Catania University, Prof. G. Rodolico, to the President of the Faculty of Sciences, Prof. G. Santagati, to the members of the Scientific Organizing Committee, to the members of the Local Organizing Commmittee and to the Local Staff, kindly placed at the Conference's disposal by the Director of the Catania Astrophysical Observatory, Prof. M. Rodonó.

We gratefully acknowledge the sponsorship of the European Physical Society - Astronomy and Astrophysics Division - and the friendly hospitality of the City of Noto and the Centro di Studi Superiori e di Ricerche, an young institution that is going to achieve an important position in the panorama of Sicilian and Italian culture.

We are very indebted to the Mayor of the City of Noto, Dr. Lawyer G. Passarello for his warm hospitality and generous sponsorship, to the President of the Province of Syracuse, Dr. S. Aparo for his kind contribution, to Prof. V. Miceli for his active interest in the organization.

We express our gratitude for the sponsorship of the Catania University, the Catania Astrophysical Observatory, the Italian Astronomical Society and the National Research Council.

We look for an interesting and successful Conference.

Thank you.

Gaetano Belvedere

FOREWORD

The present volume collects the review and contributed papers presented at the Europhysics Study Conference "Accretion Disks and Magnetic Fields in Astrophysics" held in Noto, Sicily, Italy, from 16 to 21 June 1988, under the High Patronage of the President of the Italian Republic and the local patronage of the Centro di Studi Superiori e di Ricerche (Center for Advanced Study and Research), a scientific institution supported by the City of Noto and the University of Catania.

The attendance to the Conference was limited to about fifty selected participants from Canada, China, Denmark, France, Germany F.R., Italy, Japan, the Netherlands, Spain, Switzerland, United Kingdom and United States, in order to give the meeting the character and appearance of a real workshop, whose main aim was to highlight a varied but somewhat unified picture of the interaction, in different astrophysical systems, of rotation, turbulence and magnetism, with specific reference to accretion disks.

To this regard, we made the experiment - which revealed successful, according to the favourable comments of the participants - of bringing together colleagues actively working on different aspects of the problem, with relation to different fields of Astrophysics. In fact we gathered for six days eminent scientists with various personal expertises, such as cloud dynamics, star formation, solar and stellar activity, binary X-ray sources, white dwarfs and neutron stars, and active galactic nuclei (AGN), but with a common framework of knowledge of magnetohydrodynamics (MHD), turbulence and non - thermal high energy processes.

This led to a careful analysis of all aspects of the matter and a cross fire of questions and comments during and after the scientific sessions, which reached its climax in the course of two topical joint discussions on the subjects of Stellar disks and AGN disks. Of course, all this favoured a thoughtful and profitable exchange of knowledge and ideas, and the active and valuable contribution of everybody.

As mentioned by all colleagues attending the Conference, the atmosphere was friendly and enthusiastic. Not inessential to this regard was perhaps the location of the meeting, a nice hotel on a beatiful beach not far from the pretty baroque town of Noto, in the province of Syracuse.

As to the structure of the Conference, three basic but informal sections may be outlined: (i) solar and stellar activity, (ii) protostellar and binary system disks, (iii) AGN disks.

The first section was devoted to the physics of convection and magnetism in rotating stellar shells and to dynamo models of stellar activity.

The second section included topics such as basic stellar accretion disk theory, MHD processes in clouds, disks and magnetospheres, and turbulent generation of high velocity flows and jets in protostellar systems.

The third section comprised subjects such as accretion onto supermassive objects, the central powerhouse of AGN, and the physics of AGN outflows and jets.

This does not mean a real splitting up, as there were several intersections and overlaps, specially of sections (ii) and (iii), according to the spirit of the meeting.

The order of the papers published here follows the above outlined sections and the Scientific Programme. The book opens with a review of the topics of the Conference and a comment on its results, and closes with a summary, which also takes into account the outcome of the two joint discussions.

We want to express our friendly and warm thanks to the colleagues attending the meeting, who all did their best for its success.

Gaetano Belvedere

SCIENTIFIC ORGANIZING COMMITTEE

G. BELVEDERE	University of Catania, Italy
A. O. BENZ	ETH, Zürich, Switzerland
R. N. HENRIKSEN	Queen's University, Kingston, Canada
R. V. LOVELACE	Cornell University, Ithaca, New York, USA
F. PACINI	Arcetri Astrophysical Observatory, Florence, Italy
S. N. SHORE	New Mexico Tech., Socorro, N.Mexico, USA
H. C. SPRUIT	Max Planck Institute for Astrophysics, Garching, FRG
W. M. TSCHARNUTER	University of Wien, Austria
N. O. WEISS	University of Cambridge, UK

LOCAL ORGANIZING COMMITTEE

G. BELVEDERE	University of Catania
A. M. ANILE	University of Catania
O. MUSCATO	University of Catania
M. RODONO'	Director, Catania Astrophysical Observatory

LIST OF PARTICIPANTS

A. ACHTERBERG	Sterrekundig Instituut, Utrecht The Netherlands
M. ANILE	Dipartimento di Matematica, Catania Italy
V. ANTONUCCIO	Nordita, Copenhagen Denmark
D. AVERNA	Istituto di Astronomia, Catania Italy
M. BEGELMAN	JILA, Boulder, Colorado U.S.A.
S. BELLMER	Sternwarte Göttingen F.R.G.
G. BELVEDERE	Istituto di Astronomia, Catania Italy
A. O. BENZ	Institut für Astronomie - ETH, Zürich Switzerland
M. CAMENZIND	Landessternwarte Königsstuhl, Heidelberg F.R.G.
S. CATALANO	Istituto di Astronomia, Catania Italy
D. S. DIXON	Astrophys. Research Center, Socorro, N.Mexico U.S.A.
A. FERRARI	Istituto di Astronomia, Torino Italy
F. FIORE	Istituto di Astronomia, Roma Italy
G. GEERTSEMA	Sterrekundig Instituut, Utrecht The Netherlands
A. I. GOMEZ DE CASTRO	Observatorio Astronómico Nacional Spain

P. HARDEE	Dept. Physics and Astronomy, Tuscaloosa, AL U.S.A.
R. N. HENRIKSEN	Dept. Physics, Kingston, Ontario Canada
D. W. HUGHES	Dept. Appl. Mathem. Theor. Phys., Cambridge U.K.
A. R. KING	Dept. Astronomy, Leicester U.K.
A. KÖNIGL	Dept. Astron. Astrophys., Chicago, IL U.S.A.
A. LANZAFAME	Dipartimento di Matematica, Catania Italy
J. P. LASOTA	Observatoire de Meudon France
G. LETO	Istituto di Astronomia, Catania Italy
Y. Q. LOU	NCAR-HAO, Boulder, Colorado U.S.A.
R. V. LOVELACE	Dept. Appl. Phys., Ithaca, New York U.S.A.
A. MAJORANA	Dipartimento di Matematica, Catania Italy
L. MESTEL	Astronomy Center, Brighton U.K.
E. MEYER	Max Plank Institut - Astrophysik, Garching F.R.G.
E. MEYER-HOFMEISTER	Max Planck Institut - Astrophysik, Garching F.R.G.
D. MOLTENI	Dipartimento di Fisica, Palermo Italy

P. MONGER	Space Telescope Sci. Inst., Baltimore, U.S.A
S. MOTTA	Dipartimento di Matematica, Catania Italy
O. MUSCATO	Dipartimento di Matematica, Catania Italy
L. NOBILI	Dipartimento di Fisica, Padova Italy
M. NORMAN	Nation. Center Supercomputer Appl., Urbana, IL U.S.A.
I. PAGANO	Istituto di Astronomia, Catania Italy
L. PATERNO'	Istituto di Astronomia, Catania Italy
V. PIRRONELLO	Dipartimento di Fisica, Cosenza Italy
R. E. PUDRITZ	Dept. Physics, Hamilton, Ontario Canada
M. RODONO'	Istituto di Astronomia, Catania Italy
D. SCHMITT	Sternwarte Göttingen F.R.G.
H. C. SPRUIT	Max Planck Institut - Astrophysik, Garching F.R.G.
R. TUROLLA	Dipartimento di Fisica, Padova Italy
Y. UCHIDA	Department of Astronomy, Tokio Japan
A. A. VAN BALLEGOOIJEN	Center for Astrophysics, Cambridge, MA U.S.A.
R. VENTURA	Istituto di Astronomia, Catania Italy

N. O. WEISS Dept. Appl. Math. Theor. Phys., Cambridge
U.K.

LOCAL STAFF

A. CALI' Osservatorio Astrofisico, Catania
Italy

L. RAPISARDA Osservatorio Astrofisico, Catania
Italy

S. SARDONE Osservatorio Astrofisico, Catania
Italy

M. C. SCUDERI Istituto di Astronomia, Catania
Italy

C. SORGE Osservatorio Astrofisico, Catania
Italy

A. WANAUSEK Osservatorio Astrofisico, Catania
Italy

THE RELEVANCE OF MAGNETIC FIELDS TO STARS AND ACCRETION DISKS

G.Belvedere
Istituto di Astronomia, Università di Catania,
Italy

As clearly stated in the Introductory talk during the Opening Ceremony, the significance of the Conference was in gathering together scientists actively working in different research fields with a common background of expertises, in order to highlight a varied but in some sense unified scenario of the interaction of rotation, turbulence and magnetism in different astrophysical systems, with primary interest in stars and accretion disks and specific regard to objects and topics ranging from star formation, solar and stellar magnetic activity, and close binaries, to neutron stars, X-ray galactic sources and active galactic nuclei (AGN).

In the course of the meeting, the theory of accretion disks and the phenomenology of various astrophysical systems where accretion disks exist or are supposed to exist, were revisited and extensively discussed with ample contribution of reviews and recent original work.

The relevance of the magnetic field as to stellar activity, to structure, dynamics and energetics of disks and interstellar clouds, to high velocity flows and jets in both protostellar objects and AGN, was widely debated in the scientific sessions and during the two topical joint discussions on the subjects of Stars and AGN. In fact, the role of the magnetic field was one of the outstanding pieces of the Conference.

The choice of topics was intended to put into relief the unitary nature of the physical processes and mechanisms in spite of different masses, sizes, energy contents and outputs, and characteristic timescales involved in the phenomenology of active stars and accretion disks onto various objects, though carefully analyzing the peculiarity of different astrophysical contexts.

As mentioned in the Foreword, three basic but informal sections (or sequential items) may be outlined in the structure of the meeting: (a) solar and stellar activity, (b) protostellar and binary system disks, (c) AGN disks.

G. Belvedere (ed.), Accretion Disks and Magnetic Fields in Astrophysics, 1–10.

However, this is not a real splitting up into distinct subjects, as cross intersection and overlap of topics was largely frequent, according to the unitary spirit of the meeting.

The first section was devoted to the physics of convection and magnetism in rotating stellar shells, to dynamo theory of stellar activity and to energetic phenomena occurring in stellar flares.

N.O.Weiss thoroughly reviewed linear and non-linear dynamo models, stressing the positive results obtained in reproducing the observed phenomenology of activity and the open problems still existing in the theoretical approach. To this regard it is to be remarked that no real agreement has been reached by the helioseismology experts as to the dependence of angular velocity on radial distance and latitude in the convection zone and in the deep interior of the Sun, which is essential to correctly modelling the stellar dynamo. The latter should operate in a relatively stable layer just at the boundary between the convection zone and the radiative interior.

The stability of the toroidal magnetic field stored there was investigated by Hughes and Cattaneo, in order to understand how a large scale diffuse field can give rise to isolated flux tubes bursting through the photosphere. In fact a non-linear analysis shows that the breakup of the magnetic layer may occur, driven by the Rayleigh-Taylor instability. It is relevant the fact that this conclusion may apply to any astrophysical context were a strong field is trapped beneath a non-magnetic region.

Coming from the interior to the surface, Lou discussed the non-linear interaction of rotation and zonal magnetic field, and showed that, in the magnetohydrodynamical (MHD) regime with comparable Rossby and Alfvén wave speeds, two kinds of non-linear finite amplitude MHD waves propagate in the azimuthal direction. If detected, these waves may shed light on the background uniform solar rotation and the zonal magnetic field strength at the equator.

Now, what is the relevance of stellar MHD studies to accretion disks? In a way, one may simply consider the fact that several uncertainties still remain in dynamo theory and its application to stars, and deduce that investigating the role of a large scale magnetic field in different (also geometrically) structures like accreting systems may be premature. On the other hand, however, we firmly believe that the study of the interaction of plasma and magnetic field in stellar rotating shells may serve as a landmark for the theory of rotating accretion disks, if a large scale field is allowed to develop, specially in the case of geometrically and optically thick disks (accretion tori).

Unfortunately, up to date the role of the magnetic field in the structure and dynamics of disks has scarcely been emphasized in the literature, perhaps because considering also the MHD aspect of the problem makes it by far more difficult. What we look for, anyway, is the simultaneous and rigorous solution of the relevant equations, including the MHD ones, which should led, among other things, to a self consistent determination of the angular velocity field inside the disk, this appearing more convincing than the widely assumed keplerian rotation.

Other important phenomena largely investigated in stellar activity, which may shed light on similar events occurring in accreting sources and AGN are stellar flares. The study of the pre-flare and post-flare topology of magnetic fields in a close astrophysical laboratory like the Sun, of the emission of high energy particles and their interaction with the magnetic field, of the radio counterpart, may give insight on the onset of instabilities in much more distant objects, leading to brightness variability or explosive processes, in which the role of the magnetic field is likely to appear relevant.

The second section included topics as basic accretion disk theory, MHD processes in clouds, disks and magnetospheres, and turbulent generation of high velocity flows and jets. As mentioned above, the second section was primarily concerned with binary system and protostellar disks, but no rigid boundaries or walls were raised between this section and the AGN one.

For instance, King gave a review on the basic ideas of disk accretion, stressing the common aspects between problems in stellar and AGN disk modelling. He clearly outlined the progress so far achieved from comparing theory to observations.

The role of small scale and large scale magnetic fields in disks and even the interaction between the disk's intrinsic magnetic field and the collapsed companion's were extensively examined in various contributions with reference to various observed phenomene and underlying physical processes.

Meyer-Hofmeister investigated the transition from a cool state to a hot state in dwarf nova accreting system outbursts, in particular the delay observed between the optical and UV flux rises. The study of this instability in the framework of the classical thin disk model should shed light on the value of the viscosity stress tensor due to small scale magnetic fields, parametrized as $\tau_{r\varphi} \sim \alpha P$, where P is the pressure. During the transition α should remain constant.

The evidence for the existence of cool prominences,

magnetically sustained as in the stellar case, in the coronae of accretion disks was discussed by Meyer. This should be relevant to explain occultation features in high inclination systems of cataclysmic variables and also the broad line regions of AGN.

Achterberg was concerned with the dynamics of isolated magnetic flux tubes and their interaction with the surrounding medium. It was pointed out that the concept of added mass, which parametrizes the effect of the external forces as an increased inertia, becomes problematical when considering the disk vertical stratification and the Coriolis force.

The possibility of magnetic driving of cataclysmic variable observed winds was examined by Van Ballegooijen. He discussed a model of radial transport of magnetic flux in an accretion disk, taking into account both flux accretion from the white dwarf and diffusion inside the disk, and found that the field decay rates are very short (some hours). This might put severe constraints on the efficiency of a dynamo mechanism. However, a self-consistent dynamo model was not worked out, as no intrinsic mechanism of large scale field generation in a differentially rotating disk was included.

The interaction between the magnetic fields of the compact object and of the disk is one of the most interesting topics concerned with disk activity.

To this regard, Lasota, Hameury and King suggested that the intense magnetic fields of white dwarfs in short period cataclysmic variables may prevent the formation of disks. Indeed observations show that they are not present in syncronous system (polars) and in some non-syncronous systems (intermediate polars); Benz and Güdel proposed a magnetic connection between the accretion disk and the magnetic poles of the white dwarf, to explain the delay between the optical and radio outbursts of dwarf novae. They consider a maser mechanism with the source above the magnetic poles of the white dwarf and acceleration of non-thermal particles occurring at some distance from the central object, in a region near the disk.

MHD processes in disks and jets were extensively reviewed by Lovelace, with particular reference to the protosolar nebula and to disks with bipolar flows. According to this author, ordered magnetic fields in accretion disks may have a direct role in the formation of oppositely directed plasma jets along the rotation axis, in the outward transport of angular momentum and in the ohmic heating of the disk. Further, it is shown that a weak disordered field can lead to the amplification of an ordered field component, this being of great interest to dynamo generation of large scale magnetic fields and to magnetic alignment of jets. The transition from the Poynting flux jet (energy carried by the electromagnetic field) to the kinetic energy flux jet (energy carried by the matter) is also relevant to the physical description of jets and comparison to observations. Finally it is stressed that the magnetic field can have an important role in the outward transport of angular momentum, even in the absence of jets.

An alternative to classical accretion mechanisms with angular momentum transport was proposed by Spruit, who pointed out that a system of stationary shock waves, excited by moderate nonaxisymmetries at large distances from the point mass (spiral shocks) can do the job. Further, the spiral shocks may also be efficient particle accelerators. Then, the spiral flow with fast particles, which results from this model, should have the characteristics required by ion-supported tori involved in models of central engines of AGN.

The current mechanisms of turbulent generation of high velocity flows and jets in clouds, disks and magnetospheres were reviewed in detail by Henriksen, with the emphasis placed on the star forming regions. In addition some physical processes peculiar to collapsed objects for producing beams and jets were surveyed and discussed in terms of galactic (pulsars) and extragalactic (AGN) sources.

The fully relativistic problem of the interaction among rotation, dynamo magnetic field and plasma injected into the magnetosphere of rapidly rotating compact objects was discussed by Camenzind in the stationary and axisymmetric case. This problem can be reduced to a non-linear system consisting of the relativistic Grad - Schlüter - Shafranov equation for the magnetic flux and the hot wind equation for the plasma poloidal velocity. Solutions were given for the case of a millisecond pulsar or a supermassive rotator, and for a black hole. In both cases the centrifugal force induced outflows can be collimated outside of the light cylinder to give rise to relativistic jets.

The cloud dynamics and the early stages of star formation were also discussed by Mestel with special attention devoted to the MHD point of view. The topics treated included: anisotropic collapse and disk formation; fragmentation; magnetic braking; flux leakage and field topology. It is stressed the necessity for an adequate theory to predict typical masses, angular momenta and remnant magnetic fluxes for protostars and their accretion disks.

The growing observational evidence for the presence of circumstellar disks around low mass newly formed stars and for the association of such circumstellar disks with energetic bipolar outflows along their rotation axes,was the starting point of the review presented by Königl. In addition the magnetic fields around these sources are generally found to be well aligned with the direction of the outflow and the rotational axis, suggesting a relevant dynamical role of ordered (large scale) magnetic fields in the accretion and outflow processes. In Königl's picture the outflows are powered by accretion and carry away the angular momentum that is to be removed in order the accretion to proceed. The coupling between inflow and outflow is due to the disk ordered large scale magnetic field which gives rise to a centrifugally driven wind, which becomes super-Alfvénic and magnetically collimated at finite height above the disk.

The removal of angular momentum from the massive central region of a protostellar cloud may occur, according to Uchida, through the combined effect of rotation and magnetic field. In the condensation process a rotating pancake-like mass is formed and a toroidal field is generated, which pinches the object by wrapping it up. The back-reaction of the pancake induces mass and angular momentum loss, whose observational evidence is the presence of rotating streamer structures in protostellar objects. Indeed, observational studies of star forming regions in the Galaxy have demonstrated the ubiquity of mass outflows from objects in the various stages of star formation, as bipolar outflows and jets, and motivated hydromagnetic models.

To this regard, the necessity for self consistent models of magnetic disk accretion was pointed out by Stone, Norman and Mihalas, who outlined recent progress made in this field by using the Zeus code, that is an emerging multiphysics fluid dynamics code for application in Astrophysics.

The last topic of the second section was the formation of H_2 in interstellar clouds. On the basis of the experimental evidence that molecular hydrogen is released in the gas phase from icy targets, as H_2O, CH_4, NH_3, upon bombardment by fast ions, Pirronello and Averna evaluated the production rate of H_2 inside dense molecular clouds by interaction of cosmic rays with ice mantles of dust grains and found that this mechanism is more efficient than recombination of adsorbed atoms on amorphous mantles. This contribution was somewhat marginal to the outline of the Conference but intrinsically interesting to explain H_2 production in molecular clouds.

The third section was concerned with subjects as magnetic field generation in protogalaxies, accretion onto supermassive objects, the central powerhouse of AGN, and the physics of AGN outflows and jets.

Although magnetic field generation in galaxies may seem

to be a subject somewhat marginal to the main course of the Conference, nevertheless it is of great importance, as everybody is interested to know about the origin (how and when) of this basic ingredient of our meeting.

Recent observational evidence and theoretical aspects of this problem were reviewed by Pudritz. According to this author shocked gas clouds, that may have filled protogalactic haloes, are an ideal medium in which to generate magnetic fields at primeval epochs. Further, these fields seem to have played an important role in the support of primordial molecular clouds and in the formation of primordial halo stars.

The physics of outflows from AGN was reviewed by Ferrari, who discussed the formation of jets from the funnels of accretion disks and their propagation outside, the interpretation of observed morphologies in extended radio sources, the presence of emission line regions and the termination of outflows. Particular attention was given to the role of instabilities and shocks as to generation and maintenance of jets.

The formation of radio jets and lobes and its relationship with the central engine were outlined by Uchida in a model which may be considered the extension to AGN of the previous work here presented in the second section. The basic role is played by a dynamo generated toroidal field, which winds up continuously due to disk differential rotation. The Lorentz force in the unwinding twist expells out the mass in the disk atmosphere into axisymmetric helical paths, giving origin to jets and lobes. The back reaction of this process would be magnetic braking of the inner part of the disk due to angular momentum loss, causing the enhancement of accretion of magnetized gas onto the central object.

Numerical simulations of propagating weakly magnetized jets were presented by Hardee and Norman. In their picture the jet is driven by a periodic sinusoidal oscillation in time of the angle between the jet's velocity and the axial direction at the inlet. At early times the jet maintains approximate axisymmetry, whilst at later times sideway oscillation occurs which destroys directionality and collimation, so that the jets transform itself into a growing lobe. The distance at which the jet's collimation is destroyed depends on the luminosity and the physical properties of the environment. The magnetic field is introduced in closed loops, but no magnetic diffusivity is considered (ideal MHD).

The central powerhouse of AGN and the energy transport in hot thick (pressure supported) accretion tori were carefully discussed by Begelman, who stressed the importance of plasma and radiative effects in disks and their envelopes, as well as the mechanism of angular momentum transport (flywheel) from a rotating supermassive black hole. The dependence of some characteristic time scales on the accretion rate $\dot{m}$ and the central object mass M were also examined.

The latter was estimated by Fiore and Massaro as $\sim 10^7 M_\odot$, for the Schwarzschild metric, and $\sim 10^8 M_\odot$, for the extreme Kerr metric, in the case of the Seyfert NGC 4151. These values were deduced by interpreting the small amplitude X-ray time variability as the signature of the orbital time relative to the accretion of matter near the innermost stable orbit around the collapsed object.

Relativistic effects, with regard to accretion and radiation transport, were finally pointed out by some authors.

Anile presented an extensive discussion of fluid models for relativistic warm plasmas which are of great interest in connection with stellar winds, pulsar magnetospheres and collisionless accretion onto neutron stars and black holes (also supermassive). His approach is based on the entropy principle and introduces a fluid model in which the constraint equations (with reference to the moments of the relativistic Vlasov equation) are satisfied exactly.

Turolla and Nobili analyzed the Thorne structure equation governing radiation field moments in the case of stationary spherically symmetric accreting flows. Their results give a satisfactory description of the general relativistic radiative transfer and stress the fact that the moment formalism is a powerful tool in dealing with problems in which mass flows and radiation fields are coupled.

Lasota showed the existence of a new branch of equilibrium solutions for stationary accretion disks around black holes, corresponding to moderate super-Eddington accretion rates. This should be a consequence of cooling due to general relativistic Roche lobe overflow and horizontal advection of heat at the inner edge of disks.

The basic questions raised in the scientific sessions were widely and intensely debated in the course of two topical joint discussions on the subjects of Stellar disks and AGN disks.

Here we shortly outline these points and the most remarkable results come out, in our opinion, from the sessions and the joint discussions:

- the keplerian rotation has some observational evidence for thin disks (not extremely convincing, however), but does not represent the most general rotational regime, the latter point being essential when modelling thick disks

- thin disk models have some observational support as far as binary systems are concerned, but,when considering AGN,the peculiar physical scenario requires thick disk (accretion torus) models. In this regard it has been pointed out that plasma and radiative effects in AGN disks and their environments are of primary importance

- future work should deal with more sophisticated modelling of accretion disks, eventually leading to the simultaneous solution of all relevant equations, including the MHD ones, if the role of the large scale magnetic field appears to be prominent in the disk dynamics (accretion disk dynamo). Of course, this depends also on the competitivity between production and dissipation of magnetic field, and is related to the disk nature (thin or thick)

- the magnetic field of the compact object in binary systems may affect the disk structure and dynamics in a considerable way. Under some circumstances, disk formation may be inhibited by the collapsed companion's very strong magnetic field. Moreover, if disks do have a large scale dynamo regulated magnetic field, the interaction between the latter and the former should adequately be investigated

- some difficulties have been raised as to the possibility of simple scaling between disk models for galactic and extragalactic sources. Indeed, it has been stressed the strong dependence of the physical processes on the peculiar environmental scenario, specially the source of the accreting matter, the accretion rate, the energy transport inside the disk, which in turn depends on temperature, density and opacity. Thus, simple scaling according to the mass of the central object, the disk size and angular momentum, and the rate of accretion, should not be allowed, except in the case of comparable mass and size systems and, in general,for the magnetic field of the central object which should scale as its linear size

- as to protostar and AGN jet formation and dynamics, mechanisms of acceleration and collimation of different nature have been proposed or revisited, and, in this regard, the importance of the accretion rate, the jet density and opacity and the physical properties of the environment has been emphasized, as well as the balance between inflow and outflow. The magnetic field does not seem to be dynamically relevant to the jet motion, as there is no compelling observational evidence for magnetic collimation in AGN jets. Thus, mechanisms for field dissipation should be included in the numerical simulation of

MHD jets, to really test the relevance of a dynamo-like magnetic field to the jet dynamics, at least on the theoretical point of view.
As to jet formation, however, the magnetosphere of a rapidly rotating compact object could play a relevant role.

- The existing observational data in favour or contrary to spinar "clean" models (low entropy central engines) and gravar "dirty" models (high entropy central engines) have carefully been examined. In the former case the primary source of energy is rotation, in the latter is conversion of gravitational binding energy. However, no agreement was reached, as no strong argument was given against either model.

Of course, the present review was intended to give only a general, although detailed, outline of the topics discussed during the meeting and to point out some remarkable results and conclusions.

We address the reader to the specific articles and the summary of the joint discussions for a full understanding of the scientific contents and highlights of the Conference.

DYNAMO PROCESSES IN STARS

N.O. WEISS
Department of Applied Mathematics and Theoretical Physics
University of Cambridge
Cambridge CB3 9EW
UK

ABSTRACT. Our understanding of stellar dynamos rests on detailed study of the Sun and recent observations of magnetic activity in other late-type stars. The interaction between rotation and convection leads to differential rotation and cyclonic eddies (possessing net helicity) which are parametrized in mean-field dynamo models. Nonlinear solutions provide examples of chaotic time-dependent behaviour and illustrate the development of complicated spatial structure. However, it is difficult to extrapolate from slow rotators like the Sun to rapidly rotating, highly active stars.

1. Introduction

In attempting to describe the generation of magnetic fields in cool stars like the Sun, theory has been led by observations. Seventy years ago Larmor (1919) suggested that magnetic fields in sunspots were produced by dynamo action within the Sun but the development of dynamo theory has been impeded by technical difficulties which have given the subject its peculiar flavour. Although planetary, stellar and galactic magnetic fields are apparently maintained by hydromagnetic dynamos, applied mathematicians have trouble in explaining what nature so easily achieves.

The decay time for the magnetic field in the radiative interior of a cool star is comparable with its lifetime on the main sequence, so there would be no need for a dynamo if the field were observed to be steady. (In planets, on the other hand, the decay time is short and dynamos are needed to explain the presence of magnetic fields.) In fact the Sun's magnetic field reverses after 11 years and similar behaviour has been detected in other late-type stars. This solar-stellar connection means that we can combine detailed observations of magnetic activity in the Sun, where fine structure can be resolved and the record extends over several centuries, with measurements of stars with different ages, rotation rates and structure (Tayler

G. Belvedere (ed.), Accretion Disks and Magnetic Fields in Astrophysics, 11–29.

1989). The challenge to theoreticians is to construct models that reproduce the properties of the solar cycle and can be extrapolated to explain the behaviour of more active stars.

In this review I shall first describe the observational constraints. Next, in Section 3, I shall outline the basic physical processes (helicity and differential rotation) that lead to dynamo action, stressing the importance of recent helioseismological results. Mean field dynamo theory is treated in Section 4. Then I discuss nonlinear dynamo waves in Section 5. Finally I shall try to summarize our current understanding of nonlinear stellar dynamos in the concluding section.

2. Solar and Stellar Magnetic Activity

We shall be concerned with magnetic activity in lower main-sequence stars of spectral types G, K and M. In these dwarf stars, which have deep outer convective zones and are typically slow rotators, the combination of rotation and convection produces variable magnetic fields. (By contrast the magnetic field in an Ap star, though it may be much stronger, remains almost constant in a rotating frame so there is no need to postulate a dynamo in order to maintain it.) Although solar activity is relatively weak the Sun is the obvious prototype with which other cool stars can be compared.

The incidence of sunspots has been studied for many years. As is well known, solar activity varies cyclically with a mean period of about 11 years. At the beginning of a cycle spots appear in both hemispheres around latitudes of $\pm 30°$; the sunspot zones spread towards the equator and then contract, so that the last spots of the old cycle occur at low latitudes as the next cycle begins. Thus a diagram showing the incidence of spots in latitude as a function of time produces the familiar butterfly pattern (e.g. Yallop & Hohenkerk 1980). Sunspots occur in active regions formed by emerging magnetic flux tubes with a predominantly azimuthal orientation. The associated fields have directions which are opposite in the two hemispheres and reverse after 11 years, giving rise to a 22-year magnetic cycle whose statistical properties have been analysed by Bracewell (1988). This cycle is aperiodic and its amplitude is modulated, with episodes of reduced activity such as the Maunder minimum of the late seventeenth century (Eddy 1976). The envelope of the activity cycle can be extended back over the past 9000 years by using proxy evidence. The incidence of galactic cosmic rays, which lead to the formation of ^{14}C and ^{10}Be in the earth's atmosphere, is modulated by magnetic fields in the solar wind. Fluctuations in the abundance of ^{14}C in trees and of ^{10}Be in polar ice-caps therefore provide records of solar activity which agree with observations over the last 300 years, including the Maunder minimum (Stuiver & Quay 1980; Beer, Siegenthaler & Blinov 1988). The ^{14}C proxy record shows many grand minima, occurring irregularly with a characteristic timescale of 200–300 years (Stuiver, Pearson & Braziunas 1986; Stuiver & Braziunas 1988). Thus we may conclude that the solar magnetic cycle exhibits chaotic behaviour with several different timescales, corresponding to the eruption of flux in active regions, the 22-year magnetic cycle and recurrent episodes of reduced activity (Ruzmaikin 1985; Weiss 1988). (It now seems likely that the layered Precambrian sediments from South Australia, which provided tantalising evidence for a regular modulation of solar activity 6.8×10^8 yr ago, were produced by tidal processes (Williams 1988, 1989).)

Magnetic activity has been detected in many late-type stars and the observations have been reviewed by Baliunas & Vaughan (1985), Hartmann & Noyes (1987), Rodonó (1987), Giampapa (1987), Soderblom & Baliunas (1988) and Tayler (1989). Magnetic fields have now been measured in a number of stars from Zeeman splitting of spectral lines (e.g. Marcy 1984; Saar 1987; Basri & Marcy 1988). Other evidence of stellar activity is provided by observations of starspots, flares, X-ray emission and, in particular, from systematic measurements of chromospheric Ca^+ emission initiated by Wilson (1978) and continued by Baliunas et al. (1989).

In the Sun Ca H and K emission is correlated with magnetic flux; for dwarfs of a given mass the emission index R' decreases monotonically with age and with increasing rotation period (e.g. Soderblom 1985). The depth of the convective zone varies among stars of different spectral type but can be taken into account by defining an inverse Rossby number $\sigma = \Omega\tau_c$, where Ω is the angular velocity and τ_c is a characteristic convective timescale (the turnover time at the base of the convective zone). Then R' behaves like a function of σ only (Noyes et al. 1984; Baliunas & Vaughan 1985). As a star evolves on the main sequence it loses angular momentum owing to the torque exerted by a stellar wind, which depends on the magnetic field (Mestel & Spruit 1987; Kawaler 1988), and gradually spins down. Observational studies of low-mass stars in clusters such as the Hyades, the Pleiades and α Persei (Lockwood et al. 1984; Stauffer et al. 1984, 1985; Radick et al. 1986; Stauffer & Hartmann 1987; Butler et al. 1987; Hartmann & Noyes 1987) allow us to reconstruct both the rotational and the magnetic history of the Sun. The variation of the equatorial rotation rate with time is sketched in Figure 1 (Rosner & Weiss 1985; Hartmann & Noyes 1987; Kawaler 1988); note that descriptions of internal rotation depend on whether large gradients in angular velocity are able to persist (Pinsonneault et al. 1989) or magnetic fields are strong enough to enforce approximately uniform rotation in the radiative interior (Mestel & Weiss 1987). The story begins with T Tauri stars, which are slow rotators (Vogel & Kuhi 1981) but magnetically active owing, perhaps, to a trapped primeval field (Tayler 1987). As they collapse towards the main sequence, conserving their angular momentum, they spin up so that Ω attains its maximum at an age of about 3×10^7 yr. At this stage stars are very active but they rapidly spin downs unless (like BY Draconis binaries) they can draw on orbital angular momentum. Indeed, the drop in Ω over the interval of 2×10^7 yr separating G dwarfs in the Pleiades from those in α Persei may be too rapid for magnetic braking to decelerate their radiative interiors (Stauffer & Hartmann 1987; Kawaler 1988). Thereafter spindown is more gradual

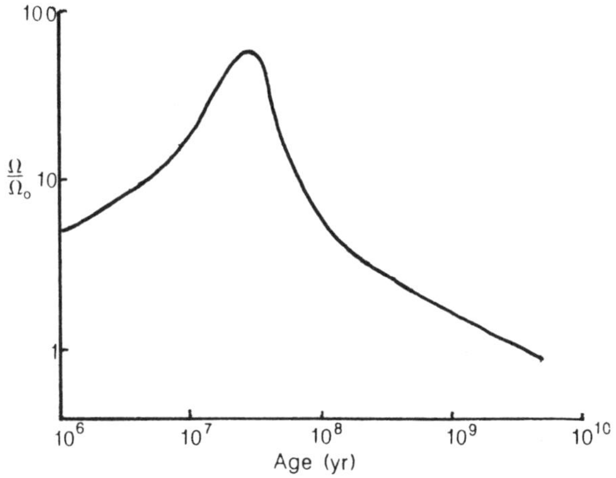

Figure 1. Rotational history of the Sun. Variation with age of the ratio of the equatorial angular velocity Ω to its present value $\Omega_o = 2.90 \times 10^{-6}$ rad s^{-1}.

as the angular velocity and magnetic activity decay together, at a rate that decreases monotonically until we reach a spindown time of 3×10^{10} yr (Pizzo et al. 1983) for the present Sun, whose feeble activity facilitates our own existence.

If we were viewing the Sun from a planet orbiting a nearby star we would be able to detect variations of about 15% in chromospheric emission over the 11-year activity cycle. Similar cyclic variations, with periods of 7 to 13 yr have been detected in a number of slowly rotating G and K dwarfs (Wilson 1978; Baliunas et al. 1989). Thus the Sun's magnetic activity can indeed by regarded as typical of a middle-aged late-type star (Soderblom & Baliunas 1988). On the other hand, young and rapidly rotating stars show much more irregular behaviour, which may include apparent periodicities. Starspots may cover a large fraction of their area, producing significant reductions in luminosity, while weaker magnetic fields extend over substantial patches on their surfaces (cf. Toner & Gray 1988), though both chromospheric Ca^+ emission and coronal X-ray emission apparently saturate as the scaled rotation rate σ increases.

3. Physical Processes: Rotation, Convection and Helicity

In a rotating star convection produces gradients in the angular velocity while Coriolis forces affect convective motion: differential rotation and cyclonic eddies are the essential ingredients of a stellar dynamo (Parker 1955). Suppose the magnetic field $\mathbf{B} = (B_r, B_\theta, B_\phi)$, referred to spherical polar coordinates (r, θ, ϕ), is initially confined to meridional planes $(B_\phi = 0)$. Then an angular velocity $\Omega(r, \theta)$ will draw the field lines out, generating an azimuthal component B_ϕ wherever $\mathbf{B}.\nabla\Omega \neq 0$. Thus differential rotation readily produces toroidal flux from poloidal flux. The reverse process is more tricky: in a stratified layer rising fluid will expand and eddies can therefore acquire a systematic sense of rotation. This cyclonic motion twists toroidal field lines to regenerate poloidal flux, at a rate determined by the locally averaged helicity $< \mathbf{u}.\ \mathrm{curl}\ \mathbf{u} >$, where $\mathbf{u}$ is the fluid velocity in a uniformly rotating frame. This effect can be represented by a parameter

$$\alpha = -\frac{1}{3}\tau_c < \mathbf{u}.\ \mathrm{curl}\ \mathbf{u} > ,$$

where τ_c is the local convective timescale. Much of hydromagnetic dynamo theory is concerned with calculating the effects of these two processes on a magnetic field in the presence of diffusion (Moffatt 1978; Parker 1979; Krause & Rädler 1980; Zel'dovich, Ruzmaikin & Sokoloff 1983). Applications to solar and stellar magnetic fields have regularly been reviewed (Cowling 1981; Weiss 1983; Schüssler 1983; Belvedere 1985; Ruzmaikin 1985; Stix 1987; Mestel 1988).

In order to apply this theory to a lower main-sequence dwarf we must first establish how Ω varies within the star. Newly obtained results from helioseismology have upset our preconceptions about the internal rotation of the Sun. Solar p-modes of degree l and order m, with periods around 5 minutes, have frequencies ν_{lm}; in the absence of rotation ν_{lm} is virtually independent of m but rotational splitting removes this degeneracy. The earliest results, for sectoral modes with $|m| = l$, supplied information about the angular velocity $\Omega(r, \pi/2)$ in the equatorial plane (Duvall & Harvey 1984; Duvall et al. 1984; Brown 1985; Libbrecht 1986). The high degree modes, which are concentrated near the surface of the sun, yield a rotation period of about 25.1 days: this is equal to the period derived from magnetic features (Snodgrass 1983; Howard 1984) which is about 2% less than the period obtained from photospheric Doppler measurements (Snodgrass 1984). The angular velocity deduced from sectoral modes remains constant through the convection zone and then drops by around 7%. The observations are consistent with uniform rotation for $\mathbf{r} \leq 0.6R_\odot$ though there is some evidence for a rapidly rotating core from modes with $l \leq 2$ (Duvall & Harvey 1984).

Recent measurements of rotational splitting for tesseral modes with $|m| \leq l$ have been inverted to give the latitudinal variation of Ω. Independent data sets obtained with the Fourier Tachometer at Sacramento Peak (Brown & Morrow 1987; Morrow 1988; Brown et al. 1989) and at Big Bear Solar Observatory (Libbrecht 1988, 1989; Christensen-Dalsgaard & Schou 1988; Dziembowski, Goode & Libbrecht 1989) yield essentially the same results: the surface angular velocity, given by an expression of the form

$$\Omega(R_\odot, \theta) = (2.90 - 0.36\cos^2\theta - 0.53\cos^4\theta) \times 10^{-6}\ \mathrm{rad\ s^{-1}}$$

persists throughout the convective zone. Between the base of the convective zone at $r \approx 0.7R_\odot$ and a level around $r = 0.65R_\odot$ the latitudinal variation disappears and the results are consistent with a uniform angular velocity $\Omega_i = 2.70$ rad s^{-1} (corresponding to a period of 26.9 days) for $0.4 < r/R_\odot < 0.65$. This is the surface rotation rate at a latitude of $\pm 37^o$: we would expect Ω_i to be equal to the surface angular velocity at some latitude around $30^o - 45^o$ so that there is no net torque exerted on the uniformly rotating interior (Gilman, Morrow & Deluca 1989). Estimates of the magnetic field in the interior of the radiative zone suggest that it is strong enough to enforce rigid rotation (Mestel & Weiss 1987). Hence the variation of angular velocity within the Sun is as sketched in Figure 2(a), with latitudinal gradients in the convective zone and radial gradients immediately below it.

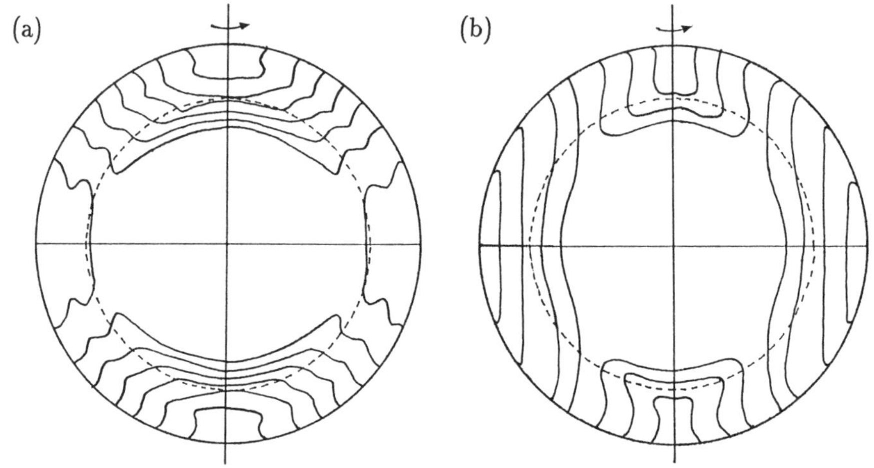

Figure 2. Differential rotation in the interior of the Sun. Sketches showing contours of constant angular velocity Ω in the convective zone and the outer part of the radiative zone. (a) Results derived from rotational splitting of p-mode oscillations with Ω constant on radii in the convective zone (after Libbrecht 1988). (b) Results from numerical simulations of convection in a rotating system (after Rosner & Weiss 1985). The broken line indicates the base of the convective zone.

This pattern owes its origin to convection. Hart et al. (1986) investigated the effect of rotation on convection in an ingenious experiment carried out under zero-gravity conditions in a spacecraft. The spinning system had spherical geometry with a radial gravitational acceleration mimicked by an electrostatic field. When Coriolis forces are dominant convection occurs in elongated banana cells, aligned with the rotation axis. As the heating rate is increased the banana pattern grows increasingly confused until there is an irregular tesselated pattern with no obviously preferred orientation. These results suggest that the pattern of convection depends on the extent to which rotation dominates, which is measured by the inverse Rossby number σ. The Sun, with a convective turnover time $\tau_i \approx 25$ days, turns out to be a marginal case ($\sigma \approx 6$) but it is clear from Figure 1 that younger, rapidly-rotating stars (or smaller stars with larger values of τ_c) will have $\sigma \gg 1$ so that rotational effects will be dominant and banana cells should be expected.

Such behaviour has been simulated numerically for convection in a rotating spherical shell in the Boussinesq (Gilman 1979) and anelastic (Glatzmaier 1985a) approximations. In these models the Proudman-Taylor constraint leads to motion in irregularly shaped cells elongated parallel to the rotation axis in the equatorial zone (and thus resembling a cartridge-belt rather than a bunch of bananas). Advection and diffusion of angular momentum are balanced in a steady state in which the equator rotates faster than the poles, with Ω constant on cylinders at low latitudes. These results imply that the distribution of angular velocity in a rapidly-rotating star should be as sketched in Figure 2(b), with $\Omega = \Omega(s)$ in the equatorial part of the convective zone, where $s = r\sin\theta$, and (as before) a constant angular velocity Ω_i below $r \approx 0.6R_\odot$ in the radiative zone (Rosner & Weiss 1985). The numerical simulations also yield a net negative (positive) helicity in the northern (southern) hemisphere; for these models we therefore have $\partial\Omega/\partial r > 0$ and $\alpha > 0$ in the lower part of the convective zone in the northern hemisphere.

Gilman (1983) and Glatzmaier (1985b) have also computed self-consistent nonlinear dynamo models in this geometry, assuming laminar diffusivities, which model turbulent diffusion in a stellar convection zone. Gilman fixed the thermal gradient (measured by a Rayleigh number) and varied both the rotation rate (measured by a Taylor number T) and the ratio ς of the magnetic to the thermal diffusivity. He found an intermediate range of T where cyclic dynamos appeared (via a subcritical Hopf bifurcation) as ς was decreased. The nonlinear solutions shared strong differential rotation, with toroidal fields of opposite sign that migrated towards the poles in the two hemispheres. As ς was further decreased the field became increasingly disordered and differential rotation was suppressed by the Lorentz force. These models confirm that oscillatory dynamos can exist. Although they were originally designed to simulate the generation of magnetic fields in the Sun the angular velocity gradients are probably more appropriate for a rapidly rotating star, while the field migrates in the poleward rather than the equatorward direction (as expected from the signs of $\partial\Omega/\partial r$ and α).

4. Mean Field Dynamo Theory

Most attempts to describe solar and stellar magnetic fields have relied on simplified dynamo models with azimuthally averaged properties. Then we can separate the magnetic field $\mathbf{B}$ into poloidal and toroidal components by writing

$$\mathbf{B} = \mathbf{B}_P + \mathbf{B}_T,$$

where

$$\mathbf{B}_P = \text{curl }(\mathrm{A}\hat{\boldsymbol{\phi}}) = \text{curl }(\frac{\chi}{\mathrm{s}}\hat{\boldsymbol{\phi}}), \qquad \mathbf{B}_{\mathrm{T}} = \mathrm{B}_\phi \hat{\boldsymbol{\phi}} \ ,$$

with (s, ϕ, z) cylindrical polar coordinates and $\hat{\boldsymbol{\phi}}$ a unit vector in the ϕ-direction. Then the induction equation

$$\frac{\partial \mathbf{B}}{\partial t} = \text{curl }(\mathbf{u} \times \mathbf{B}) - \frac{1}{\sigma'} \text{curl }\mathbf{j} \ , \tag{1}$$

where σ' is the electrical conductivity and $\mathbf{j}$ the electric current, has a poloidal component given by

$$\frac{\partial A}{\partial t} = \mathbf{u} \times \mathbf{B}.\hat{\boldsymbol{\phi}} - j_\phi / \sigma' \ . \tag{2}$$

Cowling's theorem (e.g. Cowling 1976) shows that a steady axisymmetric field cannot be maintained. For consider behaviour at a neutral point of the poloidal field, where $\mathbf{B}_P = 0$ but $j_\phi \neq 0$: then $\mathbf{u}.\mathbf{B} \times \hat{\boldsymbol{\phi}} = 0$, so the current cannot be maintained by induction. This result has been extended by Hide & Palmer (1982) to cover unsteady fields in a compressible star. Let the flux function $\chi = 0$ on the axis $s = 0$ and suppose that $|\chi(s, z)|$ attains its maximum value on the circle Γ_m of radius s_m through the neutral point where $\chi = \chi_m$. Then the total magnetic flux through a surface S_m spanning Γ_m is $F = 2\pi\chi_m$. If the neutral point moves with a velocity $\mathbf{U}$ as the field evolves

$$\dot{F} = \int_{S_m} \frac{\partial \mathbf{B}}{\partial t}.d\mathbf{S} + \int_{\Gamma_m} \mathbf{B}.\mathbf{U} \times d\mathbf{l} = \int_{\Gamma_m} (\mathbf{u} - \mathbf{U}) \times \mathbf{B}.d\mathbf{l} - \int_{\Gamma_m} (\mathbf{j}/\sigma').d\mathbf{l} \qquad \text{(from(1))}$$

$$= (2\pi s_m / \sigma') j_\phi = 2\pi\eta(\frac{\partial^2 \chi}{\partial s^2} + \frac{\partial^2 \chi}{\partial z^2}) \ ,$$

where the magnetic diffusivity $\eta = (\mu_o \sigma')^{-1}$. Hence $\dot{F} < 0$ if $F > 0$ and $\dot{F} > 0$ if $F < 0$, so by following whichever neutral point of the poloidal field has the greatest value of $|F|$ we can show that the total flux decays monotonically. (If $j_\phi = 0$ on Γ_m it is necessary to integrate over a small tube enclosing Γ_m and bounded by a field-line of $\mathbf{B}_P$.)

The mean poloidal field can only be maintained if non-axisymmetric motion produces a mean azimuthal electromotive force $\mathbf{E}$. We consider parametrized $\alpha\Omega$-dynamos with $\mathbf{E} = \alpha \mathbf{B}_T$ and $\mathbf{u} = s\Omega(s, z)\hat{\boldsymbol{\phi}}$, so that we can write

$$\left.\begin{aligned} \frac{\partial A}{\partial t} &= \alpha B_\phi + \eta(\nabla^2 - \frac{1}{s^2})A \ , \\ \frac{\partial B_\phi}{\partial t} &= s\mathbf{B}_P.\nabla\Omega + \eta(\nabla^2 - \frac{1}{s^2})B_\phi \ , \end{aligned}\right\} \tag{3}$$

where α represents the effects of helicity and η is now an eddy diffusivity. This parametrization can be justified in certain circumstances which do not obtain in a stellar convective zone (Cowling 1981; Weiss 1983). For a prescribed geometry the behaviour of solutions of the linear dynamo equations (3) depends on a single stability parameter, the dynamo number

$$D = \frac{\alpha \Omega' L^4}{\eta^2} \quad , \tag{4}$$

where Ω' is a measure of the differential rotation and L is a characteristic length scale. In a stellar convective zone we expect that $D \approx \sigma^2$ (Durney & Latour 1978). Typically we find that the trivial solution ($\mathbf{B} = 0$) undergoes an oscillatory (Hopf) bifurcation and loses stability when $D = D_{crit}$; for $D > D_{crit}$ solutions of the linear problem give dynamo waves which (for $\alpha > 0$ in the northern hemisphere) propagate towards the poles if Ω decreases inwards.

For $D > D_{crit}$ exponential growth of the magnetic field should eventually be halted owing to nonlinear effects. Since magnetic activity in lower main sequence stars is observed to depend on σ, and therefore on D, it seems appropriate to include the effect of the Lorentz force on the motion within the mean field formalism. For instance we could calculate the perturbation v_ϕ in the azimuthal velocity from the equation

$$\frac{\partial v_\phi}{\partial t} = \frac{1}{\mu_o \rho}[(\text{curl }\mathbf{B}_\text{T}) \times \mathbf{B}_\text{P}.\hat{\phi}] + \nu(\nabla^2 - \frac{1}{\text{s}^2})\text{v}_\phi \quad , \tag{5}$$

where ρ is the density and ν is an eddy viscosity (cf. Weiss, Cattaneo & Jones 1984). Since the Lorentz force is quadratic in the magnetic field we would expect to find both an overall suppression of differential rotation (which occurred in Gilman's simulations) and fluctuations in angular velocity with twice the frequency of the magnetic cycle (Schüssler 1981; Yoshimura 1981). Such torsional waves, with a period of 11 years at any latitude and propagating from the poles to the equator over 22 years, have actually been detected on the Sun (Howard & LaBonte 1980; Howard 1984; Ulrich et al. 1988). They provide the strongest observational evidence available for the existence of a dynamo (rather than an oscillator) in the Sun.

Nonlinear $\alpha\Omega$-dynamo models with spherical geometry reproduce the essential features of the solar cycle for mildly supercritical values of the dynamo number. As D is further increased more complicated spatial and temporal behaviour may appear. Figure 3 shows butterfly diagrams for two cases studied by Belvedere, Pidatella & Proctor (1989). In the first the oscillations are periodic, with waves travelling towards the equator and symmetrical about it. The second case shows periodically modulated behaviour. In the next section we shall explore such transitions for models that are drastically simplified (Ruzmaikin 1985).

5. Nonlinear Dynamo Waves

We can simplify the geometry by introducing local cartesian coordinates (x, y, z) in the (θ, ϕ, r) directions and then considering dynamo waves in a flat star with

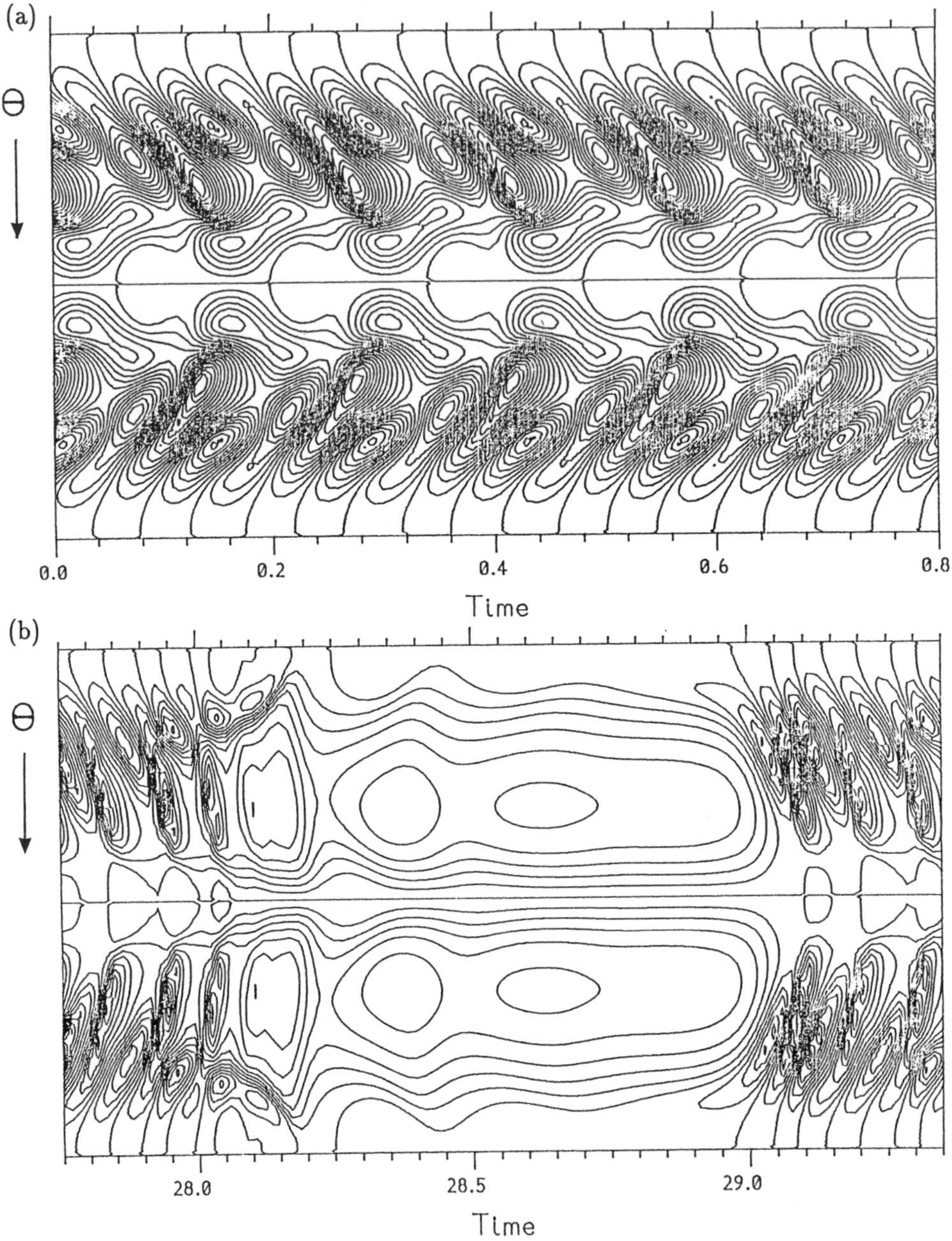

Figure 3. Butterfly diagrams for mean field $\alpha\Omega$-dynamos. Contours of the scaled azimuthal field $sB_\phi(\theta, t)$ for the nonlinear spherical dynamo model investigated by Belvedere et al. (1989). (a) $D \approx 6D_{crit}$, with periodic solutions showing complex spatiotemporal structure. (b) $D \approx 12D_{crit}$ showing a pulsed periodic solution with an episode of quasistatic activity.

$\mathbf{B} = \nabla \times (A\hat{\mathbf{y}}) + B\hat{\mathbf{y}}$ and $\mathbf{u} = V(z)\hat{\mathbf{y}}$. Then the dynamo equations (3) take the form

$$\frac{\partial A}{\partial t} = \alpha B + \eta \nabla^2 A \ , \quad \frac{\partial B}{\partial t} = \frac{\partial A}{\partial x}\frac{dV}{dz} + \eta \nabla^2 B. \tag{6}$$

If we assume $A = A'(t) \exp\,(\mathrm{i}kx)$, $B = B'(t) \exp\,(\mathrm{i}kx)$ then the system (6) becomes (after rescaling and suppressing primes)

$$\dot{A} = 2DB - A, \qquad \dot{B} = \mathrm{i}A - B, \tag{7}$$

where the dynamo number $D = \alpha V'/2\eta^2 k^3$. Thus there are unstable dynamo waves travelling in the x-direction for $D > 1$ (Parker 1955, 1979).

To describe the dynamical effects of the Lorentz force we suppose that the perturbation to the velocity shear is given by

$$v' = V'[\omega_o(t) + \omega(t) \exp\,(2\mathrm{i}kx)],$$

where ω_o is real. The variation of ω_o and ω with time can then be extracted from the cartesian analogue of (5) and included as a nonlinear contribution to the toroidal component of the induction equation. After rescaling we obtain a seventh-order nonlinear extension of (7):

$$\left.\begin{aligned} &\dot{A} = 2DB - A, \quad \dot{B} = \mathrm{i}(1+\omega_o)A - \frac{1}{2}\mathrm{i}A^*\omega - B, \\ &\dot{\omega}_o = \frac{1}{2}(A^*B - AB^*) - \nu\omega_o, \quad \dot{\omega} = -\mathrm{i}AB - \nu\omega \end{aligned}\right\} \tag{8}$$

(Weiss et al. 1984). If we arbitrarily set $\omega = 0$ this reduces to a fifth-order system with a nonlinear periodic solution

$$B = [\nu(D-1)]^{\frac{1}{2}} \exp\,(\mathrm{i}t)/D \ , \quad \omega_o = -(1 - 1/D) \tag{9}$$

which is stable for all D. Setting $\omega_o = 0$, on the other hand, yields a sixth-order system describing dynamo action limited by torsional waves. Once again there is a nonlinear periodic solution but this solution becomes unstable for $\nu < 1$; successive bifurcations lead to quasiperiodic and eventually to chaotic motion. It is possible to derive a reduced fifth-order system describing the modulation of the basic cycle (Jones, Weiss & Cattaneo 1985), which is illustrated in Figure 4 for $D = 8$. Even this toy system, with minimal spatial structure, can mimic some of the complicated temporal behaviour of the solar cycle: the irregular modulation of activity is consistent with the recurrent grand minima in the ^{14}C record.

This approach can be extended to include some spatial structure by considering a system that is periodic in x, which represents the colatitude (Stix 1972).

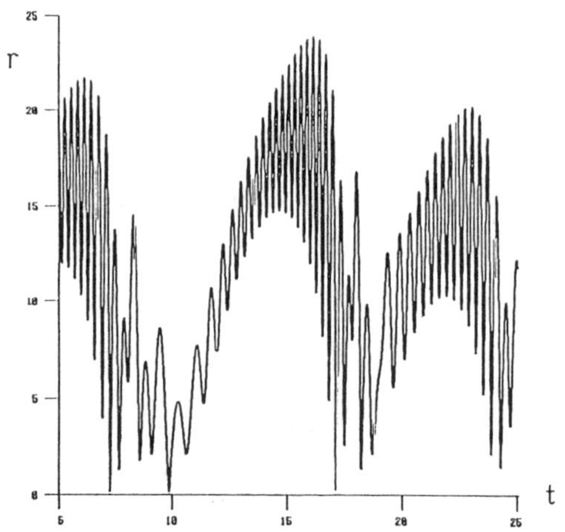

Figure 4. Chaotically modulated dynamo waves. Variation of the amplitude r of the basic cycle with time for the reduced fifth-order system of Jones et al. (1985) with $D = 8$. Note the recurrent minima associated with a "ghost" torus.

For simplicity nonlinear interactions can be parametrized by introducing parameters τ, κ, λ that describe quenching of the α-effect and of differential rotation and enhanced losses through magnetic buoyancy respectively. Thus we obtain the partial differential equations

$$\left.\begin{aligned} \frac{\partial A}{\partial t} &= (1+\tau B^2)^{-1} DB\cos x + \frac{\partial^2 A}{\partial x^2}, \\ \frac{\partial B}{\partial t} &= (1+\kappa B^2)^{-1}\frac{\partial A}{\partial x}\sin x + \frac{\partial^2 B}{\partial x^2} - \lambda B^3 \end{aligned}\right\} \qquad (10)$$

with boundary conditions $A = B = 0$ at $x = 0, \pi$ (Jennings & Weiss 1989). These equations possess two families of solutions with different symmetries at the equator: dipole solutions have $\partial A/\partial x = B = 0$ at $x = \pi/2$ while quadrupole solutions have $A = \partial B/\partial x = 0$. As $|D|$ is increased there is a sequence of pitchfork and Hopf bifurcations from the trivial solution giving rise to "multipole" solutions. Linear theory therefore suggests that as the dynamo number is increased the spatial structure of magnetic fields maintained by dynamo action becomes increasingly complicated (Parker 1979; Durney, Mihalas & Robinson 1981). To see what happens in the nonlinear domain we must establish the bifurcation structure for the problem. Branches of steady or oscillatory solutions emerge from primary bifurcations from the trivial solution and themselves undergo secondary bifurcations at which they may acquire or lose stability. Thus there is a complicated network of solution branches and it is important to discover which of these are stable. For the system (10) with $D < 0$ the first mode to become unstable as $|D|$ is increased is a steady quadrupole, followed by an oscillatory dilpole. Further primary bifurcations involve steady and oscillatory quadrupole as well as oscillatory dipole solutions. With $\kappa = \lambda$ and $\tau = 0$ a branch of stable steady solutions with quadrupole symmetry emerges from the first bifurcation. It loses stability to a branch of periodic mixed-mode solutions which lead to stable oscillations wiht dipole symmetry. These in turn eventually give way to mixed-mode oscillations which coexist with stable quadrupole oscillations (Jennings & Weiss 1989; D. Schmitt, private communication, has reported similar results). Thus secondary and tertiary bifurcations lead to solutions with complicated spatial structure; other choices of parameters allow quasiperiodic and chaotic behaviour too. Even in the Sun magnetic activity is not symmetric about the equator, indicating that mixed modes have been excited. In more rapidly rotating stars we should expect dynamo-produced fields to display much richer spatial structure (Parker 1979).

6. Solar and Stellar Dynamos

The only theories of stellar magnetic activity that have been explored in any detail are those involving dynamo action that can be represented within the mean field formalism. Although the view that such a mechanism is responsible for the solar cycle is widely shared, other opinions are expressed from time to time. Models involving torsional oscillations, hydromagnetic waves and azimuthal convective rolls have been proposed (and criticized) from different viewpoints by various authors (e.g. Cowling 1981; Snodgrass & Wilson 1987; Parker 1987a; Dicke 1988; Wilson 1988; Gough 1988; Bracewell 1988). I shall assume that variable magnetic fields in late-type stars are generated by dynamo action in or immediately below their convective zones.

The first question to be answered is: where is the dynamo located? Once flux tubes enter the convective zone the combined effects of convection and magnetic buoyancy should bring them to the surface in a time of order τ_c; hence Parker (1979) argued that the solar dynamo must be located near the base of the convective zone. This hypothesis is supported by the observation that long-lived magnetic features (originating well below the surface) show less differential rotation than shallow, short-lived features (Golub et al 1981; Stenflo 1989). Studies of nonlinear magnetoconvection show that the motion and magnetic fields are segregated from each other (Galloway & Weiss 1981). Isolated flux tubes cast thermal shadows which may generate convective downflows that counteract magnetic buoyancy (Parker 1987b,c) while topological asymmetries in the convective pattern may help to keep magnetic flux submerged (Moffatt 1983; Arter 1983; Cattaneo, Hughes & Proctor 1988). Hence it has been suggested that stellar magnetic fields originate in a magnetic layer beneath the base of the covective zone, in a region of weak convective overshoot (Spiegel & Weiss 1980; van Ballegooijen 1982; Schmitt, Rosner & Bohn 1984; Pidatella & Stix 1986). Within this shell toroidal flux is created by differential rotation and the magnetic field increases until the layer becomes unstable to modes driven by magnetic buoyancy (Parker 1979; Acheson 1979; Schmitt & Rosner 1983; Hughes 1985a,b, 1987; Hughes & Proctor 1988) and breaks up, as discussed by Hughes & Cattaneo in these Proceedings. As a result flux tubes are liberated and enter the convective zone, where they erupt in active regions (Moreno-Insertis 1987; Parker 1988). Further support for the shell dynamo hypothesis comes from considering the sign of the dynamo number: beneath the convective zone gas converges onto rising plumes and density stratification is less important; hence the helicity changes sign (Glatzmaier 1985b) so that $\alpha < 0$ in the northern hemisphere. With $\partial\Omega/\partial r > 0$ at low latitudes it is then possible to generate dynamo waves that propagate towards the equator. Simple models of linear and nonlinear mean field dynamos with plane geometry are consistent with this picture (Parker 1979; DeLuca & Gilman 1986, 1988; Gilman, Morrow & DeLuca 1989). It should be noted, however, that the arguments in favour of a shell dynamo rely on detailed observations of the Sun. We might expect to find analogous behaviour in stars with convective zones of similar depth and comparable rotation rates but it is not obvious that the same dynamo process operates in F stars, with shallow convective zones, or in M stars that are almost fully convective. It may be significant that active regions cannot be recognized in the emission from early F stars (Giampapa & Rosner 1984) while short cycle periods (2–5 yr) have only been detected in late F stars (Baliunas et al 1989). The M stars include the rapidly rotating and highly active dMe (or BY Draconis) stars with chromospheric Hα emission but older, more sedate M dwarfs still show signs of magnetic activity. Since late M stars are fully convective any dynamo must operate within their convective zones. Yet there is no convincing evidence for an abrupt change in their magnetic properties (Giampapa & Liebert 1986; Giampapa 1987).

Next we can attempt to identify the processes responsible for equilibration in nonlinear stellar dynamos. The simple models discussed in the previous section included quenching of the α-effect or of differential rotation, torsional waves and enhanced losses through magnetic buoyancy. Observations indicate that activity increases with increasing angular velocity and we expect that $D \propto \Omega^2$: from (9), quenching of differential rotation leads to fields that decrease as $D^{-1/2}$ for large D so that mechanism can be eliminated. If we restrict our attention to slow rotators with activity cycles like the Sun's some further process can be made by investigating the variation of cycle period with rotation rate (Noyes, Weiss & Vaughan 1984). Although the available sample is limited to 12 stars in addition to the Sun there are clear indications that for a star of given spectral type the cycle period decreases as Ω increases. This is consistent with the behaviour of simple nonlinear models where dynamo action is limited by buoyancy losses or torsional waves but not with quenching of the α-effect or of differential rotation, which leave the cycle period unchanged. This conclusion has been criticized on the grounds that the sample is unduly limited (Baliunas & Vaughan 1985; Soderblom & Baliunas 1988) but we should not expect to find analogous behaviour in rapidly rotating stars.

On theoretical grounds the convective pattern should change as Ω is increased: as we have seen, the dominant Coriolis force leads (through the Proudman-Taylor constraint) to a cartridge-belt pattern of rolls when σ is large. Indeed, the pattern of large-scale convection may be affected by magnetic fields too (Zel'dovich et al 1983; Jones & Galloway 1988). Such a change must alter the pattern of differential rotation and dynamo action should therefore be profoundly different in a rapidly rotating star (Knobloch, Rosner & Weiss 1981). Thus it is risky to assume that other more active stars have magnetic cycles that are simply more vigorous versions of what we can see on the Sun. In particular, it is not obvious which way the dynamo waves will travel or even that the dynamo will produce fields that regularly reverse.

Observationally, it is clear that young and active stars are indeed very different from the Sun. They have strong fields over a substantial fraction of their surfaces in addition to starspots. While sunspots only produce luminosity variations of order 0.1% starspots may cause reductions of 30% in luminosity as a star rotates. Moreover dMe stars show variation on a variety of timescales ranging from 10 to 80 years. Finally, activity patterns cannot be related to a single parameter, such as the inverse Rossby number σ: even the solar cycle is interrupted by grand minima and stars of the same age, mass and rotation rate may exhibit very different patterns of activity (Rodonó 1987). For example, components of visual binaries with similar spectral type can show different magnetic behaviour (Baliunas et al 1989). Such lack of uniqueness is not surprising, for dynamos are highly nonlinear and multiple solutions with different basins of attraction are a common feature of nonlinear systems.

What we can do is to classify different patterns of behaviour that might be found in early G-type stars, ordered by increasing angular velocity. This corresponds to looking at samples of younger stars or to following the Sun's evolution backwards. We begin with slow rotators, with periods of 20 days or more. This is the one class that we know about: its members show weak activity with aperiodic cycles like the Sun's. As the rotation period is decreased to around 10 days stars become moderately active and we can hope to describe their behaviour by extrapolating from the Sun. Thus we expect to find rich spatial structure as well as more complicated time-dependence. Rapid rotators with periods around 5 days exhibit wildly irregular activity while the ultra-rapid rotators with periods of 0.5 days are hyperactive with convection dominated by magnetic fields.

As I have explained, persistent observations have given us some understanding of solar activity and it seems safe to extrapolate from the Sun to other slow rotators. As the rotation rate increases, or the depth of the convection zone changes, such extrapolations grow more risky. Dynamo action seems ubiquitous but it is hard to predict just how the dynamo in an accretion disc should behave.

References

Acheson, D.J. 1979 Solar Phys. **62**, 23.

Arter, W. 1983 J. Fluid Mech. **132**, 25.

Baliunas, S.L., Donahue, R.A., Horne, J.H., Noyes, R.W., Porter, A., Gilliland, R.L., Duncan, D.K., Frazer, J., Lanning, H., Misch, A., Mueller, J., Soyumer, D., Vaughan, A.H., Wilson, O.C. & Woodard, L.A 1989. Astrophys. J., submitted.

Baliunas, S.L. & Vaughan, A.H. 1985. Ann. Rev. Astron. Astrophys. **23**, 379.

van Ballegooijen, A.A. 1982. Astron. Astrophys. **113**, 99.

Basri, G. & Marcy, G.W. 1988, Astrophys. J. **330**, 274.

Beer, J., Siegenthaler, U. & Blinov, A. 1988. In Secular Solar and Geomagnetic Variations in the Last 10,000 Years (ed. F.R. Stephenson & A.W. Wolfendale), p. 297. Kluwer; Dordrecht.

Belvedere, G. 1985. Solar Phys. **100**, 363.

Belvedere, G., Pidatella, R.M. & Proctor, M.R.E. 1989. Geophys. Astrophys. Fluid Dyn., submitted.

Bracewell, R.N. 1988. Mon. Not. Roy. Astron. Soc. **230**, 535.

Brown, T.M. 1985. Nature **317**, 591.

Brown, T.M., Christensen-Dalsgaard, J., Dziembowski, W., Goode, P., Gough, D.O. & Morrow, C.A. 1989. Astrophys. J., in press.

Brown, T.M. & Morrow, C.A. 1987. Astrophys. J. **314**, L21.

Butler, R.P., Cohen, R.D., Duncan, D.K. & Marcy, G.W. 1987. Astrophys. J. **319**, L19.

Cattaneo, F., Hughes, D.W. & Proctor, M.R.E. 1988. Geophys. Astrophys. Fluid. Dyn. **41**, 335.

Christensen-Dalsgaard, J. & Schou, J. 1988. In Seismology of the Sun and Sun-like Stars (ed. E.J. Rolfe), p. 149. ESA-SP286, Noordwijk,

Cowling, T.G. 1976. Magnetohydrodynamics (2nd ed.). Hilger, Bristol.

Cowling, T.G. 1981. Ann. Rev. Astron. Astrophys. **19**, 115.

DeLuca, E.E. & Gilman, P.A. 1986. Geophys. Astrophys. Fluid Dyn. **37**, 85.

DeLuca, E.E. & Gilman, P.A. 1988. Geophys. Astrophys. Fluid Dyn. **43**, 119.

Dicke, R.H. 1988. Solar Phys. **115**, 171.

Durney, B.R. & Latour, J. 1978. Geophys. Astrophys. Fluid Dyn. **9**, 241.

Durney, B.R., Mihalas, D. & Robinson, R.D. 1981. Publ. Astron. Soc. Pac. **93**, 537.

Duvall, T.L., Dziembowski, W.A., Goode, P.R., Gough, D.O., Harvey, J.W. & Leibacher, J.W. 1984. Nature **310**, 22.

Duvall, T.L., & Harvey, J.W. 1984. Nature **310**, 19.

Dziembowski, W.A., Goode, P.R. & Libbrecht, K.G. 1989. Astrophys. J. **337**, L53.

Eddy, J.A. 1976. Science **286**, 1189.

Galloway, D.J. & Weiss, N.O. 1981. Astrophys. J. **243**, 945.

Giampapa, M.S. 1987. in Cool Stars, Stellar Systems and the Sun (ed. J.L. Linsky & R.E. Stencel), p.236. Springer, New York.

Giampapa, M.S. & Liebert, J. 1986 Astrophys. J. **305**, 784.

Giampapa, M.S. & Rosner, R. 1984. Astrophys. J. **286**, L19.

Gilman, P.A. 1979. Astrophys. J. **231**, 284.

Gilman, P.A. 1983. Astrophys. J. Suppl. Ser. **53**, 243.

Gilman, P.A., Morrow, C.A. & DeLuca, E.E. 1989. Astrophys. J., in press.

Glatzmaier, G.A. 1985a. Astrophys. J. **291**, 300.

Glatzmaier, G.A. 1985b. Geophys. Astrophys. Fluid Dyn. **31**, 137.

Golub, L., Rosner, R., Vaiana, G.S. & Weiss, N.O. 1981. Astrophys. J. **243**, 309.

Gough, D.O. 1989. In Solar-Terrestrial Relationships and the Earth Environment in the Last Millennia (ed. G. Cini-Castagnoli), p.90. Elsevier, Amsterdam.

Hart, J.E., Toomre, J., Deane, A.E., Hurlburt, N.E., Glatzmaier, G.A., Fichtl, G.H., Leslie, F., Fowlis, W.W. & Gilman, P.A. 1986. Science **234**, 61.

Hartmann, L.W. & Noyes, R.W. 1987. Ann. Rev. Astron. Astrophys. **25**, 271.

Hide, R. & Palmer, T.N. 1982. Geophys. Astrophys. Fluid Dyn. **19**, 301.

Howard, R. 1984. Ann. Rev. Astron. Astrophys. **22**, 131.

Howard, R. & LaBonte, B.J. 1980. Astrophys. J. **239**, L33.

Hughes, D.W. 1985a. Geophys. Astrophys. Fluid Dyn. **32**, 273.

Hughes, D.W. 1985b. Geophys. Astrophys. Fluid Dyn. **34**, 99.

Hughes, D.W. 1987. Geophys. Astrophys. Fluid Dyn. **37**, 297.

Hughes, D.W. & Proctor, M.R.E. 1988. Ann. Rev. Fluid Mech. **20**, 187.

Jennings, R.L. & Weiss, N.O. 1989. Mon. Not. Roy. Astron. Soc., to be submitted.

Jones, C.A. & Galloway, D.J. 1988. In Secular Solar and Geomagnetic Variations over the last 10,000 Years (ed. F.R. Stephenson & A.W. Wolfendale), p.97. Kluwer, Dordrecht.

Jones, C.A., Weiss, N.O. & Cattaneo, F. 1985. Physica **14D**, 161.

Kawaler, S.D. 1988. Astrophys. J. **333**, 236.

Knobloch, E., Rosner, R. & Weiss, N.O. 1981. Mon. Not. Roy. Astron. Soc. **197**, 45P.

Krause, F. & Rädler, K-H. 1980. Mean-Field Magnetohydrodynamics and Dynamo Theory. Pergamon, Oxford.

Larmor, J. 1919. Brit. Assoc. Report, p.159.

Libbrecht, K.G. 1986. Nature **319**, 753.

Libbrecht, K.G. 1988. In Seismology of the Sun and Sun-like Stars (ed. E.J. Rolfe), p.131. ESA-SP 286, Noordwijk.

Libbrecht, K.G. 1989. Astrophys. J. **336**, 1092.

Lockwood, G.W., Thompson, D.T., Radick, R.R., Osborn, W.H., Baggett, W.E., Duncan, D.K. & Hartmann, L.W. 1984. Publ. Astron. Soc. Pac. **96**, 714.

Marcy, G.W. 1984. Astrophys. J. **276**, 286.

Mestel, L. 1988. In Activity in Cool Star Envelopes (ed. O. Havnes et al.), p.1. Kluwer, Dordrecht.

Mestel, L. & Spruit, H.C. 1987. Mon. Not. Roy. Astron. Soc. **226**, 57.

Mestel, L. & Weiss, N.O. 1987. Mon. Not. Roy. Astron. Soc. **226**, 123.

Moffatt, H.K. 1978. Magnetic Field Generation in Electrically Conducting Fluids. Cambridge.

Moffatt, H.K. 1983. Rep. Prog. Phys. **46**, 621.

Moreno-Insertis, F. 1987. In The Role of Fine-Scale Magnetic Fields on the Structure of the Solar Atmosphere (ed. E.-H. Schröter, M. Vázquez & A.A. Wyller), p.167. Cambridge.

Morrow, C.A. 1988. In Seismology of the Sun and Sun-like Stars (ed. E.H. Rolfe), p.91. ESA-SP 286, Noordwijk.

Noyes, R.W., Hartmann, L.W., Baliunas, S.L., Duncan, D.K. & Vaughan, A.H. 1984. Astrophys. J. **279**, 763.

Noyes, R.W., Weiss, N.O. & Vaughan, A.H. 1984. Astrophys. J. **287**, 769.

Parker, E.N. 1955. Astrophys. J. **122**, 293.
Parker, E.N. 1979. Cosmical Magnetic Fields. Oxford.
Parker, E.N. 1987a. Solar Phys. **110**, 11.
Parker, E.N. 1987b. Astrophys. J. **312**, 868.
Parker, E.N. 1987c. Astrophys. J. **321**, 984.
Parker, E.N. 1987d. Astrophys. J. **321**, 1004.
Parker, E.N. 1988. Astrophys. J. **325**, 880.
Pinsonneault, M.H., Kawaler, S.D., Sofia, S. & Demarque, P. 1989. Astrophys. J., in press.
Pidatella, R.M. & Stix, M. 1986. Astron. Astrophys. **157**, 338.
Pizzo, V., Schwenn, R., Marsh, E., Rosenbauer, H., Mühlhäuser, K.H. & Neubauer, F.M. 1983. Astrophys. J. **271**, 335.
Radick, R., Thompson, D.T., Lockwood, G.W., Duncan, D.K. & Baggett, W.E. 1987. Astrophys. J. **321**, 459.
Rosner, R. & Weiss, N.O. 1985. Nature **317**, 790.
Rodonó, M. 1987. In Solar and Stellar Physics (ed. E.-H. Schröter & M. Schüssler), p.39. Springer, Berlin.
Ruzmaikin, A.A. 1985. Solar Phys. **100**, 125.
Saar, S.H. 1987. In Cool Stars, Stellar Systems and the Sun (ed. J.L. Linsky & R.E. Stencel), p.10. Springer, New York.
Schmitt, J.H.M.M. & Rosner, R. 1983. Astrophys. J. **265**, 901.
Schmitt, J.H.M.M., Rosner, R. & Bohn, H.U. 1984. Astrophys. J. **282**, 316.
Schüssler, M. 1981. Astron. Astrophys. **94**, L17.
Schüssler, M. 1983. In Solar and Stellar Magnetic Fields (ed. J.O. Stenflo), p. 213. Reidel, Dordrecht.
Snodgrass, H.B. 1983. Astrophys. J. **270**, 288.
Snodgrass, H.B. 1984. Solar Phys. **94**, 13.
Snodgrass, H.B. & Wilson, P.R. 1987. Nature **328**, 1987.
Soderblom, D.R. 1985. Astron. J. **90**, 2103.
Soderblom, D.R. & Baliunas, S.L. 1988. In Secular Solar and Geomagnetic Variations over the last 10,000 Years (ed. F.R. Stephenson & A.W. Wolfendale), p.25. Kluwer, Dordrecht.
Spiegel, E.A. & Weiss, N.O. 1980. Nature **287**, 616.
Stauffer, J.R. & Hartmann, L.W. 1987. Astrophys. J. **318**, 337.
Stauffer, J.R., Hartmann, L.W., Burnham, J.N. & Jones, B.F. 1985. Astrophys. J. **289**, 247.
Stauffer, J.R., Hartmann, L.W., Soderblom, D.R. & Burnham, N. 1984. Astrophys. J. **280**, 202.
Stenflo, J.O. 1989. Nature, submitted.
Stix, M. 1972. Astron. Astrophys. **20**, 9.
Stix, M. 1987. In Solar and Stellar Physics (ed. E.-H. Schröter & M. Schüssler), p.15. Springer, Berlin.
Stuiver, M. & Braziunas, T.F. 1988. In Secular Solar and Geomagnetic Variations over the last 10,000 Years (ed. F.R. Stephenson & A.W. Wolfendale), p.245. Kluwer, Dordrecht.
Stuiver, M., Pearson, G.W. & Braziunas, T. 1986. Radiocarbon **28**, 980.
Stuiver, M. & Quay, P.D. 1980, Science **207**, 11.
Tayler, R.J. 1987. Mon. Not. Roy. Astron. Soc. **227**, 533.

Tayler, R.J. 1989. Q. J. Roy. Astron. Soc., in press.
Toner, C.G. & Gray, D.F. 1988. Astrophys. J. **334**, 1008.
Ulrich, R.K., Boyden, J.E., Webster, L., Snodgrass, H.B., Padilla, S.P., Gilman, P. & Shieber, T. 1988. Solar Phys. **117**, 291.
Vogel, S.N. & Kuhi, L.V. 1981. Astrophys. J. **245**, 960.
Weiss, N.O. 1983. In Stellar and Planetary Magnetism (ed. A.M. Soward), p.115. Gordon & Breach, London.
Weiss, N.O. 1988. In Secular Solar and Geomagnetic Variations over the last 10,000 Years (ed. F.R. Stephenson & A.W. Wolfendale), p.69. Kluwer, Dordrecht.
Weiss, N.O., Cattaneo, F. & Jones, C.A. 1984. Geophys. Astrophys. Fluid Dyn. **30**, 305.
Williams, G.E. 1988. Climatic Change **13**, 117.
Williams, G.E. 1989. J. Geol. Soc. **146**, 97.
Wilson, O.C. 1978. Astrophys. J. **226**, 379.
Wilson, P.R. 1988. Solar Phys. **117**, 217.
Yallop, B.D. & Hohenkerk, C.Y. 1980. Solar Phys. **68**, 303.
Zel'dovich, Ya. B., Ruzmaikin, A.A. & Sokoloff, D.D. 1983. Magnetic Fields in Astrophysics. Gordon & Breach, London.

THE NONLINEAR BREAKUP OF THE SUN'S TOROIDAL FIELD

D.W. Hughes	and	F. Cattaneo
D.A.M.T.P.		J.I.L.A.
University of Cambridge		University of Colorado
Silver Street		Boulder
Cambridge CB3 9EW		CO 80309-0440
U.K.		U.S.A

ABSTRACT. There are good reasons for believing that the sun has a strong toroidal magnetic field in the stably-stratified region of convective overshoot sandwiched between the radiative zone and convective zone proper. We have modelled the magnetic field in this region by studying the behaviour of a layer of uniform field embedded in a sub-adiabatic atmosphere. Since the field can support extra mass, such a configuration is top-heavy and instabilities of the Rayleigh-Taylor type can occur. By numerical integration of the 2-dimensional compressible MHD equations we have followed the evolution of this instability into the nonlinear regime. The initial buoyancy-driven instability of the magnetic field gives rise to strong shearing motions, thereby exciting secondary Kelvin-Helmholtz instabilities which wrap the gas into regions of intense vorticity. The somewhat surprising subsequent motions are determined primarily by the strong interactions between vortices.

1. Introduction

Although high-resolution observations from both the ground and from space have enabled the behaviour of the magnetic field at the solar surface to be studied in some detail, exactly what is going on in the convective and radiative zones remains somehat of a mystery. In particular, the dynamo process which keeps the field going through its approximately 11-year cycles is not well understood and, moreover, even the site of the sun's large toroidal magnetic field is a topic for debate.

As pointed out by Parker (1975), the intrinsic buoyancy of isolated magnetic flux tubes poses a major problem in keeping the field stored within the convection zone for periods comparable to that of the solar cycle. Several authors (e.g. Spiegel & Weiss 1980; van Ballegooijen 1982) have suggested that such a problem may not arise if the bulk of the toroidal field is stored below the convection zone proper, in the subadiabatic region of convective overshoot. This of course does not preclude a convection-zone dynamo - field may be generated in the convection zone itself and then be pushed downwards by turbulent transport of some kind (see, for example, Moffatt 1983; Spiegel 1987).

The work described in this paper was motivated by considerations of the evolution of a magnetic field trapped within such a convectively stable region. After all, wherever the large-scale toroidal field in the sun actually is, we know that at least part of it escapes to the surface where it is observed in the form of sunspots

G. Belvedere (ed.), Accretion Disks and Magnetic Fields in Astrophysics, 31–36.

and other active regions. We were therefore interested in considering instabilities of the magnetic field, one of our aims being to discover whether a diffuse toroidal field could give rise, through such instabilities, to the isolated flux tubes observed at the surface. Our model, which is described in detail in §2, has as its basic state a layer of magnetic field embedded in a subadiabatic atmosphere. The magnetic field, by virtue of its pressure, supports extra weight against gravity and consequently Rayleigh-Taylor instabilities ensue. We have pursued these instabilities into the fully nonlinear regime and have unearthed several interesting features (see §3).

So far we have restricted our attention to two-dimensional modes of instability. The undular modes, with motion along the field and in the vertical direction, were found to be always stable due to the prohibitive effects of the magnetic pressure. (The complicated issue of the preferred mode in magnetic buoyancy instabilities is discussed in detail by Hughes & Cattaneo 1987). Interchange modes on the other hand, for which the magnetic field lines remain straight with motion perpendicular to them, were readily destabilised and in this paper it is on these modes that we shall concentrate. A comprehensive account of this work, including the linear theory and the effects of rotation, can be found in glorious technicolour in Cattaneo & Hughes (1988).

2. Mathematical Formulation

As explained above, we have considered an initial state consisting of a region of uniform horizontal magnetic field embedded in a field-free, convectively stable atmosphere. This of course is not a true equilibrium since, for any finite value of the magnetic diffusivity η, the magnetic diffusion equation ($\eta\nabla^2\mathbf{B} = 0$) is not satisfied across the interfaces of the magnetic field. However, this is not a serious problem provided that the instabilities grow on a time scale which is much shorter than that on which the field diffuses appreciably — this may be checked *a posteriori* and, fortunately, does indeed turn out to be true.

Our computational domain is between $z = 0$ (top) and $z = d$ (bottom). Initially the atmosphere is piecewise polytropic with a temperature distribution of the form $T = T_0 + \Delta z$ and a uniform horizontal magnetic field $B_0\hat{\mathbf{y}}$ confined to the region $z_1 < z < z_2$. The total (gas + magnetic) pressure is everywhere continuous, thereby giving rise to discontinuities in the density at $z = z_1$ (where the gas is top-heavy) and at $z = z_2$ (botttom-heavy). It is worthwhile pointing out the differences between a magnetic layer as described here and an isolated magnetic flux tube. A magnetic layer has no horizontal inhomogeneities and (in the absence of magnetic diffusion) can be in hydrostatic equilibrium whatever the temperature distribution. An isolated flux tube, on the other hand, can only be static if its interior temperature is cooler than the surroundings by a certain amount — otherwise it will rise or fall. The stability properties of magnetic layers and isolated flux tubes are markedly different (see Cattaneo & Hughes 1988).

For reasons explained in the introduction we shall concentrate only on interchange modes in this paper. All variables are independent of y and the fluid flow is in the xz-plane only, perpendicular to the magnetic field $B\hat{\mathbf{y}}$. Our computational domain may then be thought of as a meridional slice near the equator with a toroidal field and with axisymmetric flows (see figure 1). To simplify the problem, the thermal conductivity K, the shear viscosity μ and the magnetic diffusivity η are assumed to be constants; furthermore the plasma obeys the perfect gas law with constant specific heats. By choosing the layer depth d and the sound crossing time $d/\sqrt{RT_0}$

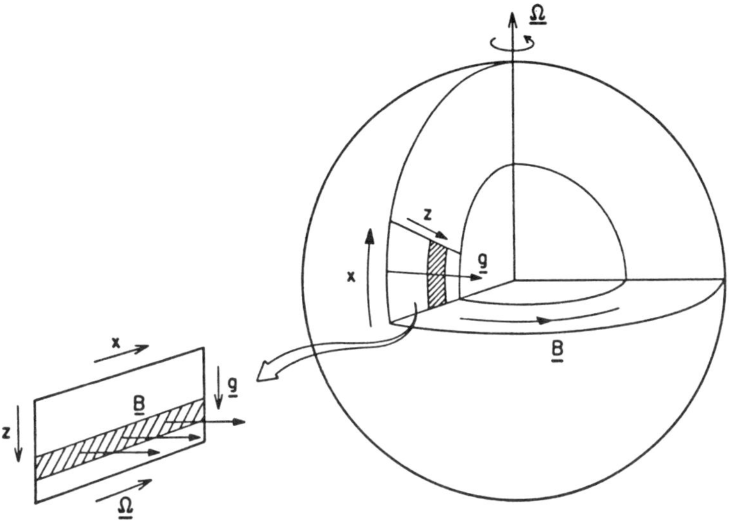

Figure 1 Schematic representation of the computational domain. The shaded region depicts the undisturbed magnetic layer (rotational effects are neglected in this paper).

as the units of length and time respectively (R is the gas constant) the evolution equations can be expressed in dimensionless form as follows:

$$p = \rho T, \tag{2.1}$$

$$\partial_t \rho + \nabla \cdot \rho \mathbf{u} = 0, \tag{2.2}$$

$$\partial_t B + \nabla \cdot (B\mathbf{u}) = \tau C_k \nabla^2 B, \tag{2.3}$$

$$\partial_t \rho \mathbf{u} + \nabla \cdot \rho \mathbf{u}\mathbf{u} = -\nabla(p + \frac{B^2}{2\beta}) + \theta(m+1)\rho\hat{\mathbf{z}} + C_k\sigma(\nabla^2\mathbf{u} + \frac{1}{3}\nabla(\nabla\cdot\mathbf{u})), \tag{2.4}$$

$$\rho(\partial_t T + \mathbf{u}\cdot\nabla T) + (\gamma-1)p\nabla\cdot\mathbf{u} = \gamma C_k \nabla^2 T + (\gamma-1)C_k(\sigma\phi_{ij}\partial_i u_j + \tau(\nabla B)^2/\beta), \tag{2.5}$$

where $\phi_{ij} = \partial_i u_j + \partial_j u_i - \frac{2}{3}\delta_{ij}\nabla\cdot\mathbf{u}$.

The six dimensionless parameters are defined by:

$$\sigma = \frac{\mu C_p}{K}, \qquad \tau = \frac{\eta \rho_0 C_p}{K}, \qquad \beta = \frac{\mu_0 p_0}{B_0^2},$$

$$m = (\frac{g}{R\Delta} - 1), \qquad \theta = \frac{\Delta d}{T_0}, \qquad C_k = \frac{K}{\rho_0 C_p d\sqrt{RT_0}},$$

where p_0 and ρ_0 are, respectively, the values of the pressure and density at $z = 0$, μ_0 is the magnetic permeability and C_p is the specific heat at constant pressure.

The formulation of the problem is completed by imposing boundary conditions at the top and bottom and by requiring that all variables be periodic in the horizontal direction. In the physical context of the solar overshoot zone it is unclear exactly what the boundary conditions should be and, in any case, it is certainly unrealistic to apply complicated conditions to our highly idealised model. Thus, with the understanding that any instability ceases to be of interest once the influence of the boundaries is significant, we decided on the following simple boundary conditions for the horizontal surfaces:

$$T = 1 \quad \text{at} \quad z = 0 \quad \text{and} \quad T = 1 + \theta \quad \text{at} \quad z = 1;$$

$$w = \partial_z u = \partial_z B = 0 \quad \text{at} \quad z = 0, 1.$$

To solve equations (2.1)-(2.5) we employed a pseudo-spectral method with a fourth-order finite-difference representation in the z-direction and a Fourier decomposition in x. The time-stepping of the advective terms was achieved by a partially-corrected Adams-Bashforth scheme. The diffusive terms were treated implicitly or explicitly depending on timestep considerations. If an explicit scheme was chosen it was implemented as for the advective terms above; alternatively an implicit Crank-Nicolson scheme was employed.

3. The Nonlinear Evolution

In this section we shall describe the evolution of the magnetic layer from random initial conditions — the magnetic field at various stages is portrayed in figure 2. The initial phase is characterised by the exponential growth of the kinetic energy and by the formation of small undulations on the upper interface (figure 2a). These

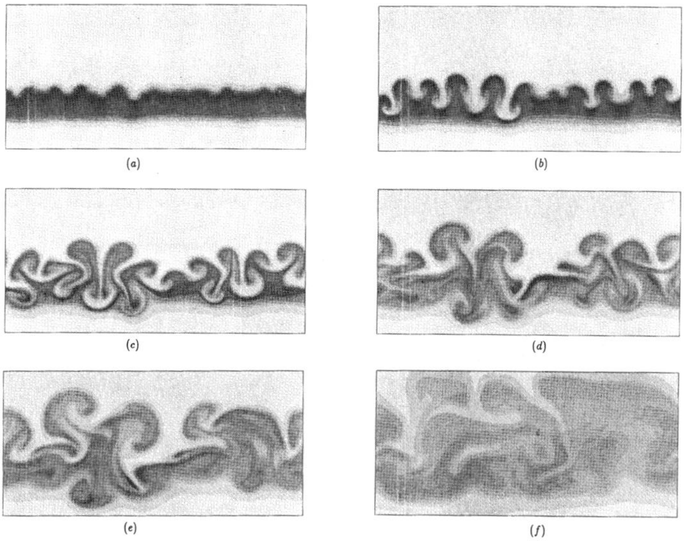

Figure 2 Evolution of the magnetic field from random initial conditions. $\gamma = 5/3$, $m = 1.6$, $\theta = 2$, $\beta = 1$, $\sigma = 0.1$, $\tau = 0.01$, $C_k = 0.01$, $z_1 = 0.6$, $z_2 = 0.8$.

then grow into the mushrooms typical of Rayleigh-Taylor instabilities driven by only a modest density contrast (see, for example, Daly 1967) (figure 2b). Associated with the formation of the mushrooms is a strong, almost horizontal, shear flow which leads via the Kelvin-Helmholtz mechanism to concentrations of vorticity in the wings of the mushrooms. The subsequent evolution of the magnetic field is governed primarily by the interactions between these vortices. The dominant interactions turn out to be those between neighbouring vortices on adjacent mushrooms — these can be so strong as to pull down regions of strong field that are actually lighter than their surroundings (figures 2c and 2d). The random initial conditions lead to a wide range both in the size of the mushrooms and their associated vorticity - consequently the symmetry of the mushrooms is quickly destroyed and the layer rapidly distorted. The final stages of the instability are controlled primarily by diffusion, spreading the magnetic field throughout the entire domain, and by a slow rundown of the motions by viscous processes. For times long compared to the growth time of the instability but short in comparison with the Ohmic diffusion time (figure 2f), the magnetic field is heavily fragmented but, because of the effects of vorticity, the peak magnetic energy density occurs in the neighbourhood of z_2, the base of the initial magnetic layer.

4. Discussion

It is of course difficult to make quantitative predictions about the sun's magnetic field since the dimensionless parameters of §2 take rather extreme values in the overshoot zone (for example, β is very large, σ and τ very small) and it is just not possible to perform our numerical calculations with these values. However, we do feel that we have captured the essential physics since, although we have overestimated the density contrast between the magnetic and field-free regions, thereby exaggerating the instability, we also have considerably more stabilising dissipation. Thus it certainly appears that any strong toroidal field generated beneath a field-free region will be drastically distorted by magnetic Rayleigh-Taylor instabilities and that a key rôle is played by the vorticity. Whether isolated flux tubes are the natural upshot of these instabilities is still unclear and will probably remain so without a fully 3-dimensional calculation.

Besides the study of 3-dimensional modes, a formidable task, there are other important extensions of the work described here, such as the inclusion of an overlying convectively unstable region or a less simplistic initial field configuration. In collaboration with T. Chiueh we are currently studying this latter problem by considering how the instability is affected by a twist in the magnetic field.

Acknowledgements

It is a pleasure to thank Prof. Juri Toomre for many stimulating discussions. This work was supported by the National Aeronautics and Space Administration through grants NSG-7511, NAGW-91 and NAG5-513 and contract NAS8-31958.

References

CATTANEO, F. & HUGHES, D.W. 1988 The nonlinear breakup of a magnetic layer: instability to interchange modes. *J. Fluid Mech.* **196**, 323-344.

DALY, B.J. 1967 Numerical study of two fluid Rayleigh-Taylor instability. *Phys. Fluids* **10**, 297-307.

HUGHES, D.W. & CATTANEO, F. 1987 A new look at the instability of a stratified horizontal magnetic field. *Geophys. Astrophys. Fluid Dyn.* **39**, 65-81.

MOFFATT, H.K. 1983 Transport effects associated with turbulence with particular attention to the influence of helicity. *Rep. Prog. Phys.* **46**, 621-664.

PARKER, E.N. 1975 The generation of magnetic fields in astrophysical bodies. X. Magnetic buoyancy and the solar dynamo. *Astrophys. J.* **198**, 205-209.

SPIEGEL, E.A. 1987 Hydrostatic adjustment time for the solar subconvective layer. In *The Internal Solar Angular Velocity*, eds. B.R. Durney and S. Sofia (D.Reidel), 321-327.

SPIEGEL, E.A. & WEISS, N.O. 1980 Magnetic activity and variations in solar luminosity. *Nature* **287**, 616-617.

VAN BALLEGOOIJEN, A.A. 1982 The overshoot layer at the base of the solar convective zone and the problem of magnetic flux storage. *Astron. Astrophys.* **113**, 99-112.

NONLINEAR ROSSBY-ALFVÉN GLOBAL WAVES ON THE SURFACE OF A ROTATING SPHERE

YU-QING LOU
Advanced Study Program
and
High Altitude Observatory
National Center for Atmospheric Research[1]
P.O. Box 3000
Boulder, Colorado 80307
U.S.A.

ABSTRACT. We consider two-dimensional nonlinear magnetohydrodynamic (MHD) waves of global scale for a thin magnetofluid layer on the surface of a rotating sphere. The fluid motions are assumed to be incompressible and the radial dimension is assumed to be in magnetohydrostatic balance. We are interested in a parameter regime in which horizontal spatial scales of wave motions, rotation rate of the background atmosphere, and the background zonal magnetic field strength are such that Rossby and Alfvén wave speeds are comparable. We obtain two kinds of finite-amplitude MHD waves propagating in the azimuthal directions relative to the uniformly rotating background atmosphere in the presence of a background zonal magnetic field and a steady differential zonal flow. These two kinds of nonlinear MHD waves are fundamentally due to the joint effects of uniform rotation of the background atmosphere and the background zonal magnetic field; the first is an inertial wave of the Rossby-Haurwitz type, modified by the presence of background zonal magnetic field, while the second is a magnetic Alfvén-like wave which is modified by the uniform rotation of background atmosphere. We discuss possible applications of these results in the context of geophysical and astrophysical systems.

1. Introduction

The linear Rossby wave was first discussed in the so-called β-plane approximation (Rossby 1939) in the context of atmospheric science. A global linear Rossby wave solution was soon obtained for shallow fluid shell on the surface of a rotating sphere (Haurwitz 1940). It was later realized that the Haurwitz's solutions exactly satisfy the two-dimensional nonlinear vorticity wave equation (Ertel 1943, Craig 1945, Neamtan 1946, Thompson 1982), i.e., these waves are, in fact, finite-amplitude permanent Rossby waves. The existence of the linear Alfvén wave in the presence of uniform magnetic field was predicted by Alfvén (1942). Walén (1944) suggested finite-amplitude Alfvén wave solution for an incompressible and uniformly magnetized medium. Parker (1984) found finite-amplitude Alfvén waves in a thermally stratified fluid. Recently, Namikawa and Hamabata (1988) obtained exact solutions of magnetohydrodynamic (MHD) wave equations in a circular

[1] The National Center for Atmospheric Research is sponsored by the National Science Foundation

G. Belvedere (ed.), Accretion Disks and Magnetic Fields in Astrophysics, 37–42.

magnetic field geometry. Hide (1966) investigated linear MHD waves due to the interaction of rotation and magnetic field in the context of "geomagnetic secular variation" using the β-plane approximation. Gilman (1969) obtained Alfvén-Rossby waves which were very similar to those obtained by Hide, but which allowed for vertical structure. In this paper, we present exact solutions of two-dimensional, global finite-amplitude Rossby-Alfvén waves for incompressible thin fluid shell on the surface of a rotating sphere in the presence of background zonal magnetic field (Lou 1987). These global waves are physically due to the interaction of Coriolis and Lorentz forces. One kind of solution is an inertial wave of the Rossby-Haurwitz type, modified by the presence of zonal magnetic field, while the other is a magnetic Alfvén-like wave which is modified by the uniform rotation of background atmosphere. In the limit of zero zonal magnetic field, the solution reduces to nonlinear Rossby-Haurwitz global wave propagating in the opposite direction of background rotation; in the limit of non-rotating atmosphere, the solution reduces to nonlinear Alfvén global waves propagating in both azimuthal directions. We discuss possible applications of these results in the context of geophysical and astrophysical systems.

2. A General Formulation

The ideal MHD equations describing an incompressible thin magnetofluid layer on the surface of a rotating sphere, when written in the corotating reference frame, are

$$\frac{\partial \vec{v}}{\partial t} + (\vec{v}\cdot\nabla)\vec{v} + 2\vec{\Omega}_o \times \vec{v} = -\frac{\nabla p}{\rho} + \frac{(\nabla\times\vec{B})\times\vec{B}}{4\pi\rho} - \nabla V, \tag{2.1}$$

$$\frac{\partial \rho}{\partial t} + \nabla\cdot(\rho\vec{v}) = 0, \qquad \frac{\partial \vec{B}}{\partial t} = \nabla\times(\vec{v}\times\vec{B}), \tag{2.2, 2.3}$$

$$\nabla\cdot\vec{B} = 0, \qquad \nabla\cdot\vec{v} = 0, \tag{2.4, 2.5}$$

where $\vec{v}$, $\vec{B}$, $\vec{\Omega}_o$, p, and ρ are the flow velocity in the co-rotating frame, the magnetic field in the co-rotating frame, the angular velocity of the rotating sphere, the fluid pressure, and the fluid mass density, respectively. The total potential V in equation (2.1) is composed of two parts: the gravitational potential V_g and the centrifugal potential $V_c \equiv -|\vec{\Omega}_o \times \vec{R}|^2/2$, where $\vec{R}$ is the position vector on the surface of the rotating sphere. We shall adopt the magnetohydrostatic approximation in the radial dimension, and thus ignore the radial structure of the thin fluid shell. In the shallow fluid approximation, we obtain an equation for the free fluid surface in the spherical coordinate system (R,θ,ϕ) from condition (2.5), viz.,

$$\frac{\partial(R_S - R_B)}{\partial t} + \frac{1}{R\sin\theta}\frac{\partial}{\partial\theta}\left[\sin\theta(R_S - R_B)v_\theta\right] + \frac{1}{R\sin\theta}\frac{\partial}{\partial\phi}\left[(R_S - R_B)v_\phi\right] = 0, \tag{2.6}$$

where R_S, R_B, and R are the radial coordinates of the free surface, of the bottom, and of the mean radial distance of the thin magnetofluid shell, respectively. In the following, all physical variables are assumed to depend on θ, ϕ, and t, with R a constant parameter. Consistent with the "shallow fluid" approximation, we shall further assume that the "surface divergence" of the flow vanishes, which is satisfied by introducing a stream function $\psi(\theta,\phi,t)$ such that

$$v_\theta = -\frac{1}{R\sin\theta}\frac{\partial\psi}{\partial\phi}, \qquad v_\phi = \frac{1}{R}\frac{\partial\psi}{\partial\theta}. \tag{2.7}$$

Since the r-component of magnetic field plays no dynamic role, we assume, in addition, that B_r vanishes. Thus equation (2.4) is satisfied by introducing a magnetic flux function $A(\theta,\phi,t)$ such that

$$B_\theta = -\frac{1}{R\sin\theta}\frac{\partial A}{\partial\phi}, \qquad B_\phi = \frac{1}{R}\frac{\partial A}{\partial\theta}. \tag{2.8}$$

We orient the axis $\theta = 0$ along $\vec{\Omega}_\circ$, and introduce the Coriolis parameter $f \equiv 2\Omega_\circ \cos\theta$. We seek solutions having the functional form $\psi = \psi(\theta, \phi')$, where $\phi' \equiv \phi - \omega t$, with ω a constant angular wave speed relative to the corotating reference frame. It follows from equations (2.6) and (2.7), from equations (2.2), (2.5), and (2.7), and from the θ- and ϕ-components of the induction equation (2.3), that

$$R_S - R_B = H(\Phi), \qquad \rho = D(\Phi), \qquad \text{and} \quad A = G(\Phi), \tag{2.9, 2.10, 2.11}$$

respectively, where $H(\Phi)$, $D(\Phi)$, and $G(\Phi)$ are three free functions, and

$$\Phi \equiv \psi + \omega R^2 \cos\theta. \tag{2.12}$$

From the θ- and ϕ-components of the momentum equation (2.1) together with equations (2.7), (2.8) and (2.10), it is straightforward to obtain

$$\frac{1}{2}\left[\frac{1}{R^2 \sin^2\theta}\left(\frac{\partial\Phi}{\partial\phi'}\right)^2 + \frac{1}{R^2}\left(\frac{\partial\Phi}{\partial\theta}\right)^2\right] + \frac{p}{\rho} + V + \frac{(2\Omega_\circ + \omega)\omega R^2 \cos^2\theta}{2} = W(\Phi), \tag{2.13}$$

where $W(\Phi)$ is the fourth free function (Tsinganos 1981).

We now derive a single equation for the stream function ψ. Again from the θ- and ϕ-components of the momentum equation (2.1), with the help of equations (2.7), (2.8), (2.10), (2.11), and (2.13), for nontrivial ψ, it follows that

$$\nabla^2\psi + f - \frac{1}{4\pi\rho}\frac{dG(\Phi)}{d\Phi}\nabla^2 A = \frac{dW(\Phi)}{d\Phi} + \frac{p}{\rho^2}\frac{dD(\Phi)}{d\Phi}. \tag{2.14}$$

where ∇^2 is the two-dimensional Laplacian operator on a spherical surface. In general, when the four free functions $H(\Phi)$, $D(\Phi)$, $G(\Phi)$, and $W(\Phi)$ are prescribed, the stream function ψ can be solved by combining equations (2.12) and (2.14). Then v_θ and v_ϕ are obtained from equation (2.7); with the help of equation (2.11), B_θ and B_ϕ are obtained from equation (2.8); R_S, ρ, and p are obtained from equations (2.9), (2.10), and (2.13), respectively. Equation (2.14) describes motions of vorticity in the presence of magnetic field.

We shall seek propagating global wave solutions relative to a uniformly rotating background atmosphere in the presence of a steady differential zonal flow and zonal magnetic field; the angular wave speed relative to the uniformly rotating background atmosphere satisfies a certain condition, which we discuss next.

3. A Recipe For Solutions

3.1. A STEADY STATE SOLUTION

For a steady solution, we set $\omega = 0$ in equation (2.14) and choose the free functions as follows:

$$\frac{dD(\psi^\circ)}{d\psi^\circ} = 0, \qquad \frac{dW(\psi^\circ)}{d\psi^\circ} = -\kappa\psi^\circ, \qquad G(\psi^\circ) = \alpha\psi^\circ, \tag{3.1}$$

where α and κ are two constants, and ψ° is a steady stream function; α is the magnitude ratio of magnetic field to velocity field in the corotating reference frame. For unconstrained κ, the regular solution to equation (2.14) is

$$\psi^\circ = -\mu R^2 \cos\theta, \qquad \text{where} \qquad \mu \equiv \frac{2\Omega_\circ}{\kappa R^2 - 2(1 - \alpha^2/4\pi\rho)}. \tag{3.2}$$

Solution (3.2) corresponds to a uniformly rotating atmosphere relative to the corotating reference frame. The zonal magnetic field strength is characterized by a finite parameter $\gamma \equiv \alpha\mu$.

3.2. NONLINEAR ROSSBY-ALFVÉN WAVE SOLUTIONS

Given the steady solutions (3.1) and (3.2), we seek time-dependent ($\omega \neq 0$) solutions. We choose the free functions in equation (2.14) in the following manner:

$$\frac{dD(\Phi)}{d\Phi} = 0, \qquad \frac{dW(\Phi)}{d\Phi} = -\nu\Phi, \qquad G(\Phi) = \beta\Phi, \tag{3.3}$$

where $\beta(\neq \alpha)$ and $\nu(\neq \kappa)$ are two constants; β is the magnitude ratio of magnetic field to velocity field in the comoving frame of angular speed ω.

Since we are looking for two-dimensional solutions of equation (2.14), the parameter ν must have the following form

$$\nu = \frac{L(L+1)}{R^2}\left(1 - \frac{\beta^2}{4\pi\rho}\right), \tag{3.4}$$

where L is an integer. The regular solution to equation (2.14) is thus

$$\begin{aligned}\psi(\theta, \phi, t) = &-\frac{\{2\Omega_o + \left[L(L+1)(1-\beta^2/4\pi\rho) + \beta^2/2\pi\rho\right]\omega\}R^2}{\left[L(L+1) - 2\right](1-\beta^2/4\pi\rho)} \cos\theta \\ &+ \sum_{m=-L}^{L} a_m P_L^m(\cos\theta) \exp\left[im(\phi - \omega t)\right],\end{aligned} \tag{3.5}$$

where m is an integer, $P_L^m(\xi)$ is the associate Legendre polynomial of order (L, m), and $\{a_m\}$ are complex coefficients. We note that the $m = 0$ term in solution (3.5) represents a steady differential zonal flow superposed onto the uniformly rotating background atmosphere. By identifying background rotation and zonal magnetic field, the angular wave speed of perturbations relative to a uniformly rotating background atmosphere is

$$\omega - \mu = \frac{-(\Omega_o + \mu) \pm \sqrt{(\Omega_o + \mu)^2 + L(L+1)\left[L(L+1) - 2\right]\Omega_A^2}}{L(L+1)}, \tag{3.6}$$

where

$$\Omega_A^2 \equiv \frac{\gamma^2}{4\pi\rho} \tag{3.7}$$

is the square of angular Alfvén speed along the zonal magnetic field. The lower and upper signs correspond to an inertial Rossby-Haurwitz wave modified by zonal magnetic field, and to an Alfvén wave modified by uniform background rotation, respectively. Various limits can be easily obtained from condition (3.6) for relative angular speed $\omega - \mu$. For example,

(i) For small spatial scales $L \gg 1$, the Coriolis effect is unimportant and

$$\omega - \mu \approx \pm\Omega_A, \tag{3.8}$$

i.e., the global waves propagate in both azimuthal directions with Alfvén angular speed.

(ii) For norotating atmosphere $\Omega_o + \mu = 0$,

$$\omega - \mu = \pm\sqrt{\frac{L(L+1) - 2}{L(L+1)}}\Omega_A. \tag{3.9}$$

The global waves are purely magnetic in nature. The factor in front of Ω_A is due to the curved background zonal magnetic field which varies with latitude.
(iii) For nonmagnetic atmosphere $\Omega_A = 0$,

$$\omega - \mu = -\frac{2(\Omega_0 + \mu)}{L(L+1)}, \tag{3.10}$$

and

$$\omega - \mu = 0. \tag{3.11}$$

The wave associated with equation (3.10) is the well-known Rossby-Haurwitz inertial wave. Condition (3.11) is a degenerate case; it represents a stationary MHD flow pattern in an inertial reference frame.

In general, a structured steady differential zonal flow can be superposed onto a uniformly rotating background atmosphere (see equation 3.5). Furthermore, the axis of the spherical harmonics which make up the perturbation part of the solution can be oriented in an arbitrary direction (Thompson 1982). If only one of the two types of MHD waves is excited, then for given L, such waves of arbitrary amplitude can be superposed for different m, but waves of different L cannot be superposed. If both kinds of MHD waves are excited, then such waves of arbitrary amplitude can be superposed for single set of L and m. Of course, in the linear regime, all the component waves of both types are superposable.

4. Applications

One possible application of our results is the wave motions in the photospheric layers of the Sun. The premise of this application is the existence of a global scale zonal magnetic field, which is admittedly of speculative nature. If such zonal field has a strength of 10 G at the equator, it is possible to find a parameter regime such that both sound speed and surface gravity wave speed is much larger than both Alfvén speed and Rossby wave speed. In such a regime, our result is applicable.

It is well known that the solar atmosphere exhibits large-scale steady differential rotation, which is observed by both the tracer method (Newton and Nunn 1951; Snodgrass 1983) and the spectral method (Dupree and Henze 1972; Scherrer, Wilcox, and Svalgaard 1980). We shall use the observed solar angular differential rotation rate in the sidereal reference frame in the form:

$$\Omega_{sidereal}(\theta) = a - b\cos^2\theta, \tag{4.1}$$

where a and b are two positive coefficients and θ is the heliographic colatitude. The results of Section 3 suggest that we use the first term and $(L = 3, m = 0)$ part of the second term in equation (3.5) to represent the observed steady solar atmospheric differential rotation. Based on the observations, the uniform rotation rate $\Omega_0 + \mu$ of the background is 13.33 degree per day. For a photospheric mass density $\rho = 10^{-7}$ g cm^{-3}, condition (3.6) gives an inertial angular wave speed of -2.36 degree per day and a magnetic angular wave speed of 0.138 degree per day. Note that if such waves were observed, then we could work backward, and use equation (3.6) to determine the angular Alfvén frequency Ω_A, and thus the zonal magnetic field strength at the equator.

Another application of our results arises in geophysics, where it has been suggested that the observed "geomagnetic secular variation" might be interpreted as the free hydromagnetic oscillation of the Earth's molten core (Hide 1966). The liquid, metallic core of the Earth occupies a very nearly spherical shell of outer and inner radii $\sim$ 3500 km and $\sim$ 1400 km, respectively. If the mean magnetic field in the Earth's core is mainly toroidal, then our simple MHD model is a fairly reasonable approximation. Since this application to "geomagnetic secular variation" is discussed extensively elsewhere, we refer readers to Hide (1966).

In summary, we suggest a simple model to illustrate a possible nonlinear interaction of rotation and magnetic field. In a broader sense, this model may find its applications in other astrophysical objects.

ACKNOWLEDGMENTS: Discussions with B. C. Low, R. Rosner, and T. Bogdan are gratefully acknowledged.

References

Alfvén, H. 1942, *Ark. Mat. Astr. Fys.*, **29B**, No. 2.
———— 1942, *Nature*, **150**, 405.
Craig, R. A. 1945, *J. Meteor.*, **2**, 173.
Dupree, A. K., and Henze, W., Jr. 1972, *Solar Phys.*, **27**, 271.
Ertel, H. 1943, *Z. Meteor.*, **60**, 332.
Gilman, P. A. 1969, *J. Atmos. Sci.*, **26**, 1003.
Haurwitz, B. 1940, *J. Marine Res.*, **3**, 254.
Hide, R. 1966, *Phil. Trans. Roy. Soc. London, A*, **259**, 615.
Lou, Y. Q. 1987, *Ap. J.*, **322**, 862.
Namikawa, T., and Hamabata, H. 1988, *Phys. Fluids*, **31** (4), 887.
Neamtan, S. M. 1946, *J. Meteor.*, **3**, 254.
Newton, H. W., and Nunn, M. L. 1951, *M. N. R. A. S.*, **111**, 413.
Parker, E. N. 1984, *Geophys. Astrophys. Fluid Dyn.*, **29**, 1.
Rossby, C.-G., and Collaborators 1939, *J. Marine Res.*, **2**, 38.
Scherrer, P. H., Wilcox, J. M., and Svalgaard L. 1980, *Ap. J.*, **241**, 811.
Snodgrass, H. B. 1983, *Ap. J.*, **270**, 288.
Thompson, P. D. 1982, *Mon. Weather. Rev.*, **110**, 1321.
Tsinganos, K. C. 1981, *Ap. J.*, **245**, 764.
Walén, C. 1944, *Ark. Mat. Astr. Fys.*, **30A**, 1.
———— 1944, *Ark. Mat. Astr. Fys.*, **31B**, 1.

ACCRETION DISCS IN THEORY AND PRACTICE

A. R. King
Astronomy Department
The University
Leicester LE1 7RH
U.K.

ABSTRACT. I review the current status of accretion disc theory and its relation to observations.

1. INTRODUCTION

Accretion discs are widely thought to be present in many types of close binary system, and are frequently invoked in models of active galactic nuclei and quasars. As we shall see, the formation of a disc is natural if the accreting matter possesses significant angular momentum. The detailed nature of the resulting flow is still in many cases unclear however.

2. ACCRETION GEOMETRY

Let ℓ = the average specific angular momentum of matter accreting on to an object of mass M and radius R. Then the closest distance the matter can approach the object without losing angular momentum is

$$R_{min} \sim \frac{\ell^2}{GM} \quad (1)$$

Then we have two cases:
(i) If $R > R_{min}$, matter can accrete directly.
(ii) If $R < R_{min}$, matter must lose angular momentum to accrete.

Case (ii) is far more common, and leads to the formation of accretion discs.

Estimates of R_{min}: (a) Binaries

In a binary of period P and total mass M ($\sim M_\odot$) we have $\ell \sim a^2/P$, where the separation $a^3 = GMP^2/4\pi^2$ (Kepler's law), giving

$$R_{min} \sim 3 \times 10^9 \, (P/hr)^{2/3} \text{ cm}.$$

G. Belvedere (ed.), Accretion Disks and Magnetic Fields in Astrophysics, 43–53.

Since $R \lesssim 10^9$ cm for any stellar-mass compact object, we see that case (ii) above (leading to disc formation) must occur in all compact binaries. The only exception to this is when the compact object possesses a magnetic field strong enough to be dynamically important at radii $> R_{min}$. This can only occur for neutron stars or white dwarfs; the most notable class are the AM Herculis systems (a subset of CV's), where the white dwarf has a magnetic moment $\mu \simeq 10^{33} - 10^{34}$ Gcm^3, and prevents the formation of a disc. (See King, 1987, for a review of magnetic CV's.)

(b) AGN's

The typical specific angular momentum in an elliptical galaxy (approximated as e.g. a Jacobi ellipsoid) is $\ell \sim {}^1/_3\, \epsilon r_C v_C$

where ϵ = ellipticity = $1 - b/a \sim 0.1$

r_C = ½-mass radius, $\gtrsim$ 1 kpc

v_C = velocity dispersion at r_C, $\sim$ 200 km s^{-1}

Thus $\ell \sim 10^{27}$ cm^2 s^{-1}, and we estimate

$$R_{min} \sim \ell^2/GM \sim 10^{20} M_8^{-1} \text{ cm}$$

for a central black hole of mass $M = 10^8 M_8 M_\odot$. This is far larger than its Schwarzschild radius

$$R_{Schw} = \frac{2GM}{c^2} = 3 \times 10^{13} M_8 \text{ cm} .$$

Again, we can conclude that in general, if AGN's are powered by accretion, angular momentum must be lost.

3. ACCRETION DISCS

The angular momentum loss process required for accretion in case (ii) above is usually called a <u>viscosity</u>. It is important to realise that this does <u>not</u> mean that its origins are understood; it is clear for example that atomic (or "molecular") viscosity is inadequate (see below). Most conceivable processes extract angular momentum on a timescale t_{visc} larger than that for energy loss, t_{th}, and the dynamical (orbit) timescale t_{dyn}. The accreting matter thus settles into a sequence of orbits of the lowest energy consistent with a given angular momentum, i.e. <u>circular orbits</u> lying in a plane. Matter drifts slowly inwards through this sequence, losing angular momentum and energy as it does so. This is an <u>accretion disc</u>. The discussion shows that the energy loss rate is actually controlled by the angular momentum losses.

Usually the mass of the matter in the disc at any one time is much less than that of the central object, and self-gravity is negligible (although it may be important in limiting the size of AGN discs). If, in addition, centrifugal forces are much stronger than pressure gradients (i.e. the disc matter can cool, see below), the circular disc orbits must be just Kepler orbits, with angular velocity $\Omega = \Omega_K =$

$(GM/R^3)^{1/2}$ at cylindrical radius R. Note that Kepler orbits shear, i.e. $\Omega' = d\Omega/dR = -3\Omega/2R < 0$. Random motions with average velocity $\tilde{v}$ and lengthscale λ across the circular streamlines transfer angular momentum and lead to a torque between rings of matter each side of radius R. The magnitude of the torque of the outer ring on the inner is

$$G(R) = 2\pi R\nu\Sigma R^2\Omega' \tag{2}$$

where

$$\Sigma = \int_{-\infty}^{\infty} \rho dz$$

is the surface density of the disc (ρ is the mass density and z a cylindrical coordinate), and

$$\nu \sim \lambda\tilde{v} \tag{3}$$

is the kinematic viscosity. Note that $G(R) < 0$, so that angular momentum is transferred from small to large R, in agreement with our discussion above. Clearly the net torque on a ring between R, R + ΔR is given by the difference $G(R + \Delta R) - G(R) = (\partial G/\partial R)\Delta R$. the torque does work at a rate

$$\Omega \frac{\partial G}{\partial R} \Delta R = \left[\frac{\partial}{\partial R} (G\Omega) - G\Omega'\right]\Delta R \quad .$$

Here the first term just gives the rate of advection of rotational energy: the second term represents the rate of viscous dissipation by the torques. Frequently this energy is radiated over the "upper" and "lower" disc faces (a thin disc), leading to the standard expression for the dissipation rate per unit surface area of a thin disc

$$D(R) = \frac{G\Omega'}{4\pi R} = \tfrac{1}{2}\nu\Sigma(R\Omega')^2 \quad . \tag{4}$$

As expected, $D(R) \geqslant 0$, and $D = 0$ only if the disc rotates rigidly ($\Omega' = 0$). If the disc radiates thermally, its effective temperature $T_{eff}(R)$ at each radius is given by $\sigma T_{eff}^4 = D(R)$. Note that if the disc is not thin, we cannot in general assume that energy generated between radius R, R + ΔR is released there - it will in general be transported and released elsewhere, so that (4) is valid only for thin discs.

4. THE THIN DISC APPROXIMATION

A disc is "thin" if its extent H in the z-direction is << its radial position R. The flow in the z-direction must be very slow, implying near-hydrostatic balance:

$$\frac{\partial P}{\partial z} \simeq -\frac{GMz}{R^3}\rho \quad .$$

(P = pressure), so that

$$\frac{P}{H} \sim \frac{GMH}{{}^3}\rho$$

Defining the sound speed c_s by $P = \rho c_s^2$, we find

$$H \sim c_s\left[\frac{R}{GM}\right]^{1/2} R$$

i.e.

$$\frac{H}{R} \sim \frac{c_s}{v_K} \tag{5}$$

with $v_K = (GM/R)^{1/2}$ the local Kepler velocity. Clearly, for consistency of the thin disc approximation we require $c_s \ll v_K$. As c_s will depend mainly on temperature this is equivalent to demanding efficient cooling of the disc.

If $c_s \ll v_K$, the disc is automatically Keplerian as well as thin: in a steady state, the radial momentum equation is

$$v_R \frac{dv_R}{dR} - \frac{v_\Phi^2}{R} + \frac{1}{\rho}\frac{dP}{dR} + \frac{GM}{R^2} = 0 \quad .$$

We expect $|v_R| \sim R/t_{visc} \ll v_\Phi$ (see below), and $(1/\rho)(dP/dR) \sim c_s^2/R \ll v_K^2/R = GM/R^2$, so that only the terms in v_Φ^2 and GM/R^2 survive; thus

$$v_\Phi = \left[\frac{GM}{R}\right]^{1/2} = v_K \quad .$$

Thus the conditions thin, Keplerian, and efficiently cooled (i.e. $c_s \ll v_K$) are all equivalent for a disc. In particular, if any one of them does not hold, neither do the other two.

5. VISCOSITY

We shall see below that the viscous timescale $t_{visc} \sim R^2/\nu \sim (R/\lambda)(R/\tilde{v})$ for any viscosity. For atomic viscosity, λ is the mean free path to collisions ($\lesssim 10^{-2}$ cm) and $\tilde{v} \sim c_s \sim 10^6$ cm s^{-1}. But $R \sim 10^{10}$ for binaries, and is larger for AGN's, giving $t_{visc} > 3 \times 10^8$y. Clearly larger values for ν are required. Turbulence is indeed suggested by the small value of ν (atomic), which implies Reynolds numbers $\gtrsim 10^{14}$. However we know very little about how turbulence might develop in a disc, and hence cannot find λ or $\tilde{v}$. The best we can do is to place limits on them. For reasonably isotropic motions, we clearly must have $\lambda < H$; we probably have $\tilde{v} < c_s$ in most parts of the disc, as supersonic turbulence is likely to be dissipated by shocks except in restricted regions. Thus we can set

$$\nu = \alpha c_s H \qquad (6)$$

and hope that $\alpha \lesssim 1$. This Shakura-Sunyaev viscosity presciption is widely adopted: note however that there is no reason to suppose α constant with R (or z). Apart from the expectation $\alpha \lesssim 1$ little is gained by the parametrization (6).

6 THIN DISCS: RADIAL STRUCTURE

For thin discs, the radial (R) and "vertical" (z) structures decouple. The vertical structure at each R is like that of a one-dimensional star, with viscosity playing the role of nuclear burning (see e.g. Frank *et al.*, 1985). The radial structure is considered here. Since the thin disc equations are fully derived in many reviews (e.g. Pringle, 1981) and in the book by Frank *et al.* (1985), Ch. 5, I shall not present the derivations here.

Conservation of mass and angular momentum in a thin disc are expressed as

$$R \frac{\partial \Sigma}{\partial t} + \frac{\partial}{\partial R} (R \Sigma v_R) = 0 \qquad (7)$$

and

$$R \frac{\partial}{\partial t} (\Sigma R^2 \Omega) + \frac{\partial}{\partial R} (R \Sigma v_R R^2 \Omega) = \frac{1}{2\pi} \frac{\partial G}{\partial R} \qquad (8)$$

with G(R) given by (2). Eliminating v_R and setting $\Omega = \Omega_K = (GM/R^3)^{½}$ gives

$$\frac{\partial \Sigma}{\partial t} = \frac{3}{R} \frac{\partial}{\partial R} \left\{ R^{½} \frac{\partial}{\partial R} \left[\nu \Sigma R^{½} \right] \right\} . \qquad (9)$$

Given the dependence of ν on Σ (generally from the vertical structure) (14) is the basic non-linear diffusion equation determining $\Sigma(R,t)$. Solutions for simple forms of ν are given by Lynden-Bell and Pringle (1974). Given $\Sigma(R,t)$, v_R follows from

$$v_R = - \frac{3}{\Sigma R^{½}} \frac{\partial}{\partial R} \left[\nu \Sigma R^{½} \right] \qquad (10)$$

deduced from (12), (13). Dimensionally we see from (14), (15) that Σ generally changes on the viscous timescale

$$t_{visc} \sim R^2/\nu \sim R/|v_R| \qquad (11)$$

as anticipated above. From (12–14) we see that, under the action of viscosity, angular momentum diffuses outwards (to larger R) and mass inwards.

7. STEADY THIN DISCS

In the steady case (12) gives

$$2\pi R\,\Sigma(-v_R) \quad = \quad \text{constant} \quad = \quad \dot{M} \qquad (12)$$

and (8) implies

$$\nu\Sigma \quad = \quad \frac{\dot{M}}{3\pi}\left[1 - \left[\frac{R_*}{R}\right]^{1/2}\right] \qquad (13)$$

where the central star (radius R_*) is assumed non-rotating (see the discussion in e.g. Frank *et al.*, 1985). From (4) we find the simple result

$$D \quad = \quad \tfrac{1}{2}\nu\Sigma(R\Omega')^2 \quad = \quad \frac{9}{8}\,\nu\,\Sigma\,\frac{GH}{R^3}$$

$$= \quad \frac{3GM\dot{M}}{8\pi R^3}\left[1 - \left[\frac{R_*}{R}\right]^{1/2}\right] \quad . \qquad (14)$$

Strikingly, in a steady thin disc, the surface dissipation rate, and hence the local effective temperature ($\propto D^{1/4}$) is independent of the viscosity. This means that disc models may produce useful predictions even though we know almost nothing about the viscosity which actually drives them.

From (14) the total emission from a disc with outer radius $\gg R_*$ is

$$L_{disc} \quad \simeq \quad 2\int_{R_*}^{\infty} D(R)2\pi R dR$$

$$= \quad \frac{GM\dot{M}}{2R_*} \quad ,$$

i.e. exactly one-half of the total accretion luminosity L_{acc}. The other half of L_{acc} has gone into the kinetic energy of matter orbiting at R_* with the Kepler velocity (virial theorem), which is only half as tightly bound as the non-rotating stellar surface. If this matter joins the surface via a boundary layer, this must release the other half of L_{acc}. Boundary layers are briefly considered below.

8. TIME-DEPENDENT DISCS

By analogy with stellar structure we can define three local timescales for an accretion disc. The viscous timescale t_{visc} was defined in (11) and is the analogue of a star's nuclear timescale.

The dynamical timescale

$$t_{dyn} = r/v_{\Phi} \tag{15}$$

is the orbital timescale, and also that for restoring vertical hydrostatic equilibrium, since

$$\frac{H}{c_s} = \frac{H}{v_{\Phi}} \cdot \frac{R}{H} = \frac{R}{v_{\Phi}} = t_{dyn} \ .$$

The thermal timescale

$$t_{th} = \frac{\text{heat content}}{\text{dissipation}} = \frac{\Sigma c_s^2}{D}$$

$$\sim \frac{R^3 c_s^2}{GM\nu} \sim \frac{c_s^2}{v_K^2} \cdot \frac{R^2}{\nu} \qquad \text{(using 4)}$$

i.e.

$$t_{th} \sim \left[\frac{H}{R}\right]^2 t_{visc} \tag{16}$$

using (5), (11). For an alpha-disc, we have from (6)

$$t_{visc} \sim \frac{R^2}{\alpha c_s H} \sim \frac{1}{\alpha}\left[\frac{R}{H}\right]\left[\frac{R}{v_K}\right]\left[\frac{v_K}{c_s}\right]$$

but $c_s/v_K = H/R$, so from (15)

$$t_{visc} = \alpha^{-1}(R/H)^2\, t_{dyn} \tag{17}$$

giving the hierarchical relations

$$t_{dyn} \sim \alpha\, t_{th} \sim \alpha(H/R)^2\, t_{visc} \tag{18}$$

so that

$$t_{dyn} < t_{th} \ll t_{visc}$$

as assumed above.

Both thermal and viscous instabilities can occur in discs (see e.g. Frank *et al.*, 1985). Because of the hierarchy (18) these can be handled separately: for a viscous instability, we can assume dynamical and thermal equilibrium throughout, as these are established on timescales much shorter than that on which the instability grows. Thus we need to consider only the diffusion equation (9). If $\Sigma_0(R)$ is a steady solution we perturb it as $\Sigma = \Sigma_0 + \Delta\Sigma$. Writing $\mu = \nu\Sigma$ for convenience, we have

$$\Delta\mu = \frac{\partial\mu}{\partial\Sigma}\Delta\Sigma$$

and $\Delta\mu$ obeys the diffusion equation

$$\frac{\partial}{\partial t}(\Delta\mu) = \frac{\partial\mu}{\partial\Sigma}\frac{3}{R}\frac{\partial}{\partial R}\left[R^{1/2}\frac{\partial}{\partial R}(R^{1/2}\Delta\mu)\right] ,$$

where the "diffusion coefficient" is now $\partial\mu/\partial\Sigma$. Clearly if $\partial\mu/\partial\Sigma > 0$ we will get mass diffusing away from regions of high density towards low-density ones, as expected. However, if $\partial\mu/\partial\Sigma < 0$ matter tends to clump together, regions of high density attracting more mass from their surroundings. Clearly such behaviour must lead to the break-up of the disc. Hence if for a given disc structure the μ-Σ relation contains regions with $\partial\mu/\partial\Sigma < 0$, these must be avoided by the disc structure by rapid adjustments on shorter (thermal) timescales. This fact can be used to construct local limit-cycle behaviour in regions of the disc where the presence of ionization zones gives rise to S-shaped μ-Σ curves. Many attempts have been made to turn this local behaviour into global disc instabilites which could serve as a model for dwarf nova outbursts. At present there is no consensus as to the success of such models (see Verbunt, 1987, for a recent review).

9. THE INNER EDGE OF THE DISC: WHERE DOES THE ENERGY GO?

We saw above that a boundary layer between a disc and a slowly-rotating star can release up to half of L_{acc}. Studies of this process are complicated, but generally agree that the radiation temperature T_{rad} is of order

$$T_{eff} \simeq \left[\frac{GM\dot{M}}{4\pi\sigma R^3}\right]^{1/4}$$

if the boundary layer remains optically thick (Pringle, 1977). Optically thin parts of the flow will tend to reach $T_V \sim GMm_H/kR$ (Pringle and Savonije, 1979; King and Shaviv, 1984) but are likely to spread around the star and form a corona (see also King, 1986). This is confirmed by numerical hydrodynamical calculations (Robertson and

Frank, 1986; Kley and Hensler, 1987). For white dwarfs this leads to hard X-ray emission. For neutron stars, only the ions are likely to reach T_V, Compton cooling holding the coronal electrons to ~ 10^9K (King and Lasota, 1987). For black holes there is obviously no boundary layer as such, though the stability of the flow near the horizon is still under discussion.

In principle one might expect that accretion on to magnetic stars would be simpler to understand, as the field makes the flow essentially one-dimensional near the surface. However, even in the simplest white dwarf case there is as yet no agreement (for an overview see Ch. 6 of Frank *et al.*, 1985).

10. THE OUTER EDGE OF THE DISC: WHERE DOES THE ANGULAR MOMENTUM GO?

For binaries, tidal effects of the companion on the disc provide a convincing answer to this question. Because of viscosity, there is a phase lag in the tides: the resultant force between the tidal bulges and the companion returns the excess angular momentum to the binary orbit. This process truncates the disc at about 80% of the Roche lobe radius.

For AGN's there is as yet no generally accepted answer to the question. If a disc forms, it will spread under the action of viscous torques until something removes the angular momentum. There are various possibilities, involving winds, jets, tides from stars or globular clusters, etc. If none of these mechanisms is effective, the disc will ultimately reach a radius

$$R_{SG} \sim (M/H)\Sigma^{1/3}$$

where its self-gravity becomes important. This may lead to the formation of a "shepherding" star, which truncates the disc tidally, as in the binary case above (or in Saturn's rings). Alternatively, self-gravity may alter the effective viscosity (Lin and Pringle, 1987), with consequences which have not yet been investigated.

11. OBSERVATIONS

Despite our ignorance of the form of the viscosity, ν, thin disc theory is widely adopted. The main reason is the simple form (14) for the surface dissipation, which implies an effective temperature

$$T(R) \propto R^{-3/4}$$

varying strongly with radius. In a cataclysmic variable, T(R) typically ranges from ~ 6000K at the outer edge of the disc up to ~ 10^5K near the white dwarf. The disc will thus radiate in the IR-visible-UV regions: in the UV, only the central part of the disc contributes, while at longer wavelengths the brightness distribution is more uniform. This very basic prediction of thin disc theory can be tested directly by observing CVs where the companion star eclipses the disc. The eclipses are deep and narrow at short wavelengths and broad and shallow at long wavelengths, as expected. Maximum-entropy methods indeed show that the effective temperature in such discs varies approximately as $R^{-3/4}$ (Horne and Marsh, 1987).

12. THICK DISCS

The last 7 sections have concentrated on the thin disc approximation, with good reason: a tolerably coherent theory exists (despite our ignorance of viscosity) and there is good observational support for it. If the thin disc approximation is abandoned, we cannot make either statement. Without the thin disc approximation one has to solve the equations governing a radiating, dissipative inflow with only the assumption of axisymmetry (and stationarity, if imposed). This is a rather more difficult problem than that of a rapidly rotating star: not surprisingly there has been little progress. In order to learn at least something, resort is often made to ad hoc prescriptions which replace one or more governing equations. For example the shapes of the constant-pressure surfaces in a thick disc are given if we know the rotation law on them (e.g. Frank et al., 1985, Ch. 10). However, arbitrarily choosing some simple rotation law to give an appealing shape for the disc is an act of lèse majesté which must be paid for: in general such configurations will involve inconsistencies, such as negative pressures or densities, if the full set of physical equations is applied to them. While useful as illustrations of what may happen, they cannot be regarded as self-consistent solutions even in the limited sense applying to thin discs. A further set of difficulties is that such configurations appear particularly prone to dynamical instabilities (Papaloizou and Pringle, 1984).

REFERENCES

Frank, J., King, A.R. and Raine, D.J., 1985. *Accretion Power in Astrophysics*, Cambridge University Press.

Horne, K. and Marsh, T.R., in *The Physics of Accretion on to Compact Objects*, Springer, p.1.

King, A.R., 1987. in *The Physics of Accretion on to Compact Objects*, Springer, p.137.

King, A.R., 1986. *Proc. N.Y. Acad. Sci.* **470**, 320.

King, A.R. and Lasota, J.P., 1987. *Astron. Ap.* **185**, 155.

King. A.R. and Shaviv, G., 1984. *Nature* **308**, 519.

Kley, W. and Hensler, G., 1987. *Astron. Ap.* **172**, 124.

Lin, D.N.C. and Pringle, J.E., 1987. *M.N.R.A.S.* **225**, 607.

Lynden-Bell, D. and Pringle, J.E., 1974. *M.N.R.A.S.* **288**, 721.

Papaloizou, J.C.B. and Pringle, J.E., 1984. *M.N.R.A.S.* **208**, 721.

Pringle, J.E., 1977. *M.N.R.A.S.* **178**, 195.
Pringle, J.E., 1981. *Ann. Rev. Astr. Ap.* **19**, 137.
Pringle, J.E. and Savonije, G.J., 1979. *M.N.R.A.S.* **187**, 777.
Pringle, J.E. and Wade, R.A. (eds) 1985. *Interacting Binary Stars*, Cambridge University Press.
Robertson, J.A. and Frank, J., 1986. *M.N.R.A.S.* **221**, 179.
Savonije, G.J., 1983. Ch. 9 of *Accretion-Driven Stellar X-ray Sources*, eds. W.H.G. Lewin and E.P.J. van den Heuvel, Cambridge University Press.
Verbunt, F., 1987. in *The Physics of Accretion on to Compact Objects*, Springer, p.59.
Williams, G.A., King, A.R. and Brooker, J.R.E., 1987. *M.N.R.A.S.*, **226**, 725.

Instabilities in dwarf nova accretion disks.

E. Meyer-Hofmeister
Max-Planck-Institut für Physik und Astrophysik
Institut für Astrophysik
D-8046 Garching, FRG

The dwarf novae are close binary stars where the Roche lobe filling companion transfers mass to a white dwarf. Due to the high angular momentum of the overflowing material the gas cannot fall directly onto the compact object and an accretion disk is formed. The companion star is usually a low mass main sequence star. These binary systems are very common, well documented by observations in the optical wavelength region since a long time. These objects are therefore particularly well suited to study properties of accretion disks.

Magnetic fields might influence these disks in two different ways.

(1) External magnetic fields. Different orientations are possible for a magnetic field of the white dwarf. If the axis lies in the equatorial plane of the disk the magnetic pressure acts on the surface of the disk and results in a squeezing of the innermost disk, so that the disk is even thinner at its inner edge and also effective temperature and density are changed (see Anzer et al. 1987). If the axis lies vertical to the equatioral plane the magnetic field pushes the inner edge of the disk away from the central object so that - compared to the situation without magnetic field - the innermost part or an even sizeable part of the disk is missing (Liebert and Stockman (1985)). These effects are the same in disks around neutron stars.

(2) Internal magnetic fields. Small scale magnetic fields in the disk could be the cause of or contribute to the viscosity which is clearly non-molecular in these accretion disks. Usually the viscosity is parametrized by a dimensionless number α (frictional stress $\tau_{r\varphi} = \alpha P$, P pressure): A semiquantative model of frictional stresses by small scale magnetic fields yields α prop. to $(H/r)^{\frac{3}{2}}$, H scale height and r distance from central object (Meyer and Meyer-Hofmeister 1983, Appendix).

The structure of the accretion disk is determined by the mass overflow rate $\dot{M}$. The effective temperature T_e at a certain distance r from the central star (except close to the inner and outer edge of the disk) is related to $\dot{M}$ according to $\sigma T_e^4 = 3GM\dot{M}/8\pi r^3$ (σ Stefan Boltzmann constant, G gravitation constant, M mass of central star). In a stationary disk the effective temperature therefore decreases with the distance from the center prop. to $r^{-\frac{3}{4}}$. In stationary disks also $\dot{M} = 3\pi$f holds, with the viscosity integral $f = \int_{-\infty}^{+\infty} \mu dz, \mu$ viscosity. Both relations are only exactly valid for disks with zero net angular momentum flow. For rates $\dot{M}$ as occurring in dwarf nova systems ($\dot{M} \approx 10^{-11} - 10^{-9} M_{\odot}/y$) the disk structure can be described in a good approximation as geometrically thin and azimuthally symmetric around a point mass. In hot disk systems ($T_e > 10000$ k) the material is fully ionized, in cool regions the material is only partially or not ionized. At temperatures below 4000 K the uncertainties in opacity values and the approximations in the computation of the disk atmosphere influence the results.

G. Belvedere (ed.), Accretion Disks and Magnetic Fields in Astrophysics, 55–57.

The relation between the viscosity (or T_e or $\dot{M}$) and the surface density $\Sigma = \int_{-\infty}^{+\infty} \rho dz$ (z vertical height above midplane) shows whether the disk is stable. For mass overflow rates $\dot{M}$ with a slope $\partial f / \partial \Sigma < 0$ in the f-Σ relation the disk is unstable for viscous radial diffusion and for vertical thermal energy transport. No stationary state can exist, the disk changes between a hot and a cool state in a limit cycle behaviour (for a review see Smak 1984). The models for the dwarf nova outbursts describe these changes from the cool state (quiescence) to the hot state (outburst). The change-over spreads in the disk with a heating and afterwards a cooling wave.

The observations of dwarf novae outbursts for some systems display a particular feature which could not yet be modelled in a convincing way. During the rise to an outburst the ultraviolet flux shows a delay of 0.5 to 1 day behind the rise of the optical flux. This became clear after IUE observations were available, especially data for VW Hydri (Schwarzenberg-Czerny et al. 1985, Verbunt 1987). The delay of the UV flux demands that higher temperatures are reached only after 0.5 to 1 day after the onset of the instability, but also that there is already a high optical flux during this time. This especially means that the hot state does not spread to the innermost disk where rapid diffusion would lead to high temperatures within a transition wave propagation timescale. These severe contraints are discussed in detail in Meyer and Meyer-Hofmeister 1988.

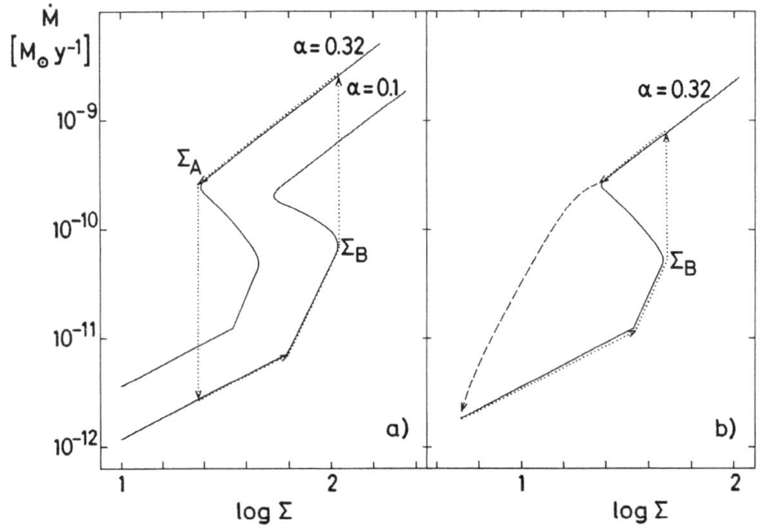

Figure 1: Schematic drawing of the limit cycle. a) usually chosen cycle with two different values α, b) modified cycle. Σ_A and Σ_B critical surface density.

We discuss a modified limit cycle (see Figure 1) with (1) a transition to only moderately high temperatures (10000 - 13000 K) during the onset of the hot state and (2) a more gradual decrease of the viscosity instead of the sudden transition to the low temperatures during the decline from the outburst. Such a gradual decrease would be supported by a model of slowed-down dynamos of the small scale magnetic field. This decrease would then be connected with a deep depletion of the disk. Such a process could explain a small UV flux remaining for some time after the decline from the outburst. On the other hand the depletion of mass leads to a steep surface density profile in the disk when new material is accumulated. The steep profile can cause a delay in the transition to the hot state in the later following next outburst.

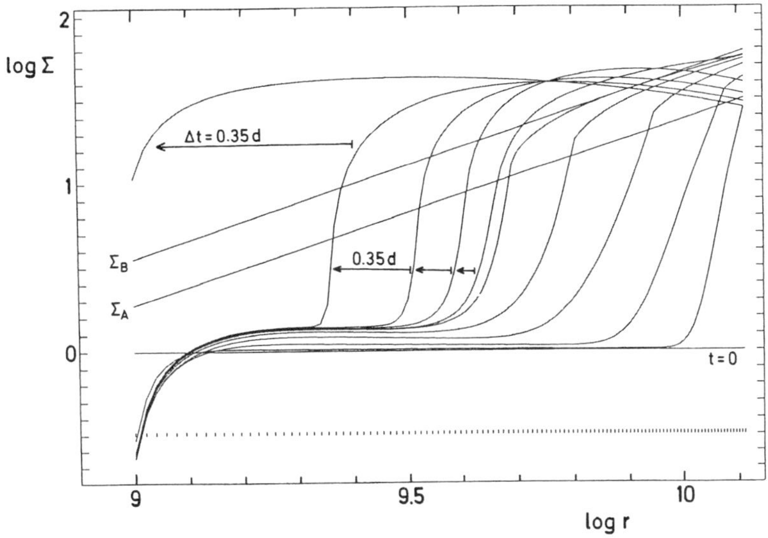

Figure 2: Accumulations of mass in the disk during the evolution, Σ surface density, Σ_A and Σ_B critical values, r distance from white dwarf, dots: spatial grid used for computation.

Computation of the disk evolution based on the proposed modified limit cycle shows such a delay. The evolution of the surface density in the disk is given in Figure 2 (for detailed results see Meyer and Meyer- Hofmeister 1988). Unfortunately there is still a discrepancy between the observed rise spectrum and the spectra computed for the disk model with a Kurucz atmosphere model code. The code for the spectra would require a temperature as low as 9000 K for the observed rise spectrum. The question then arises whether the Kurucz model atmospheres are applicable to accretion disk atmospheres where presumably strong local heating occurs. A test of synthetic accretion disk spectra by Wade (1988) also leads to the conclusion that new models for disk atmospheres need to be built. Despite this fact one might hope to get information about the viscosity in accretion disks especially from dwarf nova observations.

References:

Anzer, U. Börner G., Meyer-Hofmeister E.: 1987 Astron. Astrophys. 188, 85

Liebert J., Stockman H. S.: 1985, in Cataclysmic Variables and Low-Mass X-Ray Binaries, eds. D. Q. Lamb and J. Patterson, Astrophysics and Space Science Library, Vol. 113, p. 151

Meyer F., Meyer-Hofmeister E.: 1983, Astron. Astrophys 132, 143.

Meyer F., Meyer-Hofmeister E.: 1988, submitted to Astron. Astrophys.

Schwarzenberg-Czerny A., Ward M., Hanes D. A., Jones D. H. P., Pringle J. E., Verbunt F., Wade R. A.: 1985, Mon. Not. Roy. Astr. Soc. 212, 645

Smak J.: 1984: Publ. Astron. Soc. Pacific 96, 5

Verbunt F.: 1987, Astron. Astrophys. Suppl. 71, 339

Wade R. A.: 1988, to appear in Ap. J. Dec. 1.

ACCRETION AND PARTICLE ACCELERATION BY SPIRAL SHOCK WAVES

H.C. SPRUIT
Max Planck Institut für Physik und Astrophysik
Karl-Schwarzschildstr. 1, 8046 Garching West Germany,
and
Center for High Energy Astrophysics,
P.O. Box 41882, 1009 DB Amsterdam, Netherlands

ABSTRACT. The physics of accretion through a system of stationary spiral shaped shock waves is reviewed. The angular momentum transport by such waves can be described in terms of the negative angular momentum carried by trailing waves. Excitation processes are discussed. In 'large' disks (inner radius $r_i <<<$ outer radius r_o) there is a range of distances $r_i << r << r_o$ where the shock-accretion solution is insensitive to conditions at both r_i and r_o. For a simple equation of state and opacity this part of the solution has a selfsimilar (scale invariant) form. These properties make spiral shocked accretion especially attractive in protostellar, neutron star and AGN disks, where the traditional 'angular momentum problem' is most severe. It is shown that spiral shocks are also good locations for producing energetic protons, and I propose that this is the way in which the energetic particles inferred in many AGN's are produced.

1. Introduction

Spiral waves in disks have been studied extensively in the context of galactic dynamics (e.g. Binney and Tremaine, 1987), and to some extent that of protostellar disks (spiral waves generated by proto-Jupiters: Goldreich and Tremaine, 1980, Lin and Papaloizou, 1979, 1986; Larson, 1988). It has been known for a long time that such waves can carry a negative angular momentum, so that their dissipation would lead to accretion of the fluid supporting the waves onto the central object (Donner, 1979; Lin and Papaloizou, 1986). Yet the idea that such waves provide a natural means by which any accretion disk may solve its 'angular momentum problem' without resorting to an enhanced viscosity has developed only recently. Both Donner (1979) and Lin and Papaloizou (1986) for example still include explicitly a viscosity in their formulations, apparently under the impression (cf. Lynden-Bell, 1974) that this is somehow essential for the accretion process. (Nearly) inviscid numerical simulations by Sawada *et al.* (1985) showed accretion through spiral shaped shock waves. The existence of consistent accretion solutions of this kind was shown by more analytical means by Spruit (1987), and their properties were compared with those of viscous disks in Spruit *et al.* (1987). Applications to the protoplanetary case were made by Larson (1988). In the next section we discuss the physics of the process by mostly heuristic arguments.

G. Belvedere (ed.), Accretion Disks and Magnetic Fields in Astrophysics, 59–69.

2. Origin and angular momentum transport

Consider a thin disk of large extent, in the sense that the outer radius r_o is very much larger than the inner radius r_i. Near both boundaries plausible mechanisms exist for producing perturbations that transport angular momentum. At the inner boundary for example an asymmetric magnetosphere may excite shock waves (Michel, 1984). A sound wave or surface wave instability of the kind considered by Drury(1979, 1980, 1985) and susequently by Papaloizou and Pringle (1984, 1985), Narayan *et al.* (1987), Hawley and Blaes, (1988) can develop at either boundary. Such mechanisms may easily produce angular momentum transport within some distance of these boundaries. It is not clear however that they also suffice to produce a significant transport at distances $r << r_o$ but $>> r_i$, and it is in this range that the 'angular momentum problem' in cool (thin) disks is most acute.

Let us consider the effect of nonaxisymmetries at the outer disk boundary. For example, the mass distribution in the outer parts of a star forming disk might be irregular, depending on the history of the disk formation process. The orbital frequency at this distance is small and for simplicity we consider these irregularities just as stationary in an inertial frame. Gas orbiting at a somewhat smaller distance however sees the nonaxisymmetries as time dependent, hence waves are excited in the disk. Being excited at the outer edge, they are inward propagating waves, wound up by the differential rotation into a *trailing* spiral pattern. As the wave propagates inward, its frequency as measured in a comoving frame increases. Such waves have the interesting property of possessing a 'negative angular momentum', that is, its sign is opposite to that of the gas. Let m be the azimuthal order of the wave, Ω the Kepler frequency, so that the wave frequency in a comoving frame, σ, is

$$\sigma = \omega - m\Omega = -m\Omega \qquad 1$$

since we have assumed that the pattern frequency ω is negligible. For waves propagating in a differentially moving fluid a conserved quantity, the wave action s_w can be defined; in the short wavelength (WKB) approximation it is (Lighthill, 1978)

$$s_w = \int_V \frac{\frac{1}{2}\rho v_w^2}{\sigma} dV, \qquad 2$$

where ρ is the gas density, v_w the velocity amplitude of the wave, and the integral is over the volume V of the wave packet. This quantity is conserved along the path taken by the wave packet, as long as there are no (linear or nonlinear) dissipative processes affecting the wave. Note that the *energy* of the packet is not conserved even in the absence of dissipative processes; this expresses the fact that in a shear flow a wave exchanges energy with the mean flow. For a wave of azimuthal order m in a disk, the quantity

$$j = ms_w \qquad 3$$

is equally conserved, and can be identified with the angular momentum of the wave (Narayan *et al.* 1987). With (1)

$$j = -\int_V \frac{\frac{1}{2}\rho v_w^2}{\Omega} \quad < 0. \qquad 4$$

The minus sign turns out to have actual physical significance. It means that the disk with the wave in it has a lower angular angular momentum than without the wave (the total angular momentum being conserved in the absence of external torques). When and where the wave dissipates, it reduces the angular momentum of the gas, causing the gas to sink towards the central object. When and where will the wave dissipate? Suppose the wave starts at the outer edge with a small amplitude. As it propagates in, Ω increases inward, and to conserve j, the wave amplitude v_w in general has to increase. To see this consider the volume occupied by the wave packet, which is

$$V = 2\pi r H \Delta r \qquad 5$$

where H is the disk thickness and Δr is the radial extent of the packet. Since the front and back of the packet move in r with the radial component of the local group speed, which is proportional to the sound speed c_s, the distance between them increases at a rate $d\Delta r/dt \propto \Delta r dc_s/dr$, which can be integrated to yield

$$\Delta r \propto c_s \qquad 6$$

With (5) this yields

$$v_w \propto [\Omega j/(rHc_s\rho)]^{1/2}.$$

For the selfsimilar disk solution discussed below, $H \propto r$, $c_s \propto r^{-1/2}$, $\rho \propto r^{-3/2}$, so that

$$v_w \propto r^{-3/4}, \qquad 7$$

though this will be different for disks with a different radial structure. The amplitude of the wave *increases* inward, and eventually the wave turns into a shock. From then on the wave's angular momentum is no longer conserved, but slowly decreases (in absolute value) as the wave dissipates energy in the shock. The negative angular momentum transferred in the process to the fluid makes the fluid accrete. As the wave's angular momentum decreases, the specific angular momentum of the gas it encounters also decreases (as $r^{1/2}$). In the consistent selfsimilar solutions, the two vary exactly in parallel, so that accretion goes on throughout the entire disk.

Where the wave meets the inner edge, a part of it will reflect. The reflected wave propagates outward, and again by angular momentum conservation *decreases* in amplitude as it propagates outward. At a distance $r >> r_i$ the reflected wave therefore has no influence. It follows that there is a range in radius $r_i << r << r_o$ where the wave depends *neither on the outer nor on the inner boundary condition.* See Fig. 1. What does depend on these conditions is the extent of this range (being small or nonexistent if the initial amplitude of the wave at large radius is very small). Since the time scale for variations in the conditions at r_o ($\propto$ the orbital time scale there) is very long, the wave is nearly *stationary* in the range $r_i << r << r_o$ (compared with the orbital time scale). It follows that it must be a special kind of stationary solution of the hydrodynamic equations, namely one that contains no reference to inner or outer boundaries. Such solutions are known (Donner, 1979; Spruit, 1987): they are selfsimilar (scale invariant), spiral shaped shock waves. This kind of solution looks the same when seen at different magnifications, except for a rotation in azimuth over some angle; the spiral is logarithmic. An example is shown in Fig. 2.

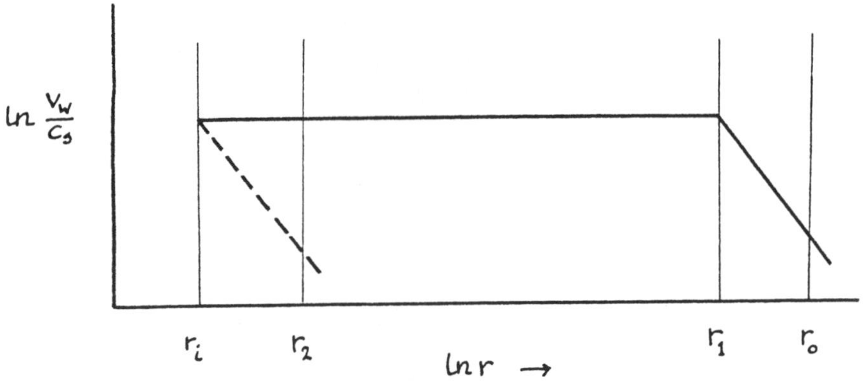

Figure 1: Amplitude of a wave generated at the outer disk boundary r_o by a stationary nonaxisymmetry there, in units of the local sound speed (schematic). Inside r_1 the wave is a shock of moderate strength. The reflected wave produced at the inner edge of the disk decreases outward (dotted line). Between r_2 and r_1 the accretion flow set up by the wave is selfsimilar

Figure 2: Example of a disk formed by a spiral shocked accretion flow. The picture shows the gas density, on a logarithmic scale.

The heuristic arguments above demonstrate (apart from mathematical concerns such as the unicity of solutions of nonlinear equations) that the selfsimilar spiral shocks found previously should occur generically in cases where the inner radius of the disk is very much smaller than the outer edge, and where a (small) amount of nonaxisymmetry exists at the outer edge. That is, the condition for their occurence is that the disk must be 'live', there must be something that adds mass nonaxisymmetrically at the the outer edge, or there must be another source of perturbation. Whereas other accretion mechanisms work plausibly near the disk edges, for example Michel's (1984) spiral shocks generated by an asymmetric magnetosphere of the central object, the present process is especially interesting because it works in its purest form at places that are neither close to the inner nor to the outer boundary.

The picture drawn here suggests that spiral shaped shock waves would be a generic feature in disks, contributing to (or even dominating) the accretion process. It does not exclude the possibility of 'dead', nonaccreting disks, however. Conceivably there might be 'fossil' disks which receive no mass, and are far from external perturbers (*e.g.* Kafka and Meyer, 1984). The level of perturbation in such disks could be too low to grow into a shock before reaching the inner edge of the disk.

Properties of spiral-shocked accretion

The properties of the solutions calculated in Spruit (1987) are illustrated in Figures 3 and 4 (see also Spruit et al. 1987). At low accretion rates the disks are cool, and the spiral shocks tightly wound. This is because the Mach number of the shocks is never very large, so that the shock always moves roughly at the speed of sound. The condition that the spiral pattern be stationary in an inertial frame then implies that the angle between the shock and the flow is of the order $c_s/\Omega r$, which is small in a cool disk. The strength of the shocks also decreases with decreasing temperature of the disk. Since the accretion rate is related to the dissipation in the shocks, accretion is less efficient in cool disks (those with low accretion rate). This can be expressed in terms of an 'effective α-parameter' for the process. This is defined as the value of the α-parameter in a standard (Shakura and Sunyaev) viscous disk model that would yield the same accretion rate at the same disk temperature. At low temperature, the numerical results in Spruit (1987) yield

$$\alpha_{eff} = 0.013\,(H/r)^{1.5}, \qquad 8$$

where H/r is the aspect ratio of the disk. Larson (1988) has derived this dependence on H/r by a heuristic argument which shows that 1.5 should indeed be read as 3/2. The argument equates the angular momentum flux in the wave to (minus) the angular momentum flux carried by the gas due to accretion, and the energy dissipated in the shock to the rate of change of orbital energy of the gas (since the shock is the way by which the orbiting gas loses its energy). This yields two relations that combine to yield a $(H/r)^{3/2}$ dependence. Larson has also reproduced the numerical factor in (8) using a weak-shock model for the waves.

Important aspects of the theory are the possibility of *adiabatic* accretion, and the role played by the ratio of specific heats γ. By adiabatic we mean here (in the usual thermo-

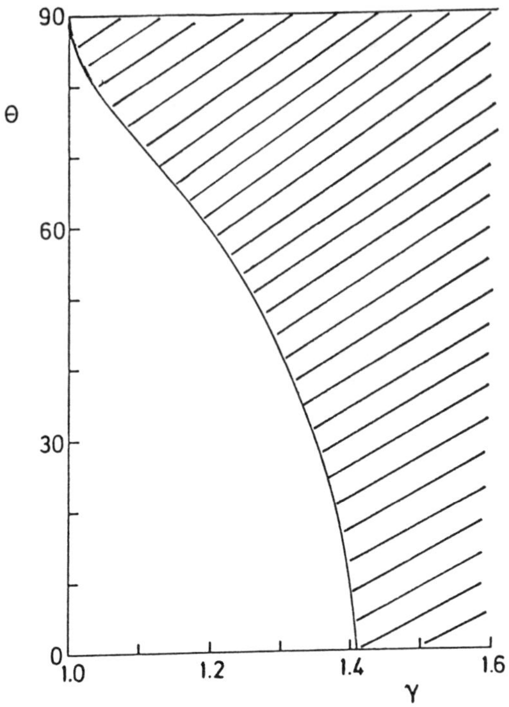

Figure 3: Spiral shocked accretion solutions. γ is the ratio of specific heats, θ the angle of the spiral. Cool solutions are at the top, hot ones at the bottom. The solid line represents adiabatically accreting flows (asymptotically reached at high accretion rates). Solutions with radiative loss occupy the shaded area.

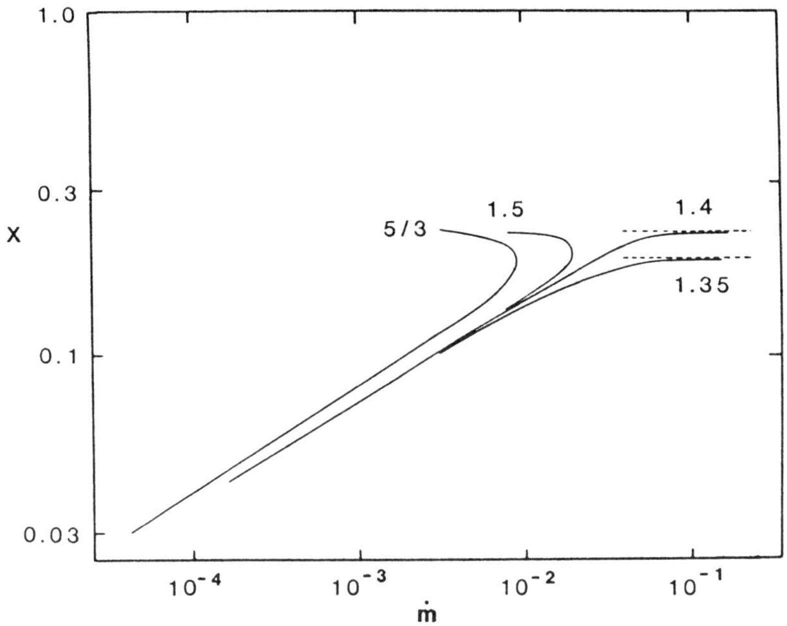

Figure 4: Temperature in units of the virial temperature as a function of dimensionless accretion rate. Parameter is the ratio of specific heats.

dynamic sense) the absence of heat losses by radiative loss or thermal conduction. It does not imply absence of dissipation. Indeed, the entropy of the gas increases on each passage through the spiral shock. When radiative losses are thus neglected, all the energy dissipated in the flow ends up as internal energy of the gas and is carried in with the accreting flow. This tends to produce high temperatures, and adiabatic accretion is possible only if the resulting temperature is not so high that the sound speed p/ρ exceeds the escape speed $(GM/r)^{1/2}$. For a gas starting at a large distance from the accreting object with negligible internal energy, the internal energy after adiabatic accretion to a circular orbit at distance r is

$$e = \frac{p}{\rho(\gamma - 1)} = \frac{1}{2}\frac{GM}{r}. \qquad 9$$

The accretion therefore proceeds at the fraction $1/2(\gamma - 1)$ of the virial temperature $\mu GM/(r\mathcal{R})$. For γ close to 1 this fraction is small, and adiabatic accretion is possible. The temperature is low in this case because the dissipated energy is distributed mainly among the many internal degrees of freedom that a low γ implies. For large γ, the resulting temperature would be too high for the accreting gas to stay bound to the central object, and hence accretion is not possible. Thus there exists a critical value of γ below which adiabatic accretion is possible, and above which accretion requires radiative loss. For spherical accretion, this critical value is 5/3 (Bondi, 1952). For accretion with angular momentum, the critical γ may well be somewhat different (but not larger than 5/3).

Fig. 3 shows the spiral shocked accretion solutions as a function of γ and the spiral angle θ (angle between the shock surface and the radial direction). The solid curve shows the relation between γ and θ for adiabatic accretion. As shown above on heuristic grounds, the solution is cool (tightly wound spiral, $\theta \approx \pi/2$) for $\gamma - 1 << 1$. The critical value of γ is about 1.4 for the twofold spiral pattern. This value increases to $\sqrt{6} - 1 \approx 1.45$ in the limit of an n-fold pattern of high n. Radiating solutions occupy the shaded area in Figure 3. The accretion rate increases downward in this diagram (line of constant) γ. If $\gamma < \gamma_c \approx 1.4$, the adiabatic solution curve is approached asymptotically as the mass accretion rate $\dot{M} \to \infty$. That is, the radiative losses become less and less important as the disk becomes denser at high accretion rate, but the temperature stays low enough for accretion to take place. This is different for $\gamma > 1.4$. There, the accretion rate has a maximum value as a function of θ (*cf.* Spruit, 1987, Fig. 6), and no solutions exist for accretion rates above this maximum. The physical reason is that accretion is now possible only with radiative losses, so that there is a maximum to the optical thickness (and hence the accretion rate) for which accretion is possible. This maximum accretion rate depends on the way in which energy loss from the disk by radiation is treated. For a radiative disk structure, the maximum accretion rate for the spiral shocked solutions is

$$\dot{M}_{max} = \dot{m}_{max}(\gamma)K \qquad (\gamma > \gamma_c), \tag{10}$$

where the dimensionless rate $\dot{m}_{max}$ depends only on γ and is about 10^{-2} for $\gamma = 5/3$, and the characteristic value K is given by

$$K = 4[\pi(\gamma-1)\sigma]^{1/2}(\mu/\mathcal{R})^2\kappa^{-1/2}(GM)^{7/4}r^{-1/4} \approx 0.67\kappa^{-1/2}(M/M_\odot)^{7/4}r_{10}^{-1/4} \quad M_\odot \mathrm{yr}^{-1}, \tag{11}$$

where σ is the Stefan-Boltzmann constant, μ the mean atomic weight, $\mathcal{R}$ the gas constant. Fig. 4 shows that the characteristic rate K plays a dual role: for $\gamma > \gamma_c$ there is a maximum accretion rate around $0.01K$, while for $\gamma < \gamma_c$ the accretion process becomes adiabatic for $\dot{M} \gtrsim 0.01K$.

The value of the critical γ given above is subject to the uncertainty introduced by the two dimensional approximation made (which is inaccurate especially near these critical values). The *existence* of a critical value of γ is demonstrated however by the existence of adiabatic solutions near $\gamma = 1$. These solutions are so cool that the thin disk approximation is justified.

The characteristic accretion rate K (Eq. 11) is physically relevant only if it is smaller than the Eddington rate $\dot{M}_E = 4\pi Rc/\kappa$, since near $\dot{M}_E$ the character of the accretion process is modified by radiation pressure. With (10,11) evaluated at the surface $r = R$ of the accreting object we have $\dot{M}_{max} < \dot{M}_E$ if

$$\dot{m}_{max}c^{-1}[\sigma(\gamma-1)/\pi]^{1/2}(\mu/\mathcal{R})^2(GM)^{7/4}\kappa^{1/2}R^{-5/4} < 1, \tag{12}$$

or

$$\kappa^{1/2}(M/M_\odot)^{7/4}(R/R_\odot)^{-5/4} \lesssim 0.1 \quad (\gamma = 5/3, \dot{m}_{max} \approx 10^{-2}). \tag{13}$$

This shows, for example, that the Eddington rate is smaller than $\dot{M}_{max}$ for accretion onto white dwarfs, neutron stars and main sequence stars. For black holes at R = last stable orbit the same holds, except for masses less than $10^{-16}M_\odot/\kappa$. For planets however, $\dot{M}_{max} < \dot{M}_E$. This means that $\dot{M}_{max}$ could be relevant for the gas accretion process that produced the large planets.

3. Particle acceleration

A favorite model for the core of an active galactic nucleus is that of a black hole accreting mass from an ion supported torus (Rees *et al.*, 1982). The ion supported torus consists of a hot proton gas that provides the pressure to keep the torus thick, and an electron gas, kept at a much lower temperature by strong synchrotron losses. The reason why such a construction is necessary is that in order for accretion on a reasonable time scale ($< 10^8$ yr, say) to be possible from a significant distance (1 pc, say), the accretion disk must be thick. Thin disks have much too long accretion time scales even for α's near unity. But with the inferred accretion rates (of the order $10^{-2} M_\odot \mathrm{yr}^{-1}$) a thick AGN disk is optically thin. The large difference in radiative loss rate between protons and electrons then leads to the suggested ion supported torus. The protons in this picture are the component of the fluid that is the primary receiver of the gravitational energy dissipated in the accretion process but the nature of this process is not specified in detail.

The power law spectra observed are interpreted to imply a nonthermal distribution of the radiating electrons. A promising candidate process for producing such a distribution is the first order Fermi acceleration in a shock wave. Though this accelerates primarily the protons (*e.g.* Blandford and Eichler, 1987) the required electron distributions are plausibly produced by loss processes acting on the protons (Zdiarski, 1986; Sikora *et al.*, 1987).

These two lines of thought can be connected in an aesthetically pleasing way by assuming that the accelerated protons inferred from the spectra are the same as those that support the ion torus. Since we need shock waves to accelerate the protons, we go on and assume that the accretion takes place by spiral shocks.

How can this idea be tested? First, the accretion must be efficient (fast) enough. For the (thick) conditions to which it is applied here, spiral shock accretion is indeed efficient (in terms of the effective α, Eq.8). Second, we must demonstrate that the relatively weak spiral shocks can accelerate protons to a sufficiently flat spectrum, such that most of the energy is put into the fastest protons (otherwise it is difficult to produce observed (photon) spectra that have most of the luminosity in the Mev range, as is the case for several sources). If the density of particles (which are assumed to be relativistic) with momentum p is a power law distribution $N \propto p^{-s}$, with some cutoff at high momentum due to loss processes, this condition is equivalent to $s \leq 2$. Finally, we must show that most of the accretion energy can be channeled into fast protons by the process. Demonstrating this last requirement is the most difficult part, and we will not address it here. Let us consider the second requirement then. Particle acceleration in a selfsimilar spiral shocked flow was calculated in Spruit (1988). The answer is that such shocks can indeed produce very flat particle spectra, in spite of their relatively low strength. The reason is that the particles pass the same shocks many times (of the order several hundred) while drifting radially from r to $r/2$ with the accretion flow. These calculations use a diffusion coefficient of the form

$$D = d\Omega r^2, \qquad 14$$

where Ω is the Kepler rotation rate $(GM/r^3)^{1/2}$, and d is taken constant. In addition, the loss of particles by diffusing vertically out of the disk is taken into account through an escape probability Q:

$$Q = D/H^2 = d(r/H)^2\Omega, \qquad 15$$

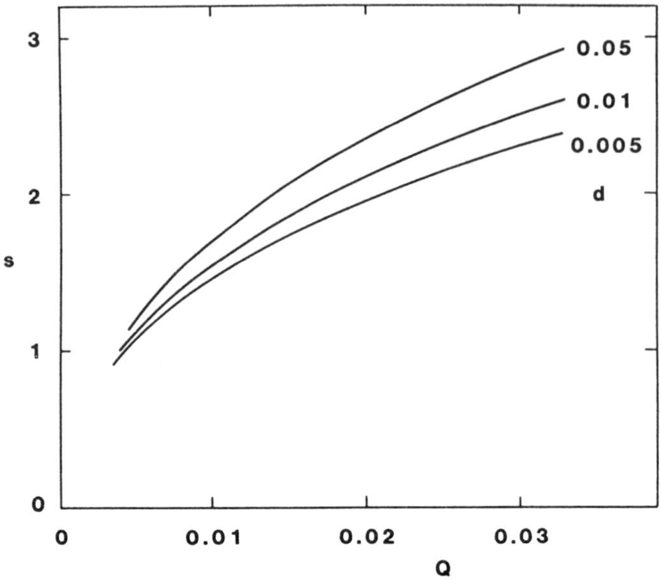

Figure 5: Fig. 5. Index s of the particle spectrum produced by an adiabatic spiral shock with $\gamma = 4/3$, as a function of the dimensionless diffusion coefficient d and the dimensionless loss parameter $Q = (r/H)^2 d$

where H is the disk thickness as before. With these radial dependences the particle distributions are power laws with index independent of position in the disk:

$$N(p, \psi, r) = r^{\nu} p^{-s} G(\psi), \quad 16$$

where p is the particle's momentum (or energy since it is assumed to be relativistic), and ψ the azimuthal coordinate (measured from the position of the shock). The radial and momentum indices ν and s are functions of the diffusion and loss parameters d and H/r, and of the flow field. For a typical flow with $\gamma = 4/3$ the momentum index is shown in Fig. 5 as a function of d and H/r. The Figure shows that a flat particle spectrum, $s \leq 2$ results when the diffusion coefficient is sufficiently low and the disk thickness sufficiently high. For $H/r = 1/3$ for example, $s < 2$ for $d < 0.005$. Abstracting a bit from the details of the numerical results, this condition can be formulated in terms of the mean free path $\lambda = 3D/c$ as

$$\lambda \lesssim 3(H/r)^2 (GMr/c^2)^{1/2} \bar{U}, \quad 176$$

where $\bar{U}$ is the radial drift speed in units of the Kepler speed, and is a few times 10^{-3} for the flows considered. We conclude that spiral shocks, in spite of their relative weakness, can do as well as or better than the strong spherically symmetric accretion shock assumed for example by Kazanas and Ellison (1986). This makes spiral shocks an attractive possibility for producing energetic particles in AGN.

References

Binney, J. and Tremaine, S.: 1987, *Galactic Dynamics*, Princeton University Press, Princeton

Blandford, R.D. and Eichler, D.: 1987, *Phys. Rep.* **154** 1

Bondi, H.: 1952, *Mon. Not. R. astron. Soc.*, **112** 195

Donner, K.J.: 1979, PhD thesis, Cambridge Univerisity, U.K.

Drury, L.O'C.: 1977, PhD thesis, Cambridge University, U.K.

Drury, L.O'C.: 1980, *Mon. Not. R. astron. Soc.*, **193** 337

Drury, L.O'C.: 1985, *Mon. Not. R. astron. Soc.*, **217** 821

Goldreich, P. and Tremaine, S.: 1980, *Ap. J.*, **241** 425

Hawley, J.F. and Blaes, O.M.: 1988, *Ap. J.*, **326** 277

Kafka, P. and Meyer, F.: 1984, in *High Energy Transients in Astrophysics*, (Proceedings of a meeting in Santa Cruz, July 11-22 1983), S.E. Woosley, ed., ISBN 0-88318-314-5

Larson, R.B.: 1988, in *The Evolution of Planetary Systems* (Proceedings of a Conference at the Space Telescope Institute, Baltimore, May, 1988), eds. H.A. Weaver, F. Paresce, L. Danly, Cambridge University Press, in press.

Lighthill, J.: 1978, *Waves in Fluids*, Cambridge University Press

Lin, D.N.C. and Papaloizou, J.: 1979, *Mon. Not. R. astron. Soc.*, **186** 799

Lin, D.N.C. and Papaloizou, J.: 1986, *Ap. J.*, **309** 846

Lynden-Bell, D.: 1974, in *Galaxies and Relativistic Astrophysics* , eds. B. Barbanis and J.D. Hadjidemetriou, Springer, Berlin, p.224, last section

Michel, F.C.: 1984, *Ap. J.*, **279** 807

Narayan, R., Goldreich, P., Goodman, J.: 1987, *Mon. Not. R. astron. Soc.*, **228** 1

Kazanas, D. and Ellison, D.C.: 1986, *Ap. J.*, **304** 178

Papaloizou, J.C.B. and Pringle, J.E.: 1984, *Mon. Not. R. astron. Soc.*, **208** 721

Papaloizou, J.C.B. and Pringle, J.E.: 1985, *Mon. Not. R. astron. Soc.*, **225** 267

Rees, M.J., Begelman, M.C., Blandford, R.D. and Phinney, E.S.: 1982, *Nature*, **295** 17

Sawada, K., Matsuda, T., Hachisu, I.: 1986, *Mon. Not. R. astron. Soc.*, **219** 75

Sikora, M., Kirk, J.G., Begelman, M.C. and Schneider, P.: 1987, *Ap. J.*, **320** L81

Spruit, H.C.: 1987, *Astr. Ap.*, **184** 173

Spruit, H.C., Matsuda, T., Inoue, M. and Sawada, K.: 1987, *Mon. Not. R. astron. Soc.*, **229** 517

Spruit, H.C.: 1988, *Astr. Ap.*, **194** 319

Zdiarski, A.A.: 1986, *Ap. J.*, **305** 45

MAGNETOHYDRODYNAMIC FLOWS IN ACCRETION DISKS AND JETS

R.V.E. Lovelace
C. M. Mobarry*
J. Contopoulos
Department of Applied Physics
Department of Physics
Department of Astronomy
Cornell University
Ithaca, New York 14853
USA

ABSTRACT. The ordered magnetic fields in accretion disks around black-holes and stars may have a direct role in the formation of the narrow jets seen in many extra-galactic radio sources and infra-red bipolar sources. An overview is given of recent theoretical work on magnetohydrodrynamics of accretion disks and on the magnetic mechanism for the formation and collimation of narrow jets.

1. INTRODUCTION

Magnetohydrodynamic (MHD) phenomena probably have an important role in many galactic and extra-galactic sources which are accreting matter. Narrow, high velocity jets of matter are observed to emanate from the compact nuclei of active galaxies and quasars (see, for example, Bridle and Eilek 1984; Begelman, Blandford, and Rees 1982). Evidence for jets on a small-scale has been found in the unusual galactic object SS 433, which exhibits Doppler-shifted emission lines due to oppositely directed high-velocity, highly collimated flows (Margon 1979, 1984). Both low-mass T-Tauri and high-mass pre-main sequence stars exhibit mass outflows with the outflows appearing bipolar or possibly unipolar (Bally and Lada 1983). In active galaxies and quasars the central source is thought to be powered by accretion onto a massive black hole with the accretion occurring through a disk of ionized plasma. Highly ordered magnetic fields are deduced to exist

*Present address: U.S. Naval Research Laboratory.

G. Belvedere (ed.), Accretion Disks and Magnetic Fields in Astrophysics, 71–90.

in some of these sources from the rapid time variations of the polarization of the optical synchrotron emission. The bipolar outflows appear to be associated with the disk accretion of gas which is only slightly ionized onto a central star which may have an appreciable magnetic moment (unlike the case of a black hole).

The origin of these astrophysical jets and outflows remains a fundamental open question. One idea for the origin of the jets is based on the occurrence of narrow, essentially empty vortex funnels in accretion flows of matter with angular momentum (Lynden-Bell 1978; Paczynski and Witta 1980; Abramowicz, Calvani, and Nobili 1980). The non-axisymmetric instabilities of disks with narrow funnels (Zurek and Benz 1986; Papaloizou and Pringle 1985) makes this idea unpromising. An alternative proposal, which is considered further here, is that the jets are magnetically driven and magnetically collimated (Lovelace 1976; Blandford 1976).

Since 1976, a variety of analytical and computer simulation models have been developed for MHD disks and jets. A non-relativistic self-similar model for jets was put forward by Blandford and Payne (1982). More recently, dynamic numerical simulations of the non- relativistic MHD equations have been undertaken to study the formation of jets (Shibata and Uchida 1985, 1986; Hanawa et al. 1988). Recent work has developed the theory of axisymmetric MHD to the point where detailed analytic calculations are feasible (Lovelace et al. 1986; Mobarry and Lovelace 1986; Camenzind 1987; Mobarry 1988). This theory has been applied to the calculation of the global electrodynamics of a viscous resistive accretion disk around a Schwarzschild black-hole with a low-density force-free plasma outside of the disk (Lovelace, Wang, and Sulkanen 1987; hereafter referred to as LWS 1987). This work showed that magnetically collimated, Poynting flux jets formed above and below the disk and acted to transport energy and angular momentum away from the disk. Time-dependent Maxwell-equations, relativistic particle-in-cell simulations of the formation of jets are being carried out by Gisler, Lovelace, and Norman (1988). An important extension of the work of LWS (1987) to the case f even field symmetry has recently been made (Wang, Sulkanen and Lovelace 1988).

In Section 2 of this work we describe some results for the even-symmetry fields inside resistive MHD disks. In Section 3, we give a brief account of a simplified model for the formation of magnetically-driven, magnetically-collimated non-relativistic MHD jets (Lovelace 1987; Mobarry 1988; Koupelis 1988; Lovelace, Mobarry and Contopoulos 1988).

2. PLASMA DYNAMICS OF ACCRETION DISKS

The basic equations for a stationary state of a rotating magnetized accretion disk are:

$$\underline{\nabla} \cdot (\rho \underline{v}) = 0 \quad , \qquad (1)$$

$$\underline{\nabla} \times \underline{B} = (4\pi/c)\underline{J} \quad , \qquad (2)$$

$$\underline{J} = \sigma(\underline{E} + \underline{v} \times \underline{B}/c) \quad , \tag{3}$$

$$\frac{\partial \underline{B}}{\partial t} = 0 = \underline{\nabla} \times (\underline{v} \times \underline{B}) + \eta \nabla^2 \underline{B} \quad , \tag{4}$$

$$\rho(\underline{v} \cdot \nabla)\underline{v} = -\underline{\nabla} p - \rho \underline{\nabla} \Phi_{grav} + \frac{1}{c} \underline{J} \times \underline{B} + \underline{f}_{vis} \quad , \tag{5}$$

$$\underline{\nabla} \cdot \underline{B} = 0 \quad . \tag{6}$$

Here, σ is the electrical conductivity, $\eta = c^2/(4\pi\sigma)$ is the magnetic diffusivity, and $\underline{f}_{vis}$ is the viscous force. Both the magnetic diffusivity and the viscosity are thought to be anomalous as a result of small-scale instabilities and turbulence (LWS 1987). Equation (1) is mass conservation, (2) is Ampère's law, (3) is Ohm's law, (4) is Faraday's law (the induction equation), (5) is the Navier-Strokes equation, and (6) is the absence of magnetic monopoles. There is, of course, an additional equation for the energy conservation which takes into account the loss of energy from the disk due to radiation and to the Poynting flux of the escaping fields (LWS 1987).

We consider axisymmetric solutions of equations (1)-(6). Axisymmetry is plausible for accretion onto a massive black hole since there is no mis-aligned magnetic dipole field and there is no nearby companion object. Therefore, the magnetic field can be written in terms of its poloidal (p) and toroidal (t) components, $\underline{B} = \underline{B}_p + B_\varphi \underline{\varphi}$ where $\underline{B}_p = (B_r, B_z)$ with $B_r = -r^{-1}(\partial\Psi/\partial z)$, with $B_z = r^{-1}(\partial\Psi/\partial r)$. The function $\Psi(r, z)$ is termed the flux-function of a magnetic field line. The poloidal projection of a magnetic field line is described by $\Psi(r, z) = \text{const}$. There are two simple field symmetries: odd z-symmetry where $\Psi(r, z) = -\Psi(r, -z)$; and even z-symmetry where $\Psi(r, z) = \Psi(r, -z)$. The odd-symmetric case is treated in LWS 1987. Figure 1 shows the nature of the fields in the even symmetry case.

For thin disks, where the vertical (z) thickness 2h is small compared with the radial distance r, the induction equation (4) can be simplified to give

$$\frac{\partial \Psi}{\partial r} + D \frac{\partial^2 \Psi}{\partial \varsigma^2} = 0 \quad , \tag{7a}$$

$$\frac{\partial}{\partial r}(uhB_\varphi) + D \frac{\partial^2 (uhB_\varphi)}{\partial \varsigma^2} = \left[\frac{d}{dr}\left[\frac{v_\varphi}{r}\right]\right]\frac{\partial \Psi}{\partial \varsigma} \quad , \tag{7b}$$

where $u \equiv -v_r > 0$ is the accretion speed, $\varsigma \equiv z/h(r)$ is a scaled vertical distance, and $D(r) \equiv \eta/(uh^2)$ has the role of a diffusion coefficient (in that dr corresponds to -dt/u). A simple example of a solution of equations (7) with even z-symmetry is:

$$\Psi = \frac{\Phi_p \cos(k\zeta)}{4\pi \cos(k)} \exp\left[-k^2 \int_r^\infty dr' D(r')\right] , \tag{8a}$$

$$B_\varphi = \frac{\Phi_p \Omega_i \sin(k\zeta)}{4\pi uh \cos(k)} \left[\delta - \frac{k\Omega(r)}{\Omega_i}\right] \exp["] . \tag{8b}$$

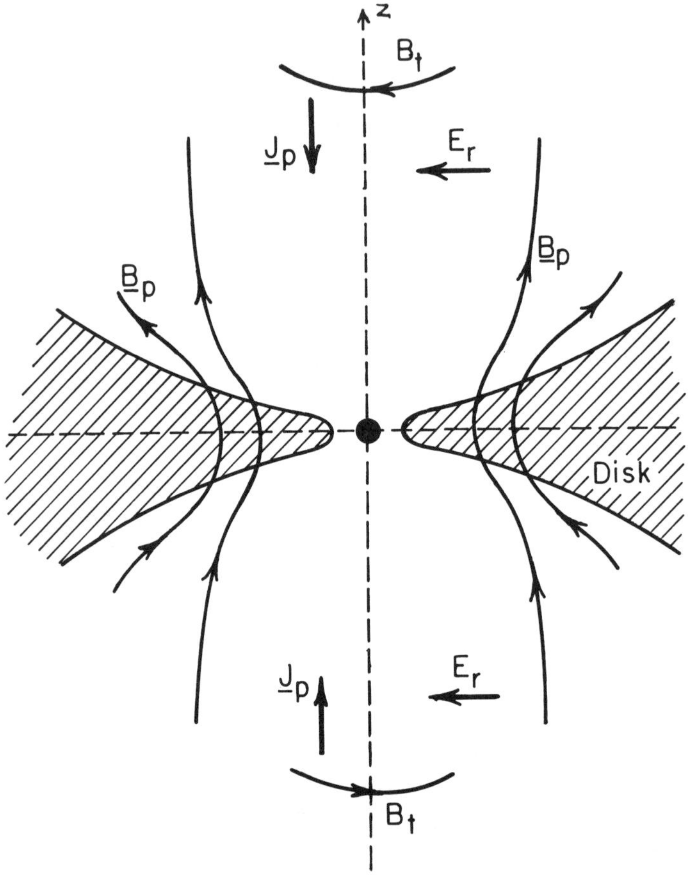

Figure 1. Magnetic field configuration with even field symmetry. The subscripts p, t, and r denote poloidal, toroidal, and radial, respectively.

(Wang, Sulkanen, and Lovelace 1988). Here, $\Phi_p/2$ is the total poloidal magnetic flux passing through the disk, k is a dimensionless constant which determines the current flow from the disk surface, and δ is a dimensionless constant which is proportional to the intrinsic component of the toroidal magnetic field (the component not due to differential rotation of the disk). Also, we have introduced $\Omega = v_\varphi/r$, which is the angular rotation rate of the disk, and $\Omega_i = \Omega(r_i)$, where r_i is the radius of the inner edge of the disk.

3. SELF-COLLIMATED MAGNETICALLY DRIVEN JETS

Here we summarize the theory of magnetically-collimated, magnetically-driven jets (Lovelace 1987; Mobarry 1988). As in Section 2 we use a cylindrical inertial coordinate system (r, φ, z) with the origin fixed at the central object. The jets are considered to be axisymmetric $(\partial/\partial\varphi = 0)$ and stationary $(\partial/\partial t = 0)$. For simplicity, the present treatment assumes non-relativistic motion in the jet and it neglects dissipative effects.

Thus, the basic equations for the jets are those of ideal magnetohydrodynamics (MHD):

$$\underline{\nabla} \cdot (\rho \underline{v}) = 0 \quad , \tag{9}$$

$$\underline{\nabla} \times \underline{B} = \frac{4\pi}{c} \underline{J} \quad , \tag{10}$$

$$\underline{E} + \underline{v} \times \underline{B}/c = 0 \quad , \tag{11}$$

$$\underline{\nabla} \times \underline{E} = 0 \quad , \tag{12}$$

$$\rho(\underline{v} \cdot \underline{\nabla})\underline{v} = -\underline{\nabla} p + \rho \underline{g} + \underline{J} \times \underline{B}/c \quad , \tag{13}$$

$$\underline{\nabla} \cdot \underline{B} = 0 \quad . \tag{14}$$

Here, $\underline{v}$ is the flow velocity in the jet, ρ is the jet mass-density, and $\underline{g}$ is the gravitational acceleration. We take $\underline{g}$ to be given approximately by the Newtonian field of the central object, that is, $\underline{g} = -\underline{\nabla}\Phi_g$, with $\Phi_g = -GM/(r^2 + z^2)^{1/2}$, where G is the universal gravitational constant, and M is the mass of the central star or black-hole.

We are primarily interested in the axial (z) variation of the jet parameters such as the axial and azimuthal flow-speeds and the magnetic field components. Therefore, we assume a simple self-similar axial variation of the transverse (r) dependences,

$$v_r(r,\ z) \approx \left[\frac{r}{a(z)}\right]\left[\frac{da(z)}{dz}\right]v_z(z) \quad , \tag{15a}$$

$$v_\varphi(r,\ z) \approx r\Omega(z) \quad , \tag{15b}$$

$$v_z(r,\ z) \approx v_z(z) \quad , \tag{15c}$$

$$\rho(r,\ z) \approx \bar{\rho}(z)f(\alpha) \quad . \tag{15d}$$

Here, $a(z)$ is the jet radius; $\Omega(z)$ is the rigid-body rotation rate of the jet at an axial distance z ; $\bar{\rho}(z)$ is the average density of the jet at the axial distance z ; $\alpha \equiv [r/a(z)]^2$ is a useful dimensionless measure of the jet radius; and $f(\alpha) \geq 0$ is a dimensionless mass-density "profile-function" normalized such that

$$\int_0^1 d\alpha\ f(\alpha) = 1 \quad .$$

The mass-flow in the jet across any z = constant surface, $\dot{M}$, is assumed constant so that

$$\dot{M} = \pi[a(z)]^2\ \bar{\rho}(z)v_z(z) = \text{const.} \tag{16}$$

Consequently, the continuity equation (9) is satisfied.

Equations (10) and (11) imply that the magnetic field obeys the "frozen-in" field equation $\underline{\nabla} \times (\underline{v} \times \underline{B}) = 0$. In turn this equation implies that

$$v_zB_r = v_rB_z \quad ,$$

or, in view of equation (7a),

$$B_r = \left[\frac{r}{a(z)}\right]\left[\frac{da(z)}{dz}\right]B_z \quad , \tag{17a}$$

and

$$\frac{\partial}{\partial r}(v_rB_\varphi) + \frac{\partial}{\partial z}(v_zB_\varphi) = r\ \frac{d\Omega}{dz}\ B_z \quad . \tag{17b}$$

Note that equation (17a) corresponds to $E_\varphi \equiv 0$, which is due to the axisymmetry. The term on the right-hand side of equation (17b) accounts for the conversion of the poloidal magnetic field into toroidal field due to the variation of the jet rotation rate with z .

In the spirit of our earlier Ansatz for the jet flow, we take

$$B_z(r, z) = B_{zi}\left[\frac{a_i}{a(z)}\right]^2 , \quad r < a(z) ,$$
$$= 0 , \quad r > a(z) , \tag{18a}$$

which satisfies $\nabla \cdot B = 0$ in view of equation (17a). Here, $B_{zi} \equiv B_z(r, z = 0)$ and $a_i \equiv a(z = 0)$ are the "initial-values" of the axial-field strength and the jet radius at the disk surface $(z = 0)$. Equation (10b) is satisfied if we take

$$B_\varphi(r, z) = \left[\frac{r}{a(z)}\right]\left[\frac{\Omega(z) - \Omega_i}{v_z(z)a(z)}\right]B_{zi}a_i^2 , \quad r < a(z) ,$$
$$= \left[\frac{a}{r}\right]\left[\frac{\Omega - \Omega_i}{v_z a}\right]B_{zi}a_i^2 , \quad r > a . \tag{18b}$$

Here, $\Omega_i \equiv \Omega(z = 0)$ is the "initial" jet rotation rate. Thus we assume in equation (18b) that $v_z(z \longrightarrow 0) \approx 0$. The poloidal field (18a) evidently corresponds to an azimuthal surface current at the jet's edge, whereas the azimuthal field (18b) corresponds to a uniform axial current density within the jet.

An equation for the conservation of axial momentum of the jet is readily derived from the general momentum conservation equation,

$$\nabla_i\left[T_p^{ij} + T_f^{ij}\right] = -\rho\nabla^j\Phi_g . \tag{19}$$

Here, $T_p^{ij} = p\delta^{ij} + \rho v^i v^j$ is the momentum flux density tensor for the particles (with p the gas pressure), $T_f^{ij} = (\delta^{ij}B^2/2 - B^iB^j)/4\pi$ is that for the magnetic field, and the right-hand side of equation (19) is the gravitational force. The contribution of the electric field is smaller by a factor $(v/c)^2 << 1$ than the magnetic terms. We integrate equation (19) over a volume of the jet contained between z and $z + dz$, as shown in Figure 2. The result is:

$$\frac{d}{d\bar{z}}\dot{P}_z = \frac{B_{\varphi *}^2}{8B_0^2}\frac{dA^2}{d\bar{z}} + \frac{F_z}{V_z} . \tag{20}$$

Here, $\bar{z} \equiv z/a_i$ is a dimensionless axial distance; $A(z) \equiv a(z)/a_i$ is a dimensionless jet radius; B_0 is a characteristic magnetic field strength taken to be $B_0 \equiv |B_\varphi(r = a_i, z \to 0)|$; $B_{\varphi *} \equiv B_\varphi(a, z)$ is the azimuthal field strength at the jet's edge; $V_z(z) \equiv v_z/v_0$ is a dimensionless axial velocity with v_0 a characteristic velocity defined below in equation (26); F_z is the axial gravitational force on the jet matter with z to $z + dz$; and $\dot{P}_z$ is a dimensionless axial momentum flow of the jet. We have

$$(a_i B_o)^2 \dot{P}_z \equiv \int_{z=\text{const.}} d^2x \left[T_p^{zz} + T_f^{zz} \right]$$

$$= \dot{M} v_z + \frac{\dot{M} kT}{m v_z} + \frac{1}{4} \int_0^a r dr \left[B_r^2 + B_\varphi^2 - B_z^2 \right] \quad . \tag{21}$$

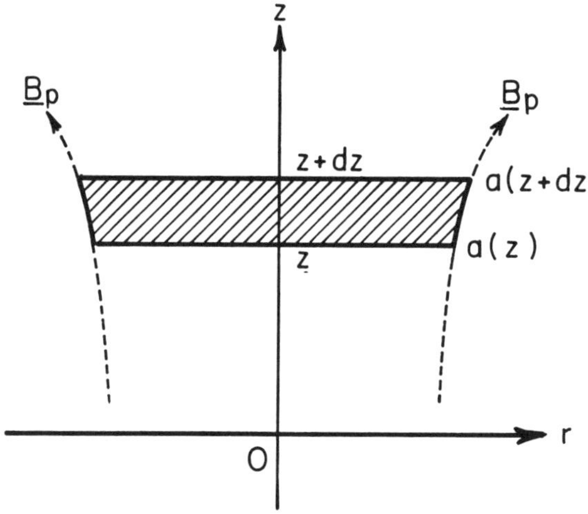

Figure 2. The integration volume used to compute momentum, angular momentum, and energy fluxes.

Here, the ideal gas law is assumed, $p = \rho kT/\mu$, where $T(z)$ is the jet temperature, k is Boltzmann's constant, and μ is the mean particle mass. $\dot{M}$ is the mass flow rate of the jet (equation 16). Notice that the gas pressure is assumed to be negligible outside of the jet (for $r > a(z)$) . However, for $r > a$ the magnetic pressure is non-zero and it contributes the first term on the right-hand side of equation (20).

From equations (18) and (21), we obtain

$$\dot{P}_z = \dot{m}\left[V_z + \frac{\tau}{V_z}\right] + \frac{b_{zi}^2}{8}\left[\frac{1}{2}\left(\frac{\omega-\omega_i}{V_z}\right)^2 + \frac{1}{2A^2}\left(\frac{dA}{dz}\right)^2 - \frac{1}{A^2}\right] , \quad (22)$$

where

$$\dot{m} \equiv \frac{\dot{M}v_0}{(a_i B_0)^2} = \text{const.}$$

is a dimensionless mass flow rate of the jet; $\tau(z) = kT(z)/(\mu v_0^2)$ is a dimensionless temperature of the jet; $b_{zi} \equiv B_{zi}/B_0$; and $\omega(z) \equiv \Omega(z)a_i/v_0$ is a dimensionless rotation rate of the jet (with $\omega_i \equiv \Omega_i a_i/v_0 = \text{const.}$).

For a jet with a uniform distribution of density as a function of r,

$$F_z = -\frac{2\delta}{A^2}\left[1 - \frac{\bar{z}}{(A^2 + \bar{z}^2)^{1/2}}\right] . \quad (23a)$$

For a jet with the density peaked at the jet's edge $(r = a(z))$,

$$F_z = -\frac{\delta\bar{z}}{(A^2 + \bar{z}^2)^{3/2}} . \quad (23b)$$

In either case,

$$\delta \equiv \frac{GM\dot{m}}{v_0^2 a_i} = \text{const.} \quad (23c)$$

is a dimensionless measure of the strength of the gravitational force.

An equation for the conservation of the total angular momentum (about the z axis) flow can be derived from the general conservation equation

$$\underline{\nabla} \cdot (\underline{r} \times (\underline{T}_p + \underline{T}_f)) = 0 . \quad (24)$$

Integration of equation (24) over the volume of the jet shown in Figure 2 gives

$$a_i^3 B_0^2 \dot{L}_z = \int d^2xr\left[T_p^{z\varphi} + T_f^{z\varphi}\right] ,$$

$$= 2\pi \int_0^a rdr\ r(\rho v_z v_\varphi - B_z B_\varphi/4\pi)$$

$$= \text{const.} ,$$

or

$$\dot{L}_z = \frac{f_p}{2} \dot{m} A^2 \omega - \frac{A^3}{4B_0^2} \int_0^1 d\alpha \alpha^{1/2} B_z B_\varphi ,$$

$$= \frac{f_p}{2} \dot{m} A^2 \omega + \frac{b_{zi}^2 (\omega_i - \omega)}{8V_z} = \text{const.} . \tag{25}$$

Here, $\dot{L}_z$ is a dimensionless angular momentum flow, and

$$f_p = 2 \int_0^1 d\alpha\ \alpha\rho \bigg/ \int_0^1 d\alpha\rho$$

is a dimensionless constant characterizing the radial distribution of density in the jet. For a uniform distribution, $f_p = 1$; for a distribution peaked near the jet edge, $f_p = 2$.

We can immediately solve equation (25) to obtain

$$\omega = \frac{\omega_0 (V_z - 1)}{A^2 V_z - y_c} . \tag{26}$$

Here, we have chosen the normalization speed v_0 so that numerator of equation (26) vanishes at $V_z = v_z/v_0 = 1$. As a result of this choice, $\dot{L}_z = (1/8)(b_{zi})^2 \omega_i$, $\omega_0 \equiv (1/4)(b_{zi})^2 \omega_i/(f_p \dot{m})$, and $y_c \equiv \omega_0/\omega_i$. Notice that the vanishing of the denominator of equation (26) at say the axial distance z_c corresponds to $v_z(z_c) = B_z(z_c)/[4\pi\rho(z_c)]^{1/2}$. This is often termed the Alfvén point of the flow (see for example, Lovelace et al. 1986). For the physical jet solutions, we find that both the numerator and the denominator of equation (26) vanish at the same axial distance.

An equation for the overall energy conservation in the jet is obtained from the general energy conservation equation

$$\underline{\nabla} \cdot (\underline{S}_p + \underline{S}_f) = 0 \quad . \tag{27}$$

Here, $\underline{S}_p = \rho\underline{v}(\underline{v}^2/2 + w + \Phi_g) + \underline{q}$ is the energy flux density carried by the fluid, including that carried by heat conduction, $\underline{q}$; $w = \int dp/\rho$ is the enthalpy; and $\underline{S}_f = (c/4\pi)\underline{E} \times \underline{B}$ is the energy flux carried by the electromagnetic field (the Poynting flux). We integrate equation (27) over the volume of the jet shown in Figure 2 to obtain

$$v_0(a_iB_0)^2\dot{E} = 2\pi \int_0^a rdr\ \rho v_z\left[\frac{1}{2}v_r^2 + \frac{1}{2}(r\Omega)^2 + \frac{1}{2}v_z^2 + w + \Phi_g\right]$$

$$+ 2\pi \int_0^a rdr\ q_z$$

$$+ \frac{c}{2}\int_0^a rdr\ E_rB_\varphi = \text{const.} \ , \tag{28a}$$

where $\dot{E}$ is the dimensionless energy flow across a $z = \text{const.}$ surface. Using the perfect conductivity, $E_r = (v_zB_\varphi - v_\varphi B_z)/c$, neglecting the heat conductivity, and using equations (18), we obtain

$$\dot{E} = \dot{m}\left[\frac{V_z^2}{2} + \frac{f_p}{4}V_z^2\left[\frac{dA}{d\bar{z}}\right]^2 + \frac{f_p}{4}(A\omega)^2 + \frac{w}{v_0^2} + \frac{\langle\Phi_g\rangle}{v_0^2}\right]$$

$$+ \frac{\omega_i(\omega_i - \omega)b_{zi}^2}{8V_z} = \text{const.} \quad . \tag{28b}$$

For a uniform density distribution,

$$\langle\Phi_g\rangle = -\frac{2GM}{a_iA}\left[\left[1 + \frac{\bar{z}^2}{A^2}\right]^{1/2} - \frac{\bar{z}}{A}\right] \quad , \tag{29a}$$

whereas for distribution peaked at the jet's edge,

$$\langle\Phi_g\rangle = -\frac{GM}{a_i}(A^2 + \bar{z}^2)^{-1/2} \quad . \tag{29b}$$

Notice that the final term in equation (28b) is positive if $\omega(z) < \omega_i$; that is, the Poynting flux is outward if the angular rotation of the jet decreases from its initial value. Notice that the magnetic field transport of angular momentum [the second term in equation (25)] is also outward if $\omega(z) < \omega_i$.

An equation for the conservation of internal energy in the jet is obtained from the first law of thermodynamics

$$\rho\left[v_z \frac{d}{dz}\right]e - \frac{p}{\rho}\left[v_z \frac{d}{dz}\right]\rho = -\frac{d}{dz} q_z ,$$

or

$$\rho T\left[v_z \frac{d}{dz}\right]s = -\frac{d}{dz} q_z . \quad (30)$$

Here, $e = (kT/m)(\gamma - 1)^{-1}$ is the internal energy per unit mass, and $s = (k/(\gamma - 1))\ell n(P/\rho^\gamma)$ is the entropy per unit mass. The heat flux-density is $q_z = -\kappa(dT/dz)$, where κ is the heat conductivity. We consider here only the limiting cases: (a) $\kappa \longrightarrow 0$ so that $q_z = 0$ (the adiabatic case); or (b) $\kappa \longrightarrow \infty$ so that T = constant (the isothermal case). In case (a), equation (28b) holds and equation (30) implies that s = const. or,

$$\tau^{1/(\gamma-1)} V_z A^2 = \text{const.} . \quad (31)$$

In the isothermal case (b), q_z is indeterminate so that equation (28b), which neglects q_z , can be expected to be only approximately correct (at best for $\tau \longrightarrow 0$) .

In the adiabatic case, the physical state of the jet at an axial distance z is completely specified by the "state vector" $\underline{U}(z) = [V_z(z), A(z), \omega(z), \tau(z)]$. The above equations (20), (26), (28b), and (31) allow the determination of $\underline{U}$ as a function of z . However, rather than using the energy conservation equation (28), we have found it useful to instead use a radial virial equation (Lovelace 1987; Mobarry 1988). The first step in deriving the virial equation is to consider the radial component of the Euler equation (13),

$$\frac{\rho r v_z}{a} \frac{d}{dz}\left[v_z \frac{da}{dz}\right] = -\frac{\partial}{\partial r}\left[p + \frac{B_z^2}{8\pi}\right]$$
$$- \frac{1}{8\pi r^2} \frac{\partial}{\partial r}(r^2 B_\varphi^2)$$
$$+ \frac{1}{4\pi} B_z \frac{\partial}{\partial z} B_r$$

$$+ \frac{\rho v_\varphi^2}{r}$$

$$- \rho \frac{\partial}{\partial r} \Phi_g \quad . \tag{32}$$

Multiplication of this equation by $2\pi r^2 dr$ and integration from $r = 0$ to the jet's edge, $r = a(z)$, gives

$$\left[\frac{f_p}{2}\right] \dot{m}A \frac{d}{d\bar{z}}\left[V_z \frac{dA}{d\bar{z}}\right] = \frac{2\dot{m}\tau}{V_z} + \frac{b_{zi}^2}{4A^2}$$

$$- \frac{A^2 B_{\varphi *}^2}{4B_0^2}$$

$$+ \frac{b_{zi}^2}{8} \frac{d}{d\bar{z}}\left[\frac{1}{A} \frac{dA}{d\bar{z}}\right]$$

$$+ \left[\frac{f_p}{2}\right] \frac{\dot{m}(A\omega)^2}{V_z}$$

$$- \frac{\dot{m}}{V_z} \left\langle r \frac{\partial \Phi_g}{\partial r} \right\rangle \quad . \tag{33}$$

Equation (33) is also valid in the isothermal case. The left-hand -side of equation (33) represents the radial force due to curvature of the jet's edge. On the right-hand-side, the different terms in the order given represent: the outward thermal pressure force; the outward force due to the poloidal magnetic field; the confining force due to toroidal magnetic field; the force due to the curvature of the poloidal magnetic lines; the outward centrifugal force due to the jet's rotation; and finally the confining force due to the gravitational field of the central object. For a jet with a uniform distribution of density as a function of r ,

$$\left\langle r \frac{\partial \Phi_g}{\partial r} \right\rangle = \frac{2GM}{v_0^2 a_i A^2} \left[(A^2+\bar{z}^2)^{1/2} + \frac{\bar{z}^2}{(A^2+\bar{z}^2)^{1/2}} - 2\bar{z} \right] \quad . \tag{34a}$$

For a density profile peaked at the jet's edge,

$$<r \frac{\partial\Phi_g}{\partial r}> = \frac{GM}{v_0^2 a_i} \frac{A^2}{(A^2+\bar{z}^2)^{3/2}} \quad . \tag{34b}$$

In the adiabatic case, we have shown (Mobarry 1988) that equation (33) can be derived directly from equations (20, (26), (28b) and (31). Thus, the approach we have taken has been that of solving equations (20), (26), (31), and (33) and then checking to verify that $\dot{E}$ of equation (28b) is a constant. Only a small number of jet solutions have been obtained in this way (with all of the terms in equation (33) retained) owing the necessity for using a shooting method to get through Alfvén point where $V_z = 1$. From these solutions, we have found that a good approximation to the 'exact' jet solutions can be obtained by dropping the z-derivative terms in the virial equation (33). The simplified virial equation is:

$$0 = \frac{2\dot{m}\tau}{V_z} + \frac{b_{zi}^2}{4A^2} - \frac{b_{zi}^2(\omega-\omega_i)^2}{4V_z^2}$$

$$+ \frac{f_p}{2}\frac{\dot{m}(A\omega)^2}{V_z} - \frac{\dot{m}}{V_z} <r \frac{\partial\Phi_g}{\partial r}> \quad . \tag{35}$$

This equation is also valid in the isothermal case.

For $z \rightarrow 0$, we assume $V_z \rightarrow 0$ and an isothermal equation of state with $\tau = \tau_i$. Notive that by definition, $A(z \rightarrow 0) \rightarrow 1$, and that $\omega \approx \omega_i[1 - V_z(1 - 1/y_c)]$ for $V_z << 1$. Thus the initial radial equilibrium (equation 35) implies that

$$0 = 2\tau_i + \left[\frac{f_p}{2}\right]\omega_i^2 - <r \frac{\partial\Phi_g}{\partial r}>_{z=0} \quad . \tag{36}$$

For a jet density profile peaked near the jet's edge and for a negligible jet temperature, $\tau_i << \omega_i^2$, equation (36) gives an initial jet rotation rate ω_i equal to the Keplerian rotation rate of the disk at the initial jet radius, a_i. Our earlier choice of the normalization field strength $B_0 = |B_\varphi(a_i, z = 0)|$ implies that

$$b_{zi} = \frac{B_{zi}}{B_0} = \left[1 + (1 + 16f_p\dot{m}\omega_i^2)^{1/2}\right] \Big/ (2\omega_i) \quad . \tag{37}$$

Thus, equation (36) can be used to determine ω_i assuming that τ_i is known. In turn, equation (37) can be used to determine b_{zi} for a given value of $\dot{m}$ (and f_p).

We solve for the jet state vector $\underline{U} = [V, A, \omega, \tau]$ as follows: for specificity, we have so far assumed that for $\bar{z} = 0$ to $\bar{z} = 1.0$ the jet is isothermal with $\tau = \tau_i$ in all cases. For $\bar{z} > 1.0$ we have analyzed both adiabatic and isothermal cases. For any value of $\bar{z}$, equation (26) gives $\omega = \omega(V_z, A)$. Therefore, in either the isothermal $(\tau = \tau_i)$ or the adiabatic case $[\tau = \tau(V_z, A)]$, the momentum conservation equation (20) can be written as

$$D_V \frac{dV_z}{d\bar{z}} = D_A \frac{dA^2}{d\bar{z}} + \frac{F_z(\bar{z}, A)}{V_z} \quad , \tag{38}$$

where D_V and D_A are algebraic functions of V_z and A which follow from equation (20). In turn, the $\bar{z}$ derivative of the simplified virial equation (35) can be written as

$$0 = B_V \frac{dV_z}{d\bar{z}} + B_A \frac{dA^2}{d\bar{z}} + C \quad , \tag{39}$$

where B_V, B_A, and C are functions of V_z, A, and $\bar{z}$. From equations (38) and (38), we obtain

$$\frac{dA^2}{d\bar{z}} = -\,(B_V F_z/V_z + D_V C)/(B_V D_A + B_A D_V) \quad . \tag{40}$$

This equation and the virial equation (35) allow us to solve for $V_z(\bar{z})$ and $A(\bar{z})$ (and thus the jet state vector) as a function of $\bar{z}$. In the isothermal case, equation (35) is a cubic in V_z and this we solve algebraically. In the adiabatic case, (35) is solved numerically for V_z using a standard root-solving technique. Thus, with V_z a known function of A and $\bar{z}$, equation (40) can be readily integrated outward from $\bar{z} = 0$.

The dimensionless parameters which enter equations (35) and (40) are $\dot{m}$, δ, and τ_i. For given $\dot{m}$ and τ_i, we find jet solutions only for $\delta \le \delta_{max}(\dot{m}, \tau_i)$. As δ decreases from δ_{max}, the asymptotic jet speed and radius, $V_z(\infty)$ and $A(\infty)$ decrease. As m increases with τ_i fixed, we find that δ_{max} decreases. In fact, we find jet solutions only for $\dot{m} \leq \dot{m}_{max}(\tau_i)$. For example, for $\tau_i = 10^{-2}$, $\dot{m}_{max} \approx 7.6 \times 10^{-2}$.

In all of the jet solutions obtained so far the magnetic field remains dynamically important as $z \longrightarrow \infty$. At $z \longrightarrow \infty$, typically 10 to 20% of the total energy flow $(\dot{E})$ is in axial kinetic energy, 15 to 30% of $\dot{E}$ is in rotary kinetic energy, and the remainder is in the Poynting flux (assuming $\tau_i << 1$). Evoluation to an entirely kinetic energy jet is of course possible in non-ideal (resistive) MHD where the magnetic flux can leak out of the jet (Mobarry 1988).

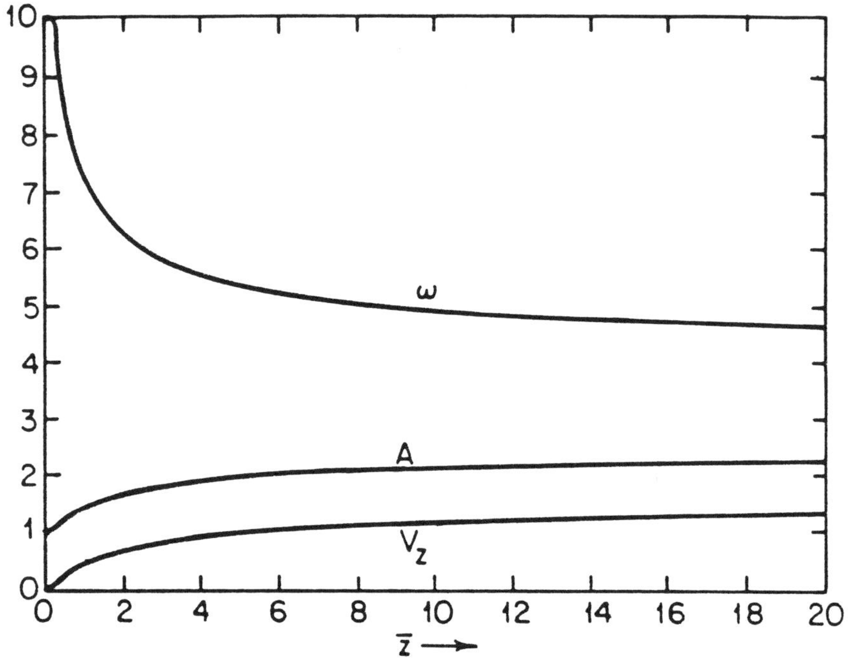

Figure 3. Axial variation of jet parameters for a sample case as discussed in the text.

Figure 3 shows the axial dependence of V_z, A, and ω for a sample case where $\dot{m} = 2.25 \times 10^{-2}$, $\delta = 6 \times 10^{-2}$, $\tau_i = 10^{-2}$ (for $\bar{z} \geq 0$), and where the density is peaked near the jet's edge. For this case, $\omega_i \approx 1.63$, $b_{zi} \approx 0.831$, $V_z(\infty) \approx 1.44$, $A(\infty) \approx 2.33$, $\omega(\infty)/\omega_i \approx 0.693$, $\dot{E}(\infty) \approx 0.143$, $\dot{E}(\infty)/\dot{L}_z \approx 1.02$, and the distance where $V_z = 1$, the Alfvén point, is $\bar{z}_c \approx 6.0$. Because τ is much less than unity, the heat flux represents only a small contribution to the total energy flow $\dot{E}$. In fact, we find that $\dot{E}$ of equation (28b) is approximately constant for $\bar{z} > 1$. The sonic point, where $V_z = \tau^{1/2}$, is at $\bar{z} \approx 0.349$. About 13% of the total energy flow at ∞ is in the axial kinetic energy, 16% is in rotational kinetic energy, and most of the remainder, 71%, is in the Poynting flux. The ratio of the axial jet velocity at ∞ to the disk velocity at a_i ,

$$\frac{v_z(\infty)}{\Omega_i a_i} = \frac{V_z(\infty)}{\omega_i} , \qquad (41)$$

is about 0.883. Figure 4 shows the three-dimensional appearance of the jet.

There are jets from both sides of the disk, and both carry away energy, angular momentum, and mass from the disk. The dimensional energy flow in say the +z jet (at ∞) can be written as

$$\dot{E}_{jet} = a_i^3 B_0^2 \Omega_i [\dot{E}(\infty)/\omega_i] , \qquad (42a)$$

where $\dot{E}$ and ω_i are our prior dimensionless variables. Notice that $a_i^3 B_0^2 \Omega_i$ is roughly the luminosity which would result if the magnetic field energy within a distance a_i of the central object were lost in a time $1/\Omega_i$. For the case of Figure 3, $\dot{E}(\infty)/\omega_i \approx 8.77 \times 10^{-2}$. With the dimensional angular momentum flow in the +z jet denoted $\dot{L}_{zjet}$, we have

$$\frac{\dot{E}_{jet}}{\dot{L}_{jet}} = \Omega_i \left[\frac{\dot{E}(\infty)}{\omega_i \dot{L}_z}\right] , \qquad (42b)$$

where the factors in the parentheses are prior dimensionless variables. For the case of Figure 3, the factor in square brackets is ≈ 0.626. The dimensional mass flow in the +z jet can be written as

$$\dot{M}_{jet} = \frac{a_i B_0^2}{\Omega_i} [\dot{m}\omega_i] . \qquad (42c)$$

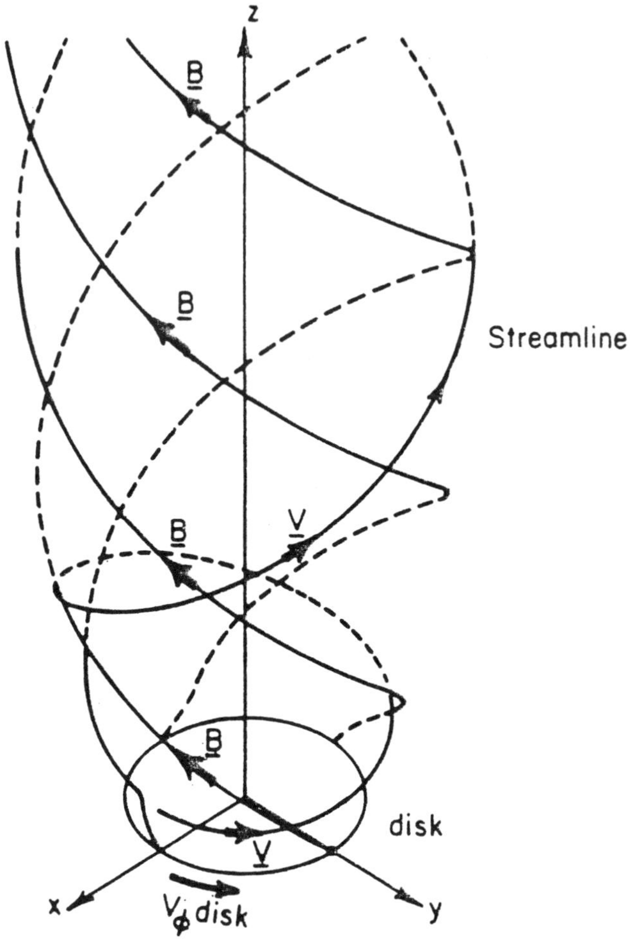

Figure 4. Three-dimensional view of four magnetic fieldlines and one streamline (all on the jet surface) for the same case as shown in Figure 3.

For the case of Figure 3, the factor in square brackets is $\approx 3{,}67 \times 10^{-2}$.

Only a brief comment is given here on the problem of matching the $\pm z$ jets to a MHD disk of Section 2 (see LWS 1987; and Lovelace,

Mobarry, and Contopooulos 1988). For a range of disk radii, say, r , to $r_2 > r_1$, the jets can remove essentially all of the excess angular momentum of a Keplerian accretion disk. Furthermore, most (if not all) of the gravitational accretion energy released in the disk (r_1 to r_2) can be carried by the jets for a particular value of $\xi \equiv 2\dot{M}_{jet}/\dot{M}_{accr}(r_2) < 1$, where $\dot{M}_{accr}(r_2)$ is the rate at which mass is accreted across $r = r_2$.

ACKNOWLEDGEMENTS

One of the authors (R.L.) is grateful to Professor Gaetano Belvedere for hospitality at the Noto Conference. We thank Carol Cook for skillful preparation of this manuscript.

REFERENCES

Abramowicz, M. A., Calvani, M., and Nobili, L. (1980). Ap. J. **242**, 772.

Bally, J., and Lada, C. J. (1983). Ap. J. **265**, 824.

Begelman, M. C., Blandford, R. D., and Rees, M. J. (1984). Rev. Mod. Phys. **56**, 255.

Blandford, R. D. (1976). M.N.R.A.S. **176**, 465.

Blandford, R. D., and Payne, D. G. (1982). M.N.R.A.S. **199**, 883.

Bridle, A. H., and Eilek, J. A. (1984). "Physics of Energy Transport in Extragalactic Radio Sources". (Greenbank: NRAO).

Camenzind, M. (9187). Astron. and Astrophys. **184**, 341.

Gisler, G., Lovelace, R.V.E., and Norman, M. L. (1988). To appear in Ap. J.

Hanawa, T., Kamahori, H., Maruyama, T., and Shibata, K. (1988). To appear in Publ. Astron. Soc. Japan, **40**, No. 6.

Koupelis, T. (1988). Ph.D. Thesis, University of Rochester.

Lovelace, R.V.E. (1976). Nature **262**, 649.

Lovelace, R.V.E. (1987). In Magnetic Fields in Extra-Galactic Objects (E. Asseo and D. Gresillon, Editors) (Editions de Physique, 91944 Les Vlis Cedex, France), p. 223.

Lovelace, R.V.E., Mehanian, C., Mobarry, C. M., and Sulkanen, M. E. (1986). Ap. J. Suppl. **62**, 1.

Lovelace, R.V.E., Mobarry, C. M., and Contopoulos, J. (1988). Preprint, Cornell University.

Lovelace, R.V.E., Wang, J.C.L., and Sulkanen, M. E. (1987). Ap. J. **315**, 504.

Lynden-Bell, D. (1978). Phys. Scripta **17**, 185.

Margon, B. (1984). Ann. Rev. Astron. and Astrophys. **22**, 507.

Margon, B., Ford, H. C., Katz, J. I., Kwitter, K. B., Ulrich, R. K., Stone, R.P.S., and Klemala, A. (1979). Ap. J. **230**, L41.

Mobarry, C. M. (1988). Ph.D. Thesis, Cornell University.

Mobarry, C. M., and Lovelace, R.V.E. (1986). Ap. J. **309**, 455.

Paczynski, B., and Witta, P. S. (1980). Astron. and Astrophys. **88**, 23.

Papaloizon, J.C.B., and Pringle, J. E. (1985). M.N.R.A.S. **213**, 899.

Shibata, K., and Uchida, Y. (1985). Publ. Astron. Soc. Japan **37**, 31.

Shibata, K., and Uchida, Y. (1986). Publ. Astron. Soc. Japan **38**, 631.

Wang, J.C.L., Sulkanen, M. E., and Lovelace, R.V.E. (1988). Preprint (for Ap. J.).

Zurek, W. H., and Benz, W. (1986). Ap. J. **308**, 123.

SLENDER FLUXTUBES IN ACCRETION DISKS

A. ACHTERBERG
Sterrekundig Instituut
Postbus 80.000
NL-3508 TA Utrecht
The Netherlands

ABSTRACT. Some of the aspects of the dynamics of slender magnetic fluxtubes in a differentially rotating thin accretion disk are considered. The effect of Coriolis- and buoyancy forces on the added mass, and of radiation pressure are considered. The results are applied to a linear stability analysis for a circular fluxtube in the equatorial plane of the disk.

1. Introduction.

Magnetic fields can possibly play an important role in the dynamics of accretion disks. Large-scale magnetic fields rooted in a disk can aid in the formation of jets or centrifugally driven winds [1,2]. Small-scale magnetic fields are probably amplified by dynamo action in the differentially rotating disk [3,4] , and can provide a mean shear stress $t_{r\varphi} = \langle B_r B_\varphi \rangle / 4\pi$ which is needed for efficient mass- and angular momentum transport in the disk [5,6].

In the case of the Sun, where the magnetic field is presumably generated in the convection zone, the magnetic field is spatially very intermittent, and is concentrated in individual thin fluxtubes or bundels of fluxtubes [7]. Numerical simulations of MHD turbulence also seem to show a tendency for the magnetic field to concentrate in fluxropes or -sheets [8 ,9].

It therefore seems relevant to consider the dynamics of slender fluxtubes (or fluxsheets) in a differentially rotating disk, in effect extending the work that has been done in the case of slender fluxtubes in the Solar convection zone [10-12]. The differences with the Solar case are twofold: first of all, the Coriolis force and the (effective) buoyancy force in a Keplerian thin disk can be of the same order-of-magnitude ; secondly, radiation pressure can be important, in particular in the inner region of the disk. This leads to a number of complications which will be considered in detail below.

2. Basic equations.

The basic equations which replace the full set of M.H.D. equations in the slender fluxtube approximation can be derived using two simplifying assumptions:

G. Belvedere (ed.), Accretion Disks and Magnetic Fields in Astrophysics, 91–97.

1) The dynamics of the fluxrope as a whole is adequately described by the motion of its axis (or a central plane in the case of a sheet)

2) The lateral force balance (perpendicular to the axis) reduces to pressure equilibrium with the surroundings.

These two assumptions limits this description to long-wavelength , low-frequency phenomena with $\lambda \gg d$ and $t \gg \max(d/a\ ,\ d/s)$. Here λ is the typical scale on which a fluxtube changes diameter or direction (measured along its axis) , d the typical transverse dimension of the tube, and t the timescale of the motion. Also a and s are the Alfvén- and sound speed respectively, defined as usual by $a \equiv B/\sqrt{(4\pi\rho)}$ and $s \equiv \sqrt{(\gamma P/\rho)}$ in terms of the magnetic fieldstrength B, density ρ and pressure P in the fluxrope.

Also one must assume a high magnetic Reynolds-number. With a magnetic diffusivity $\eta \equiv c^2/4\pi\sigma$ (σ is the conductivity), a fluxrope with diameter d will loose its identity in a timescale $\tau_D = d^2/\eta$, while the typical dynamical timescale equals $\tau_e = \lambda/V$, with V the velocity of the tube axis. One must have $\tau_D \gg \tau_e$, or equivalently $d^2V/\lambda\eta = (d/\lambda)^2 R_m \gg 1$, with $R_m \equiv \lambda V/\eta$ the magnetic Reynoldsnumber. The slender fluxtube approximation is not suitable for the description of phenomena in which involve differential motion of elements *within* the tube at a given cross-section.

In the ideal MHD limit, the magnetohydrodynamic equations can be derived from a variational principle [13] $\delta S = 0$, which can be transplanted to the slender fluxtube approximation [11,14]. The path of the fluxtube is parametrized as $\mathbf{X}(l,\ t)$ with l a coordinate along the tube axis. The unit vector along the tube axis is given by $\mathbf{t} = \partial\mathbf{X}/\partial l$ and I will assume that the magnetic field is directed along that axis: $\mathbf{B} = B(l)\mathbf{t}$.

It is convenient to define a Lagrangean "reference state" $\mathbf{X}_o(l_o, t_o)$ in which the tube has a cross-section A_o , mass per-unit-length $\mu_o = \rho_o A_o$, magnetic field B_o and magnetic flux $\Phi = B_o A_o$. The deformation of the tube from that reference state defines a displacement vector $\boldsymbol{\xi}(l,t) \equiv \mathbf{X}(l,t) - \mathbf{X}_o(l_o,t_o)$, a stretching parameter $\lambda \equiv (\partial l/\partial l_o)$ and a dimensionless density variable $\sigma \equiv \rho/\rho_o$. Since the mass in a section of the tube is conserved $\Delta m = \rho A \Delta l$ = constant, and using the conservation of the magnetic flux Φ through the tube one can derive the following relations, assuming an ideal gas in the tube with adiabatic heat ratio γ:

$$B(l,t) = B_o(l_o,t_o)\sigma\lambda \quad , \quad P(l,t) = P_o(l_o,t_o)\sigma^\gamma ,$$
$$\lambda^2 = \left(\frac{\partial\mathbf{X}}{\partial l_o}\cdot\frac{\partial\mathbf{X}}{\partial l_o}\right) = 1 + 2\mathbf{t}_o\cdot\boldsymbol{\xi}' + \boldsymbol{\xi}'\cdot\boldsymbol{\xi}' \qquad (\boldsymbol{\xi}' \equiv \partial\boldsymbol{\xi}/\partial l_o). \tag{1}$$

The pressure equilibrium between the slender fluxtube and the surrounding medium $P + B^2/8\pi = P_e$ can be written in a form which relates σ to λ and $\boldsymbol{\xi}$ *given* the external pressure distribution $P_e(\mathbf{x})$, which I will assume to be static:

$$P_o\sigma^\gamma + \frac{B_o^2}{8\pi}\sigma^2\lambda^2 = P_e(\mathbf{X}_o + \boldsymbol{\xi}). \tag{2}$$

The pressure P_e only refers to the external pressure which would also be present *in absense of the tube*. Pressure fluctuations induced by the motion of the slender fluxtube through the external medium will be dealt with separately below.

The action S which yields the equations of motion for the slender fluxtube by Hamiltons principle $\delta S = 0$ can be written, using l_o as independent variable:

$$S = \int dt \int dl_o\, \mu_o \Big(\Lambda_i (l_o, t, \boldsymbol{\xi}) + \Lambda_e (l_o, t, \boldsymbol{\xi}) \Big). \qquad (3)$$

The "internal" Lagrangean density Λ_i has the usual form of " kinetic energy - potential energy". Referring all quantities to a frame-of-reference which rotates with angular velocity $\boldsymbol{\Omega}$, Λ_i is given by (Achterberg, in preparation):

$$\Lambda_i = \frac{1}{2}\left(\frac{d\mathbf{X}}{dt}\right)^2 + \left(\boldsymbol{\Omega} \times \mathbf{X} \cdot \frac{d\mathbf{X}}{dt}\right) - \frac{P_o \sigma^{\gamma-1}}{(\gamma - 1)\rho_o} - \frac{B_o^2}{8\pi \rho_o} \sigma \lambda^2 - U_{eff}(\mathbf{X}). \qquad (4)$$

Here $U_{eff} \equiv \varphi(\mathbf{X}) - (\boldsymbol{\Omega} \times \mathbf{X})^2/2$ is the effective potential in the rotating frame due gravity and centrifugal forces. The "external" Lagrangean density Λ_e has two contributions: one due to the (static) pressure P_e [15, 16] and one due to the pressure changes in the surrounding medium induced by the motion of the tube, so-called "added mass" effects [14,17,18]:

$$\Lambda_e = - \frac{P_e(\mathbf{X})}{\rho_o \sigma} + \Lambda_{am} . \qquad (5)$$

The calculation of Λ_{am} involves solving the equation for the *induced* pressure distribution δP_e around the tube and calculating the net force on the tube per unit length $\mathbf{f}_e = -\int d\mathbf{S}\, \delta P_e$ associated with it. Here $d\mathbf{S}$ is an element of the exterior tube surface. Details of these calculations will be reported elsewhere.

In what follows I will consider only small displacements from an equilibrium for the tube in a stability analysis. One can expand the action S in powers of $|\boldsymbol{\xi}|$, i.e. $S = \sum S_n$ with $S_n = O(\xi^n)$. Each S_n is given by Eqn. (3), with $\Lambda = \Lambda_i + \Lambda_e$ replaced by a Λ_n of order ξ^n. Using $\boldsymbol{\xi}(l_o, t)$ as the dependent variable, the variation $\delta S_n = 0$ leads to Euler-Lagrange equations of order ξ^{n-1} [15]:

$$\frac{\delta \Lambda_n}{\delta \boldsymbol{\xi}} \equiv \frac{\partial \Lambda_n}{\partial \boldsymbol{\xi}} - \frac{d}{dt}\left(\frac{\partial \Lambda_n}{\partial \dot{\boldsymbol{\xi}}}\right) - \frac{d}{dl_o}\left(\frac{\partial \Lambda_n}{\partial \boldsymbol{\xi}'}\right) = 0. \qquad (6)$$

Here I have defined $\dot{\boldsymbol{\xi}} \equiv \partial \boldsymbol{\xi}/\partial t$ and $\boldsymbol{\xi}' \equiv \partial \boldsymbol{\xi}/\partial l_o$.

The zero-order density Λ_o is independent of $\boldsymbol{\xi}$ and can be discarded. The variation $\delta \Lambda_1/\delta \boldsymbol{\xi}$ yields the equations describing the equilibrium (reference-) state, while the so-called second variation $\delta \Lambda_2/\delta \boldsymbol{\xi}$ yields a set of equations linear in $\boldsymbol{\xi}$ which govern the evolution of linear perturbations.

3. Stability of an equatorial fluxring in a thin disk.

As an illustrative (but not necessarily realistic) example, I will consider the stability of a slender fluxring in the form of a torus, located in the equatorial plane of a thin disk. The major radius of the torus equals R, the minor radius equals d, which satisfies $d \ll h(R)$, where $h(R) \ll R$ is the thickness of the disk. The material in the disk itself has density ρ_e, pressure P_e and rotates with a rotation law $\Omega_e = \Omega(r/R)^{-q}$. Since the disk is assumed to be thin, it must be a Keplerian disk, and q will be put equal to $q = 3/2$ in the end. It is assumed that the accretion velocity satisfies $v_r \ll s_e \ll v_\varphi = \Omega_e r$ so that it can be neglected to lowest order. I will work in a frame which locally corotates with the disk material at $r = R$ by choosing $\mathbf{\Omega} = \Omega_e \mathbf{e}_z$. I will employ cylindrical coordinates r, φ and z with the equatorial plane at $z = 0$. The unperturbed fluxring has a constant mass per unit length $\mu_o = \pi d^2 \rho_o$, magnetic field B_o and the material in the fluxring has velocity $v_\varphi = \Omega_o R$ as seen from an *inertial* frame. The length coordinate along the unperturbed axis satisfies $dl_o = R d\varphi_o$ and $\mathbf{t}_o = \mathbf{e}_\varphi$.

I will consider perturbations which displace the axis of the fluxring from the unpertubed position $r = R$, $\varphi = \varphi_o$, $z = 0$ to $r = R + \xi_r$, $\varphi = \varphi_o + \xi_\varphi/R$, $z = \xi_z$. Expanding the Lagragian Λ in powers of ξ and considering $\delta\Lambda_1/\delta\boldsymbol{\xi}$ leads to an equation describing the force-balance of the equilibrium:

$$\Omega_o^2 = \frac{1}{r}\frac{\partial \varphi}{\partial r}\bigg|_R + \frac{B_o^2}{4\pi\rho_o R^2} + \frac{\rho_e}{\rho_o}\left(\Omega_e^2 - \frac{1}{r}\frac{\partial\varphi}{\partial r}\bigg|_R\right) \qquad (7)$$

The last term in Eqn. (7) is due to the radial pressure gradient in the disk material $\partial P_e/\partial r = \rho_e(\Omega_e^2 - \partial\varphi/r\partial r)$. In the thin-disk approximation this gradient can be neglected, and material rotates at the Keplerian frequency: $\Omega_e = \Omega_K = \partial\varphi/r\partial r = \surd(GM/r^3)$ with M the mass of the central object. Defining an Alfvén frequency by $\Omega_A^2 \equiv B_o^2/4\pi\rho_o R^2$ one finds $\Omega_o^2 = \Omega_K^2(R) + \Omega_A^2$. The material in the fluxring must rotate faster than Keplerian in order to overcome the radial magnetic tension $B^2/4\pi R$. In the thin disk approximation one must have $B^2/8\pi \ll \rho\Omega_K^2 r^2$, so seen from the corotating frame the material in the fluxring rotates with an angular velocity $\delta\Omega = \Omega_o - \Omega_K \approx \Omega_A^2/2\Omega_K \ll \Omega_A$, i.e. strongly sub-Alfvénic ($R\delta\Omega \ll a$).

Turning now to the stability of this equilibrium through the second variation $\delta\Lambda_2/\delta\boldsymbol{\xi}$, one has to take into account the induced pressure δP_e due to the motion of the fluxring perpendicular to its axis. Usual calculations of δP_e assume that the surrounding medium is not stratified, and has no (shear) flow. The influence of both flow and stratification can be seen by considering linear perturbations in the Boussinesq approximation $\nabla\cdot\mathbf{u} = 0$. Introducing local cartesian coordinates $x = r - R$, $y = R\varphi$ and $z = z$, one can derive an equation in the WJKB approximation:

$$[(\partial_t^2 + \varkappa^2)\partial_z^2 + (\partial_t^2 + N^2)(\partial_x^2 + \partial_y^2)]\partial_t \delta P_e + q\Omega_e(\partial_t^2 + N^2)\partial_x\partial_y \delta P_e = 0. \qquad (8)$$

Here $\varkappa$ is the epicyclic frequency defined by $\varkappa^2 = (4 - 2q)\Omega_e^2$ ($= \Omega_K^2$ for a thin disk), and $N^2 \equiv -g(d\ln\rho_e/dz + g/s_e^2)$ is the Brunt-Väisälä frequency with $g = \partial\varphi/\partial z$. In the thin disk approximation the effective gravitational acceleration in the corotating frame is equal to $\mathbf{g} = -g\mathbf{e}_z \approx -\Omega_K^2 z\mathbf{e}_z$. For rigid rotation ($q \equiv -d\ln\Omega_e/d\ln r = 0$) this equation describes mixed inertial-gravity waves in a rotating stratified atmosphere [19,20].

In the present case of a slender fluxtube aligned with the y-axis, one has ∂_x, $\partial_z = O(1/d) \gg \partial_y = O(1/\lambda)$. Solving Eqn.(8) for δP_e around the tube, assuming an harmonic time-dependence $\boldsymbol{\xi} \propto \exp(-i\omega t)$ with boundary condition that the external fluid velocity $\mathbf{u}$ matches the perpendicular component of the motion of the tube ($\mathbf{u}\cdot\mathbf{n} = \mathbf{n}\cdot d\boldsymbol{\xi}/dt$ where $\mathbf{n}$ is the outward normal of the exterior surface of the tube) and calculating the force per unit length on the tube associated with δP_e one finds:

$$\mathbf{f}_e \equiv \delta\Lambda_{am}/\delta\boldsymbol{\xi} = \mu_e\left((\omega^2 - N^2)(\omega^2 - \varkappa^2)\right)^{1/2}\boldsymbol{\xi}_\perp. \qquad (9)$$

Here $\mu_e \equiv \pi d^2 \rho_e$ is the mass per-unit-length of an equal volume of surrounding fluid, and $\boldsymbol{\xi}_\perp$ the component of the displacement vector perpendicular to the axis of the tube. In the high-frequency limit ($|\omega| \gg \varkappa, N$) Eqn. (9) reduces to the standard result [17,18] $\mathbf{f}_\perp = \mu_e \omega^2 \boldsymbol{\xi}_\perp = -\mu_e \partial^2 \boldsymbol{\xi}_\perp / \partial t^2$ (mass loading). The influence of the flow and stratification shows up as resonances at $|\omega| = \varkappa$, $|\omega| = N$ where the resistive force excerted on the tube by the external medium vanishes, essentially since the external medium is driven at one of its natural frequencies.

For small perturbations with $\boldsymbol{\xi} \propto \exp(im\varphi - i\omega t)$ one finds a set of equations of the form:

$$\begin{bmatrix} D_{rr} & iD_{r\varphi} & 0 \\ -iD_{r\varphi} & D_{\varphi\varphi} & 0 \\ 0 & 0 & D_{zz} \end{bmatrix} \begin{bmatrix} \xi_r \\ \xi_\varphi \\ \xi_z \end{bmatrix} = 0. \tag{10}$$

To lowest order in Ω_A/Ω_K the components of the dispersion tensor $\mathbf{D}$ read:

$$D_{rr} = \omega^2 + \eta\omega\sqrt{(\omega^2 - \Omega_K^2)} - m^2\Omega_A^2 + (1+2\eta)\Omega_K^2 + \frac{a^2}{a^2+s^2}\Omega_A^2 ,$$

$$D_{\varphi\varphi} = \omega^2 - \frac{m^2 s^2}{a^2+s^2}\Omega_A^2 , \quad D_{r\varphi} = -2\sqrt{(\Omega_K^2+\Omega_A^2)}\,\omega - \frac{m(a^2+2s^2)}{a^2+s^2}\Omega_A^2 \tag{11}$$

$$D_{zz} = \omega^2 + \eta\omega\sqrt{(\omega^2 - \Omega_K^2)} - m^2\Omega_A^2 + (\eta-1)\Omega_K^2 .$$

Here I have used that $N = 0$ at $z = 0$ and defined a density contrast $\eta \equiv \rho_e/\rho_o$. The eigenfrequencies follow from $\det\mathbf{D} = (D_{rr}D_{\varphi\varphi} - D_{r\varphi}^2)D_{zz} = 0$. The first factor describes oscillations in the disk plane. In the long-wavelength limit ($m^2\Omega_A^2 \ll \Omega_K^2$) these satisfy $\omega(\omega^2 + \eta\omega\sqrt{(\omega^2 - \Omega_K^2)} + (2\eta-3)\Omega_K^2) = 0$. This corresponds to modified epicyclic motions with a solution $|\omega| = \Omega_K$ when $\eta \approx 1$. In a typical accretion disk, heat exchange between the slender fluxtube and its surroundings is fast, so the tube is isothermal ($T_o \approx T_e$). Zero-order pressure equilibrium (Eqn. 3) then gives the density contrast as $\eta = 1 + 1/\beta$, with $\beta \equiv 8\pi P_o/B_o^2$. So $\eta \approx 1$ when the magnetic field is weak. Then the tube behaves just like the surrounding fluid. In the short-wavelength limit ($m^2\Omega_A^2 \gg \Omega_K^2$) the r and φ components of the motion decouple into radial *Alfvén waves* with dispersion relation $\omega^2 = k^2a^2/(1+\eta)$ and internal (compressive) *tube waves* [21] with dispersion relation $\omega^2 = k^2c_T^2$. Here $k \equiv m/R$ and $c_T^2 \equiv a^2s^2/(a^2+s^2)$. The Alfvén waves bend the fluxtube without changing its cross-section. The tube waves lead to a periodic bulging of the tube (like a string of sausages) leaving the tube axis in place. The mass-loading of the tube by the surrounding medium shows up in the factor $1+\eta$ in the Alfvén waves with a phase velocity $a/\sqrt{(1+\eta)} = B_o/\sqrt{(4\pi(\rho_o+\rho_e))}$.

Motions out of the disk plane (described by $D_{zz} = 0$) lead to a slow, unstable buoyant mode for long wavelengths with $\mathrm{Re}(\omega) = 0$ and $\mathrm{Im}(\omega) \equiv \Gamma \ll \Omega_K$ given by:

$$\frac{\Gamma}{\Omega_K} = -\frac{\eta}{2} + \frac{1}{2}\left(\eta^2 + 4\left(\eta - 1 - \frac{m^2\Omega_A^2}{\Omega_K^2}\right)\right)^{1/2} . \tag{12}$$

This mode is unstable when $(\eta-1)\Omega_K^2 > m^2\Omega_A^2$. For short wavelengths ($|\omega| \gg \Omega_K$) one finds an Alfvén wave modified by buoyancy with the same instability condition :

$$\omega^2 = \left(m^2\Omega_A^2 - (\eta-1)\Omega_K^2\right)/(1+\eta). \tag{13}$$

4. Influence of radiation pressure on buoyancy.

In the inner regions of accretion disks, radiation pressure becomes important and may in fact dominate. This does influence the effective buoyancy force felt by a magnetic fluxtube. One has to distinguish between the case where the fluxtube is optically thick ($\tau \equiv \sigma_T n_o d \gg 1$) and when the tube is optically thin ($\sigma_T n_o d \leq 1$). Here σ_T is the Thomson cross-section, n_o the electron density and I have assumed that electron scattering is the dominant source of opacity.

In the optically thick case the (static) buoyancy force can be calculated using pressure balance including radiation pressure between the tube and its surroundings:

$$\rho_o \mathcal{R} T_o + \frac{1}{3} a_r T_o^4 + \frac{B_o^2}{8\pi} = \rho_e \mathcal{R} T_e + \frac{1}{3} a_r T_e^4 . \quad (14)$$

Heat conduction across the field proceeds mainly through radiative conduction. The typical timescale for equilibrating the internal temperature T_o and T_e equals:

$$t_{eq} = \frac{\pi d^2 \rho_o \mathcal{R} T_o}{(\gamma-1) K_r |\Delta T|} \approx \left(\frac{\tau d}{c}\right)\left(\frac{P_g}{P_r}\right)\left(\frac{T}{\Delta T}\right) . \quad (15)$$

Here $K_r \equiv (4c/3\sigma_T n) a_r T_o^3$ is the radiative conduction coefficient, $P_g = \rho \mathcal{R} T_o$ the gas pressure, $P_r = a_r T_o{}^4/3$ the radiation pressure and $\Delta T = T_e - T_o$. Since buoyant rise of fluxtubes proceeds on a timescale of order Ω_K^{-1} or longer, the maximum temperature difference between the tube and surroundings follows from $t_{eq} \approx \Omega_K^{-1}$:

$$\frac{|\Delta T|}{T} \approx \left(\frac{\tau d \Omega_K}{c}\right)\left(\frac{P_g}{P_r}\right) \approx \tau \left(\frac{d}{h}\right)\left(\frac{s}{c}\right)\left(\frac{P_g}{P_r}\right) \quad (16)$$

Here $h \approx s\Omega_K$ is the thickness of the disk. Generally $s \ll c$ and $d \ll h$ and the temperature difference is small so that the tube can be considered as isothermal. In that case, the radiation pressure drops out of the balance equation (14) and one finds the same result for the static buoyance force per unit length as in the absense of radiation pressure:

$$\mathbf{f}_{buoy} = \pi d^2\left(-\nabla_\perp P_e + \rho_o \mathbf{g}_\perp\right) = \pi d^2 (\rho_o - \rho_e)\mathbf{g}_\perp = -\beta^{-1}\mu_o \mathbf{g}_\perp . \quad (17)$$

Here P_e is the *total* (gas- + radiation) pressure in the surrounding medium, $\beta \equiv 8\pi\rho_o \mathcal{R}T/B_o^2$ and the subscript $\perp$ denotes components perpendicular to the tube axis (i.e. $\mathbf{g}_\perp = \mathbf{g} - (\mathbf{t}\cdot\mathbf{g})\mathbf{t}$). I have used the hydrostatic balance $\nabla P_e = \rho_e \mathbf{g}$ in the corotating frame, and $\rho_e - \rho_o = \beta^{-1}\rho_o$ which follows from (14) with $T_e = T_o \equiv T$.

In the optically thin case, the effective buoyancy force consists of two separate contributions: one due to the external gas pressure P_{eg} combined with the weight of the tube, and one due to the radiation momentum intercepted per unit time by the tube from the radiation flux $\mathbf{F}$ in the external medium. Assuming that the emission from the tube itself is isotropic, and does not lead to a net force, one finds for the effective buoyancy force perpendicular to the tube axis:

$$\mathbf{f}_{buoy} = \pi d^2\left(-\nabla_\perp P_{eg} + \rho_o \mathbf{g}_\perp\right) + \pi d^2 n_o \sigma_T \left(\frac{\mathbf{F}_\perp}{c}\right) . \quad (18)$$

Assuming that the disk itself is optically thick, the diffusion approximation holds for the radiation flux: $\mathbf{F} = -(c/\sigma_T n_e)\nabla P_{er}$ in terms of the external radiation pressure. The equation of hydrostatic equilibrium in the corotating frame (where

$\mathbf{g} \approx -\Omega_K^2 z \mathbf{e}_z$) can be written as: $\nabla(P_{er} + P_{eg}) - \rho_e \mathbf{g} = 0$. These relations give a net force per unit length on the tube:

$$\mathbf{f}_{buoy} = \pi d^2 \left(\rho_o - \rho_e\right)\left(\mathbf{g}_\perp - \frac{1}{\rho_e} \nabla_\perp P_{er}\right) \approx -\frac{\alpha \mu_o}{\beta(1 + \alpha)} \mathbf{g}_\perp . \qquad (19)$$

Here I have assumed a completely ionized plasma so that $n_e/n_o = \rho_e/\rho_o$, defined $\alpha \equiv P_{eg}/P_{er}$ and used the isothermal approximation $T_o = T_e$.

Compared with the optically thick case (Eqn. 17) one sees that the net buoyancy force per unit length on an optically thin tube is strongly reduced when the radiation pressure is much larger than the gas pressure ($\alpha \ll 1$). In the opposite limit ($\alpha \gg 1$) the result coincides with that of the optically thick case. So in the inner regions of accretion disks, where radiation pressure is dominant, optically thin fluxtubes escape much slower from the disk into a disk corona then the optically thick tubes. In principle, this effect can be incorporated into the Lagrangian formalism developed above.

Acknowledgement.

This work was supported by the Netherlands Organization for Scientific Research (NWO) through a Huijgens Fellowship.

References.

1. Blandford, R.D. and Payne, D.: 1982, *Mon. Not. R. astr. Soc.* **199**, 883.
2. Pudritz, R.E. and Norman, C.A.: 1986, *Astrophys. J.* **301**, 571.
3. Pudritz, R.E.: 1981, *Mon. Not. R. astr. Soc.* **195**, 881.
4. Pudritz, R.E.: 1981, *Mon. Not. R. astr. Soc.* **195**, 897.
5. Eardly, D.M. and Lightman, A.P.: 1975, *Astrophys. J.* **200**, 187.
6. Coroniti, F.V.: 1981, *Astrophys. J.* **244**, 587.
7. Zwaan, C.: 1987, *Ann. Rev. Astron. Astrophys.* **25**, 83.
8. Meneguzzi, M., Frish, U. and Pouquet, A.: 1981, *Phys. Rev. Lett.* **47**, 1981.
9. Ting, A.C., Matthaeus, W.H. and Montgomery, D.: 1986, *Physics of Fluids* **29**, 3261.
10. Spruit, H.C.: 1981, *Astron. Astrophys.* **98**, 155.
11. Spruit, H.C.: 1981, *Astron. Astrophys.* **102**, 129.
12. Spruit, H.C. and ,Van Ballegooijen, A.A.: 1982 **106**, 158.
13. Newcomb, W.A.: 1962, *Nucl. Fusion Suppl.* part 2 , p. 451.
14. Achterberg, A.: 1988, *Astron. Astrophys.* **191**, 167.
15. Soper, D.E.: 1976, *Classical Field Theory*, J. Wiley & Sons, New York, p. 66.
16. Sakurai, T.: 1976, *Publ. Astron. Soc. Japan* **28**, 177.
17. Landau, L.D. and Lifshitz, E.M.: 1959, *Fluid Mechanics*, Pergamon Press, Oxford, p. 31.
18. Newman, J.N.: 1980, *Marine Hydrodynamics*, MIT Press, Cambridge Ma. p. 135.
19. Lesieur, M.: 1987, *Turbulence in Fluids*, Nijhoff Publ. , p. 30.
20. LeBlond, P.H. and Mysak, L.A.: 1978, *Waves in the Ocean*, Elsevier, Amsterdam, p. 45.
21. Roberts, B. and Webb, A.R.: 1978, *Solar Phys.* **56**, 5.

MAGNETIC FIELDS IN THE ACCRETION DISKS OF CATACLYSMIC VARIABLES

A. A. VAN BALLEGOOIJEN
Harvard-Smithsonian Center for Astrophysics
60 Garden Street
Cambridge, MA 02138
U.S.A.

ABSTRACT. The radial transport of magnetic flux in CV accretion disks is considered. Turbulent diffusion in a disk with finite outer radius R_d leads to decay of large-scale magnetic fields, as magnetic flux can leak out at the outer edge of the disk. Numerical computations of the decay rate and magnetic structure are presented. It is shown that the decay time is of order $R_d/(\alpha c_s)$, significantly shorter than the accretion time, $R_d/|v_r|$ (here c_s is the sound speed at the outer edge of the disk and α is the viscosity parameter). Centrifugally driven winds may be possible from the outer parts of the disk provided there exist efficient dynamo processes which regenerate the field.

1. Introduction

IUE observations of cataclysmic variables (CVs) show that the resonance lines of ions such as C IV, Si IV and N V exhibit emission lines and/or P Cygni profiles, indicative of mass loss in the form of a high-velocity wind (e.g. Krautter et al. 1981; Cordova and Mason 1982, 1985; Guinan and Sion 1982; King et al. 1983; Greenstein and Oke 1982; Verbunt et al. 1984; Mauche and Raymond 1985, 1987). The wind is believed to originate from the accretion disk which surrounds the white dwarf in these binary systems. Interpretation of the observations in terms of spherically symmetric wind models (Drew and Verbunt 1985; Drew 1987; Mauche and Raymond 1987) suggest that the wind is accelerated rather gradually, reaching its terminal velocity, $v_\infty \sim$ 5000 km s^{-1}, at $r >> 10R_{wd}$. It is not clear how the winds are accelerated. As in O-type stars, the driving force may be due to radiation pressure acting on ions with strong UV and EUV resonance lines. Alternatively, the wind may be due to magnetic fields permeating the disk: provided the field is sufficiently strong, it may cause the plasma above and below the disk to corotate with the location on the disk where the magnetic field is anchored, and the wind may be centrifugally accelerated like 'beads' on a rigid wire (Henriksen and Rayburn 1971). In the present paper we focus on the second possibility. Centrifugally driven winds in CVs were considered previously by Koen (1986) and Cannizzo and Pudritz (1988).

Centrifugal acceleration from a Keplerian disk is possible only if the field lines at the disk are tilted away from the rotation axis by an angle $\theta > 30^\circ$ (Blandford and Payne 1982). This condition arises because the outflow occurs along the field lines,

G. Belvedere (ed.), Accretion Disks and Magnetic Fields in Astrophysics, 99–106.

hence the component of centrifugal force along the field lines must be large enough to overcome the component of gravity. Whether the condition $\theta > 30°$ is satisfied in any particular case depends on the boundary conditions for the magnetic field at the disk and on the dynamics of the wind. Blandford and Payne (1982) considered a hydrodynamical model of a centrifugally driven jet, and they develop a self-similar model in which the field component perpendicular to the disk varies with radius as $B_z \sim r^{-5/4}$. The choice -5/4 for the exponent is based on the assumption that the rate of mass loss from each decade of disk radius is independent of radius. This choice is rather arbitrary: it is not a necessary consequence of the similarity assumption because the mass loss rate is not a dimensionless quantity. Instead, one must consider the question: What determines the distribution $B_z(r)$ in an accretion disk?

In this paper I consider the physical processes by which magnetic flux is transported in an accretion disk. Magnetic flux is transported radially inward by the accretion flow, while turbulent motions within the disk lead to radially outward diffusion. For simplicity I will assume that magnetic flux transport occurs *passively*, i.e., feedback of the magnetic field on the accretion flow is neglected. Furthermore, all fields are assumed to be anchored within the disk, and the magnetic fields of the white dwarf primary and the red dwarf secondary are neglected. Finally, I assume that the angular momentum loss via the wind is small compared to the 'viscous' transport of angular momentum within the disk. The structure of the disk may then be determined from the standard thin disk model (Lynden-Bell and Pringle 1974). The magnetic diffusivity η within the disk is written as the turbulent viscosity ν times a magnetic Prantl number P. Since both ν and η are due to turbulent processes, I assume that P is a constant of order unity.

In a self-similar model such as that of Blandford and Payne (1982) the disk is assumed to extend over a large range of radii. However, the disks of CVs have outer radii R_d which are only 20 - 40 times the inner radius R_{wd}. Therefore, in the development of a centrifugally driven wind model for CVs the finite size of the disk must be taken into account. From the standpoint of wind theory we are particularly interested in stationary solutions of the flux transport problem in which accretion of B_z balances magnetic diffusion. However, I will show that in a disk with finite outer radius such a balance is not possible: magnetic flux escapes at the outer edge of the disk, causing the magnetic field to decay with time. In section 2 I present a numerical model of flux transport in accretion disks, and I determine the decay rate using an eigenvalue method. The implications for centrifugally driven wind models and for dynamo models of accretion disks (e.g. Pudritz 1981a,b) are discussed in section 3.

2. Magnetic Flux Transport

The time-dependent distribution of magnetic field $\mathbf{B}(\mathbf{r},t)$ in an accretion disk is determined by the induction equation,

$$\frac{\partial \mathbf{B}}{\partial t} = \nabla \times (\mathbf{v} \times \mathbf{B} - \eta \nabla \times \mathbf{B}), \tag{1}$$

where $\mathbf{v}(\mathbf{r})$ is the velocity and $\eta(\mathbf{r})$ is the magnetic diffusivity. Introducing a cylindrical coordinate system (r, ϕ, z) centered on the white dwarf and assuming axisymmetry, the z component of equation (1) yields

$$\frac{\partial B_z}{\partial t} = \frac{1}{r}\frac{\partial}{\partial r}\left[r\left(v_z B_r - v_r B_z \right) + r\eta\left(\frac{\partial B_z}{\partial r} - \frac{\partial B_r}{\partial z} \right) \right]. \tag{2}$$

In the following I derive an equation for $B_{z,0}(r,t)$, the value of B_z at the mid-plane of the disk ($z = 0$). If the disk thickness $h(r)$ is small compared with r, $v_z/v_r \sim dh/dr << 1$; hence, unless $B_r >> B_z$ the term with v_z in equation (2) may be neglected. Then the first term in square brackets in (2) represents the influx of B_z due to accretion, while the second term represents the radial outflow due to magnetic diffusion. There are two contributions to the diffusion term. One is due to $\partial B_z/\partial r$, which gives equation (2) the character of a diffusion equation. The other is due to $\partial B_r/\partial z$, which is associated with the annihilation of oppositely directed fields B_r above and below the disk. Now assume that $B_r(r,z,t)$ varies more or less linearly with z within the disk, and let $B_{r,s}(r,t)$ denote the value of B_r at the disk surface ($z = h$). Then $\partial B_r/\partial z \approx B_{r,s}/h$.

To obtain an equation in closed form, we must express $B_{r,s}(r,t)$ in terms of $B_{z,0}(r,t)$. To achieve this, I approximate the field above and below the disk as a potential field:

$$\mathbf{B} = -\nabla\Phi. \tag{3}$$

The effect of the wind is simulated by requiring that the field is open to the interstellar medium. Such open fields may be computed by neglecting the disk thickness ($h \rightarrow 0$) and treating the disk as a monopole source with a magnetic charge per unit area equal to $B_{z,0}/2\pi$. Then the magnetic potential is given by

$$\Phi(r,z,t) = \frac{1}{2\pi}\int_0^{2\pi}\int_0^{R_d} \frac{B_{z,0}(r',t)\, r'\, dr'\, d\phi}{\sqrt{(r - r'\cos\phi)^2 + (r'\sin\phi)^2 + z^2}}, \tag{4}$$

from which a monopole-like field $\mathbf{B}(r,z,t)$ may be derived. To obtain a physically acceptable solution (with $\nabla\cdot\mathbf{B} = 0$ everywhere), we simply reverse the sign of $\mathbf{B}$ in the region $z < 0$. The component $B_{r,s}$ just above the disk ($z \downarrow 0$) may be obtained from equations (3) and (4):

$$B_{r,s}(r,t) = -\frac{\partial}{\partial r}\left[\int_0^{R_d} K\left(\frac{r'}{r}\right) B_{z,0}(r',t)\, dr' \right], \tag{5}$$

where the kernel $K(x)$ is defined by

$$K(x) = \frac{x}{2\pi}\int_0^{2\pi} \frac{d\phi}{\sqrt{1 - 2x\cos\phi + x^2}}. \tag{6}$$

Using expression (5), equation (2) yields the following integro-differential equation for $B_{z,0}(r,t)$:

$$\frac{\partial B_{z,0}}{\partial t} = \frac{1}{r}\frac{\partial}{\partial r}\left[-rv_r B_{z,0} + r\eta\frac{\partial B_{z,0}}{\partial r} + \frac{r\eta}{h}\frac{\partial}{\partial r}\left(\int_0^{R_d} K\left(\frac{r'}{r}\right) B_{z,0}(r',t)\, dr' \right) \right]. \tag{7}$$

Note that equation (7) is linear, so that the time-dependent solution is a linear superposition of modes of the following form:

$$B_{z,0}(r,t) = f(r)\, e^{-\gamma t} . \tag{8}$$

To determine the function $f(r)$ and decay rate γ for a mode, I write equation (7) in finite-difference form and treat it as an eigenvalue problem. In this paper I only consider the mode with the lowest decay rate. Once $f(r)$ is known, the potential field $\mathbf{B}(r,z)$ above the disk may be computed using equations (3) and (4). Details of the numerical method are discussed in the Appendix.

In the following I present some results obtained with a simple model of the accretion disk. The disk thickness $h(r)$ is approximated by

$$h(r) = \xi R_d \left(\frac{r}{R_d} \right)^k , \tag{9}$$

where ξ is the opening angle at the outer edge of the disk ($\xi << 1$), and k describes the shape of the disk. The turbulent viscosity is written as

$$\nu(r) = \alpha \Omega(r) h^2(r), \tag{10}$$

where $\Omega(r)$ is the Keplerian rotation rate ($\Omega \sim r^{-3/2}$) and α is the viscosity parameter. For the accretion velocity I assume:

$$v_r(r) = -1.5 \frac{\nu(r)}{r}, \tag{11}$$

which is appropriate for $r >> R_{wd}$ if tidal torques are neglected (see Lynden-Bell and Pringle 1974). The magnetic diffusivity is written as

$$\eta(r) = P\, \nu(r). \tag{12}$$

where P is the magnetic Prantl number. To avoid numerical problems in the inner part of the disk the functions $h(r)$, $\eta(r)$ and $v_r(r)$ are modified for $r < r^* = 0.1 R_d$: $h(r)$ and $\eta(r)$ are set equal to their values at r^*, while $v_r(r) = v_r(r^*)(r/r^*)$. The disk radius R_d is used as the unit of length in the model, and the decay rate γ is expressed in units of $\Omega(R_d)$, the angular velocity at the outer edge of the disk.

Table I. $c(\xi, P)$

ξ	P			
	2.0	1.0	0.5	0.25
0.10	2.75	2.61	2.36	1.98
0.05	2.53	2.46	2.32	2.10
0.02	2.38	2.35	2.29	2.19
0.01	2.32	2.30	2.28	2.22

Decay rates and magnetic field structures were computed for different combinations of the parameters ξ, P and k. Since both $v_r(r)$ and $\eta(r)$ depend linearly on α, the decay rate γ also scales linearly with α; hence, only one value of α needs to be considered (e.g., $\alpha = 1$). I find that the decay rate increases with increasing disk thickness ξ and Prantl number P. Therefore, γ is written as

$$\gamma = \alpha \xi P c(\xi, P), \tag{13}$$

where $c(\xi,P)$ is a slowly varying function. Results for the decay rates in disks with a linearly varying disk thickness ($k = 1$) are presented in Table I. For thin disks $c(\xi,P)$ is nearly independent of Prantl number, whereas for thicker disks $c(\xi,P)$ increases somewhat with increasing P; within the range $0.01 < \xi < 0.1$ and $0.25 < P < 2$ the approximation $c = 2.3$ is good to 20 %. I also computed decay rates for disks with constant thickness, $k = 0$. The results are remarkably similar, with γ only 10 % to 25 % larger than in the case $k = 1$.

The distribution of magnetic flux at the disk plane is illustrated in Figure 1, where I plot the eigenfunctions $f(r)$ for three different parameter combinations (for $k = 1$). Note that $f(r)$ vanishes at the outer edge of the disk in accordance with the imposed boundary condition (see Appendix). For $P = 1$, the transport of magnetic flux in the inner half of the disk is dominated by diffusion, causing $f(r)$ to be nearly constant; note that $f(r)$ does not depend strongly on ξ in this case. For $P = 0.25$ magnetic diffusion is reduced, and accretion produces a more peaked flux distribution.

The magnetic field structure above the disk is illustrated in Figure 2, where I plot the field lines in the (r,z) plane for the case $\xi = 0.05$, $P = 1$. The disk is located between $r = 0$ and $r = 1$ along the lower axis. Note that the magnetic field in the inner part of the disk is nearly perpendicular to the disk. In the outer part the field lines are inclined away from the rotation axis (z-axis) by an angle $\theta(r)$ which increases with increasing r; $\theta(r) = 30^\circ$ occurs at $r = 0.55$. Field-line diagrams for other values of ξ and P look qualitatively similar. For $\xi = 0.01$, $P = 1$ the point where $\theta = 30^\circ$ occurs somewhat further out ($r = 0.63$), while for $\xi = 0.05$, $P = 0.25$ it occurs further in ($r = 0.39$). The latter is a result of the more strongly peaked flux distribution at the disk mid-plane (see Fig. 1).

3. Discussion

The results on the field structure suggest that in the inner part of the disk the field lines are too closely aligned with the rotation axis for centrifugal acceleration to overcome gravity. A centrifugally driven wind may originate from the outer parts of the disk where the inclination angle θ becomes larger than 30° (Koen 1986; Cannizzo and Pudritz 1988). In this paper I did not take into account the angular momentum loss due to the wind, nor did I include tidal torques which modify the outer parts of the disk and determine where the outer edge is located (Lubow and Shu 1975; Kriz 1982). Both effects probably lead to larger accretion velocity $|v_r|$ relative to ν/r, and therefore lead to more centrally concentrated flux distributions with larger inclination angles $\theta(r)$. Hence, the radial zone over which outflow occurs would probably increase.

The present flux-transport model does not include dynamo processes by which

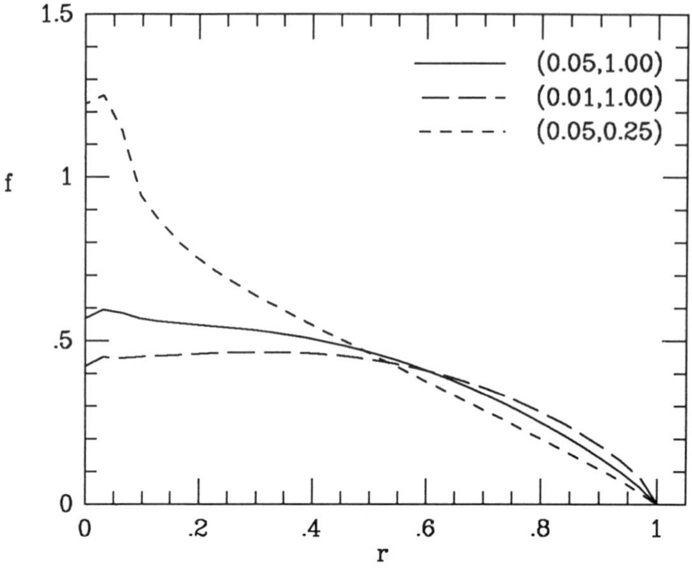

Fig. 1. Radial distribution $f\ (r\ )$ of magnetic flux in an accretion disk, for three combinations of disk thickness ξ and magnetic Prantl number P .

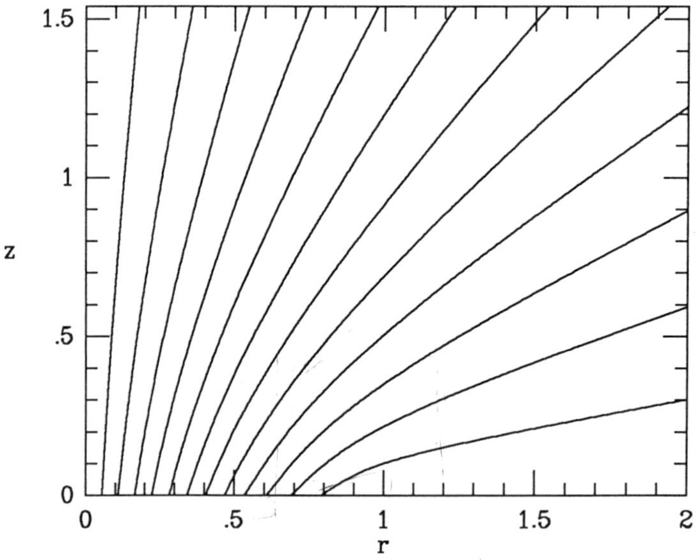

Fig. 2. Magnetic field lines for $\xi = 0.05$ and $P\ = 1$.

magnetic flux may be regenerated, and therefore the magnetic field decays with time. Using $h = c_s/\Omega$, with $c_s(r)$ the sound speed in the disk, the decay time may be written as

$$\gamma^{-1} \approx \frac{R_d}{2.3\alpha c_s(R_d)}, \tag{14}$$

where I assumed $P = 1$. This result may be compared with the accretion time scale at the outer edge,

$$\tau_a \equiv \frac{R_d}{|v_r(R_d)|} = \frac{R_d^2}{1.5\alpha h(R_d)c_s(R_d)}. \tag{15}$$

Note that the magnetic decay time is shorter than the accretion time by a factor $\sim h(R_d)/R_d$; this is a result of the fact that in a thin accretion disk the magnetic diffusion associated with $\partial B_r/\partial z$ in equation (2) is larger than that associated with $\partial B_z/\partial r$ by a factor $\sim r/h$. The decay time may increase somewhat when tidal effects and angular momentum loss via the wind are considered because this tends to concentrate the field in the inner part of the disk. Nevertheless, the relatively short decay time predicted here suggests that large-scale magnetic fields and centrifugally driven winds can be maintained only if there exists an efficient dynamo process (not included in the present model). Further work on disk dynamos is needed to determine whether such rapid regeneration is possible.

Acknowledgement: I thank Lee Hartmann and John Raymond for many stimulating discussions on accretion disk physics. This work was supported by the National Aeronautics and Space Administration under grant NAG5-87.

Appendix: Numerical Method

I introduce a uniform radial grid, $r_i = i\,\Delta r$, where $\Delta r = R_d/(N+1)$ and $i = 0, \ldots, N+1$. Let $f_i \equiv f(r_i)$. To evaluate the integral over r' in equation (7), I approximate $f(r')$ as a piecewise linear function:

$$f(r') = [(r_{j+1} - r')f_j + (r' - r_j)f_{j+1}]/\Delta r, \tag{A1}$$

for $r_j \le r' \le r_{j+1}$. Then equation (7) may be written in the following form:

$$-\gamma f_i = \frac{g_{i+1/2} - g_{i-1/2}}{r_i\,\Delta r} \qquad (i = 1, \cdots, N), \tag{A2}$$

where

$$g_{i+1/2} = -(rv_r)_{i+1/2}\frac{f_i + f_{i+1}}{2} + (r\,\eta)_{i+1/2}\frac{f_{i+1} - f_i}{\Delta r}$$
$$+ \left(\frac{r\,\eta}{h}\right)_{i+1/2} \sum_{j=0}^{N+1}\Big(W_{i+1,j} - W_{i,j}\Big)f_j. \tag{A3}$$

The matrix $W_{i,j}$ is defined by

$$W_{0,0} = W_{0,N+1} = 1/2, \tag{A4a}$$

$$W_{0,j} = 1 \quad (j=1, \cdots ,N), \tag{A4b}$$

$$W_{i,j} \equiv \int_{r_k}^{r_l} K\left(\frac{r'}{r_i}\right)\left(1 - \frac{|r' - r_j|}{\Delta r}\right)\frac{dr'}{\Delta r} \quad (i \neq 0), \tag{A4c}$$

where $k = \max(j-1,0)$ and $l = \min(j+1,N+1)$. For the boundary conditions I assume that the field becomes radial at the outer edge of the disk, $f_{N+1} = 0$, and that the net flux of B_z vanishes near the inner edge of the disk, $g_{1/2} = 0$. The latter condition yields an expression for f_0 in terms of the other f_i, and inserting this relation into (A3), equation (A2) presents an eigenvalue problem for the N unknowns f_i with $i = 1, \ldots ,$N. This eigenvalue problem is solved using standard techniques (see Press et al. 1986). The computations presented in this paper were made with $N =$ 30.

References

Blandford, R.D., and Payne, D.G. 1982, *Mon. Not. R. astr. Soc.* **199**, 883

Cannizzo, J.K., and Pudritz, R.E. 1988, *Ap. J.* **327**, 840

Cordova, F.A., and Mason, K.O. 1982, *Ap. J.* **260**, 716

Cordova, F.A., and Mason, K.O. 1985, *Ap. J.* **290**, 671

Drew, J.E. 1987, *Mon. Not. R. astr. Soc.* **224**, 595

Drew, J.E., and Verbunt, F. 1985, *Mon. Not. R. Astr. Soc.* **213**, 191

Guinan, E.F., and Sion, E.M. 1982, *Ap. J,* **258**, 217

Greenstein, J.L., and Oke, J.B. 1982, *Ap. J.* **258**, 209

Henriksen, R.N., and Rayburn, D.R. 1971, *Mon. Not. R. astr. Soc.* **152**, 323

King, A.R., Frank, J., Jameson, R.F., and Sherrington, M.R. 1983, *Mon. Not. R. astr. Soc.* **203**, 677

Koen, C. 1986, *Mon. Not. R. astr. Soc.* **223**, 529

Krautter, J., Klare, G., Wolf, B., Duerbeck, H.W., Rahe, J., Vogt, N., and Wargau, W. 1981, *Astr. Ap.* **102**, 337

Kriz, S. 1982, *Mon. Not. R. astr. Soc.* **199**, 725

Lubow, S.H., and Shu, F.H. 1975, *Ap. J.* **198**, 383

Lynden-Bell, D., and Pringle, J.E. 1974, *Mon. Not. R. astr. Soc.* **168**, 603

Mauche, C.W., and Raymond, J.C. 1985, in *Proceedings of the Ninth North American Workshop on Cataclysmic Variables*, ed. P. Szkody (Seattle: University of Washington), p. 93

Mauche, C.W., and Raymond, J.C. 1987, *Ap. J.* **323**, 690

Press, W.H., Flannery, B.P., Teukolsky, S.A. and Vetterling, W.T. 1986, *Numerical recipes* (Cambridge Univ. Press)

Pudritz, R.E. 1981a, *Mon. Not. R. Astr. Soc.* **195**, 881

Pudritz, R.E. 1981b, *Mon. Not. R. Astr. Soc.* **195**, 897

Verbunt, F., Pringle, J.E., Wade., R.A., Echevarria, J., Jones, D.H.P., Argyle, R.W., Schwarzenberg-Czerny, A., LaDous, C. and Schoembs, R. 1984, *Mon. Not. R. Astr. Soc.* **210**, 197

ACCRETION DISCS AND MAGNETIC CATACLYSMIC VARIABLES

J.P. LASOTA[1], J.M. HAMEURY[2] and A.R. KING[3]
[1] DARC, Observatoire de Paris, Section de Meudon, France
[2] DAEC, Observatoire de Paris, Section de Meudon, France
[3] Astronomy Department, University of Leicester, U.K.

ABSTRACT. The white dwarf magnetic field is likely to prevent the formation of an accretion disc in short period cataclysmic variables. Accretion discs are not present in synchronous systems ("polars") and could be absent in non-synchronous systems ("intermediate polars") for orbital periods < 5hr. This result is a consequence of the hypothesis that white dwarf magnetic moments in polars and intermediate polars are $\mu \approx 10^{33} - 10^{34}$ Gcm^3: this hypothesis is strongly supported by recent evolutionary considerations.

1. INTRODUCTION

Cataclysmic variables (CVs) are close binary systems in which a Roche-lobe filling late type secondary transfers mass on to a white dwarf primary. Magnetic CVs (MCVs) in which the white dwarf is strongly magnetized (B $\sim 10^7$ G) can be divided into two subclasses: AM Her systems (also called "polars") in which the white dwarf spin is synchronous with the orbital motion and intermediate polars ("IPs", sometimes called DQ Her systems, although DQ Her itself is very unrepresentative of the class) where the spin period is shorter than the orbital period.

It is well known that in short period binaries (P$\leq$10 hr) the orbit (and hence the period) must decrease under the influence of angular momentum losses in order to maintain mass transfer. This results in a secular evolution from longer towards shorter periods (see e.g. King 1988). A period histogram of CVs is shown in Fig. 1., separately for MCVs and nonmagnetic systems (in total 142 objects from the September 1988 version of Ritter's catalogue). It is clear from this figure that the distribution of the *sum* of two classes of MCVs is very similar to that of nonmagnetic CVs. In particular the "period gap" between $\sim$ 2hr and $\sim$ 3hr is present in both cases and the minimum period is $\sim$ 80 min. We stress that we are dealing with distributions and statistics; for example the presence of a few objects in the middle of the period gap would change nothing in the interpretation of the period histogram. The comparison of period histograms suggests that MCVs evolve in the same way as nonmagnetic CVs, but since there is a deficiency of polars above 3hrs, and only one IP (and one controversial candidate: SW UMa) below 2hrs one may conclude (Chanmugan & Ray 1984, King, Frank & Ritter 1985) that generally IPs evolve into polars. In other words, they should be the same type of objects observed at different stages of their evolution,

G. Belvedere (ed.), Accretion Disks and Magnetic Fields in Astrophysics, 107–112.

the differences between their properties accounted for by differences in binary separations (larger at longer periods) and accretion rates (of which the *long term* ($\sim 10^8$ yr) average is higher at longer periods). In particular polars and IPs should have the same magnetic moments contrary to the "weak field hypothesis" (see e.g. Lamb & Melia 1986) according to which magnetic moments in IPs should be at least an order of magnitude lower than in polars. If this latter hypothesis were true, however, the period distribution of polars would remain unexplained: all but three polars have periods below 2hrs (the fourth object present in the histogram, V 1500 Cyg is not strictly speaking a polar because it is not completely synchronous); periods of the three above 3hrs are all below 4hrs. Any attempt to make polars to evolve faster above the period gap (Schmidt, Stockman & Grandi 1986) in order to explain their period distribution leads to a total disaster: the distribution obtained is in flagrant contradiction with observations (Hameury, King, Lasota & Ritter 1987).

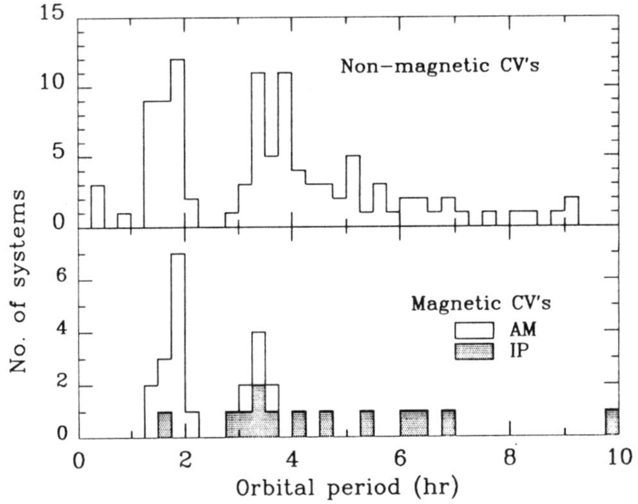

Fig. 1. *The period distribution of nonmagnetic CVs (upper figure) and magnetic CVs (lower figure).*

The period distribution puts severe constraints on possible evolutionary models and on physical properties of CVs. The fact that MCVs have essentially the same distribution as nonmagnetic systems implies that mechanisms explaining CVs evolution must apply to both type of systems. In this context it is interesting to mention evolutionary models invoking accretion discs either as affecting the angular momentum loss (Cannizzo & Pudritz 1988), or as being implicated in the creation of the period gap (Taam & McDermott 1987, but see Ritter 1988) . As we shall see accretion discs are probably absent in most MCVs, near the gap and below, so it is unlikely that they play an essential role in the secular evolution of nonmagnetic systems.

2. ACCRETION DISCS AND MAGNETIC CVs

It is generally accepted that polars do not possess accretion discs, since the magnetospheric radius of the white dwarf (for radial inflow)

$$r_\mu = 1.0 \times 10^{11} \mu_{34}^{4/7} \dot{M}_{16}^{-2/7} M_1^{-1/7} cm \tag{1}$$

(where $\mu_{33} = \mu/10^{33} Gcm^3$, $\dot{M}_{16} = \dot{M}/10^{16} gs^{-1}$ and M_1 is the primary mass in solar units) is of the same order of magnitude as the binary separation

$$a = 3.53 \times 10^{10} (M_1 + M_2)^{1/3} P_{hr}^{2/3} cm. \tag{2}$$

where P_{hr} is the orbital period in hours.

The case of IPs is less obvious because of their larger binary separations and higher accretion rates. A general criterion for disc formation is that the mass stream from the secondary can orbit freely the accreting primary and interact with itself building up a disc by spreading via viscosity. A *sufficient* condition for disc formation is that the stream's distance of closest approach R_{min} to the white dwarf should exceed the magnetospheric radius R_{mag}. Because of geometrical effects $R_{mag} \neq r_\mu$. Hameury, King & Lasota (1986, hereafter HKL1) find that $R_{mag} = 0.37 r_\mu$. Fig. 2. shows the comparison between R_{min} and R_{mag} from HKL1. Since above the period gap the secular average is $\sim 1 - 3 \times 10^{17}$ gs^{-1} it is clear that discs are unlikely to form for periods < 5hrs.

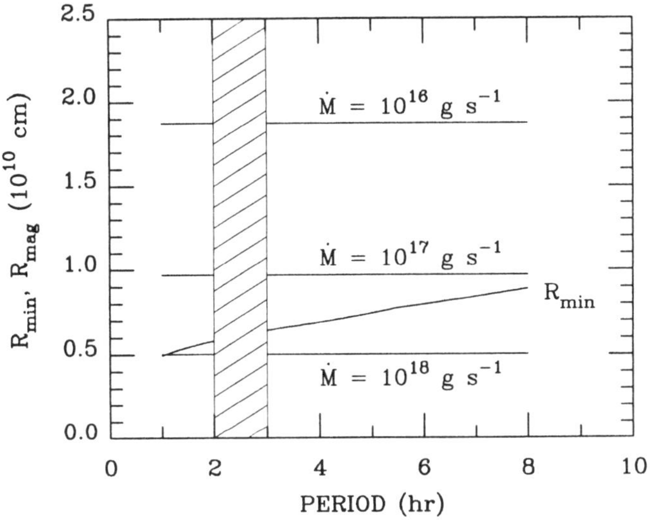

Fig. 1. *The radii R_{min}, R_{mag} plotted as function of orbital period, for magnetic moment $\mu = 3 \times 10^{33} Gcm^2$ for constant values of $\dot{M}$. For $R_{mag} > R_{min}$ no accretion disc forms.*

Lamb & Melia (1986), while accepting HKL1 criterion for disc formation, argue that the presence of discs in IPs depends on their history: systems formed at longer periods could retain their discs even if $R_{mag} > R_{min}$. This conclusion is correct only if the accretion rate remains more or less constant during the evolution. It is known however that IPs go, for extended periods, into low states, in which accretion almost stops (Warner 1986, Garnavich & Szkody 1988). It is difficult to see how those systems could retain their accretion discs. Moreover, it has been shown by Hameury, King & Lasota (1988b) that for P ≥ 3 hr, variations of $\dot{M}$ on timescales ≥ 10^5yr would move magnetic systems from the synchronous, "polar" group to the non-synchronous IP group or vice versa. One concludes therefore that it is very unlikely that accretion discs in MCVs will persist even if formed at long periods.

The condition $R_{min} > R_{mag}$ is only a sufficient condition for disc formation. Even if the accretion stream is not allowed to orbit freely around the white dwarf and is stopped by the magnetosphere, the accreting matter could in principle cross the fieldlines enough to interact with itself and create a disc. In this case finding the required criterion involves a study of relevant plasma instabilities with all the usual uncertainties connected with these problems. HKL1 conclude that $R_{min} > R_{mag}$ is probably also a necessary condition for discs formation. Finally the presence of orbital effects in the X-ray light curves of some IPs and the large size of their polar caps is easily explained if accretion discs are absent in those objects (HKL1).

3. THE POLARIZATION PROBLEM

Polars show a high degree of circular and linear polarization (hence the name "polars") modulated with the orbital period (e.g. Bailey 1985). IPs, in contrast show very low and in most cases undetectable degrees of polarization (Warner 1983, Cropper 1986). Only in BG CMi a circular polarization at a level of 0.25% in the optical (Penning, Schmidt & Liebert 1986) and 1.74% in the near infrared (West, Berriman & Schmidt 1987) has beeen detected. If one accepts that IPs have on the average the same magnetic fields as polars then the striking differences in their polarization properties have to be explained. Calculations of polarization for *homogenous* accretion on to large polar caps (Chanmugan & Frank 1987) require rather high magnetic fields in order to obtain low polarization. Preliminary calculations of the polarized emission from ring-like polar caps suggest that that these models may fit the observations better (Chanmugan & Frank 1988). However Wickramasinghe (1988) insists that the degree of polarization obtained in various geometries for fields $\simeq 10^7$ G should be observable.

To explain the "soft X-ray excess" (King & Lasota 1979, King & Watson 1987, Osborne 1988, Beuermann 1988) in polars, Kuipers & Pringle (1982) propose a model of non-uniform accretion flow consisting of plasma blobs which penetrate below the photosphere and thermalize. This idea was developed further by Frank, King and Lasota (1988) who found blob densities and sizes that produce required luminosity distribution. The same type of model should apply to accretion in IPs. Frank & Lasota (1988) suggest that the low degree of polarization in IPs is a consequence of the much higher density of the accreted plasma blobs allowing magnetic fields to be the same as in polars.

4. EVOLUTION OF MAGNETIC CVs

It is clear that the main argument in favour of the same magnetic moments for polars and IPs is based on a standard evolutionary model as described for example in King (1988). Recent studies (Hameury, King, Lasota & Ritter 1988, hereafter HKLR, Hameury, King & Lasota 1988a, hereafter HKL2, Hameury, King, Lasota & Livio 1988, hereafter HKLL, Hameury, King & Lasota 1988c, hereafter HKL3) have shown that the standard model (e.g. King 1988) finds a striking application in the description of the evolution of MCVs and predicts their detailed physical properties.

Recently HKLR proposed an explanation of the large, highly significant accumulation (6 out of 16) of polars in the period interval 113.5 – 115.5 minutes in terms of the resumption of mass transfer after these systems have crossed the period gap. When accretion resumes at a period of about 2 hr an episode of adiabatic expansion enhances the discovery probability at this period: (i) the mass transfer rate $-\dot{M}_2$ is about twice the value for a star close to the main sequence contracting in response to mass loss, and (ii) the expansion and subsequent contraction of the secondary lengthens the time spent near the initial period. To give the observed period "spike" requires most systems to have similar parameters when mass transfer resumes at the lower edge of the period gap: otherwise the initial value of P_0 would not be the same and the spike would be smeared out. At the spike, $M_2 \simeq 0.2$ M$_\odot$. Further, the period spike implies constraints on the primary masses which have been examined by HKL2. Higher masses cause a narrower period gap and hence a higher value P_0 of the initial period below the gap. HKLL show that for magnetic systems nova outbursts must remove (almost) exactly the mass accreted between explosions. The white dwarf masses therefore stay constant. Using the fact that the newly discovered AM Her system EXO033319-2554.2 has a period of 127min one concludes (HKLL) that most AM Her systems have a *unique* primary mass $\simeq 0.6 - 0.7 M_\odot$, while EXO033319-2554.2 has a significantly larger white dwarf mass ($\simeq 1.2 - 1.3 M_\odot$). This prediction (HKL2) was subsequently supported by radial–velocity measurements (Beuermann, Thomas & Schwope 1988).

The strength of the magnetic field has been measured for about 50 % of polars by observations of Zeeman splitting or cyclotron harmonics. It remarkable that with one exception (Beuermann et al 1988) all the measured values are between 20 – 30 MG (Schmidt et al 1986) whereas isolated white dwarf can have fields up to at least 5×10^8 G. Since most properties of magnetic binaries depend on the magnetic *moment*, an upper limit on the magnetic *field* has been difficult to understand. However, according to HKL2 and HKLL one may regard polars as having effectively a unique white dwarf mass. With R_1 essentially fixed at $\simeq 8 \times 10^8$ cm, the limit on B becomes a limit on moments μ_1 through $\mu_1 = 5 \times 10^{33} B_7$ G cm^3 where B_7 is B_1 in units of 10^7 G. HKL3 show that moments $\mu_1 \geq 2 \times 10^{34}$ G cm^3, (i.e. fields stronger than $B_7 \simeq 4$) probably imply catastrophic angular momentum losses $\dot{J}$ via magnetic braking involving the *white dwarf* magnetic field. These drive the secular evolution so rapidly that the systems enter a prolonged period gap at $P \sim 5$ hr, remaining as detached systems for a time of order the Hubble time and are unlikely to be observed during their MCV phase.

King (1985) proposed an explanation of the lower limit of magnetic fields observed in polars in terms of field burying by circulations in the white dwarf. This gives $B_{min} \simeq 6 \times 10^6$ G for $M_1 =$ 0.6 M$_\odot$ ($\mu \simeq 3 \times 10^{33}$ Gcm^3). The absence of *synchronous* systems with magnetic fields between 6 and 20 MG has yet to be explained.

5. CONCLUSION

From HKLL, HKL2 and HKL3 one concludes that unless 99% of polars are born at ~ 4hr, all magnetic systems have $\mu \simeq 10^{34}$Gcm^3. In this case it is unlikely that accretion discs play an essential role in the evolution of cataclysmic variables.

References

Bailey, J., 1985. 'Multiwavelength observations of AM Herculis-type binaries' in W.R. Burke(ed.), *"Recent Results in Cataclysmic Variables*, ESA-Nordwijk, p. 139

Beuermann, K., Thomas, H.C. & Schwope, A., 1988. *Astr. Astrophys.,* **195**, L15.

Beuermann, K., 1988., 'The X-ray Properties of AM Her stars' in G.Coyle et al (eds.), *Polarized Radiation of Circumstellar Origin*, Vatican Press, Vatican.

Cannizzo, J.K. & Pudritz, R.E., 1988. *Astrophys. J.,* **327**, 840.

Chanmugam, G., & Ray, A., 1984. *Astrophys. J.,* **285**, 252.

Chanmugam, G. & Frank, J., 1987. *Astrophys. J.,* **320**, 746.

Chanmugam, G. & Frank, J., 1988. *Astrophys. J.,* , . in preparation

Cannizzo, J.K. & Pudritz, R.E., 1988. *Astrophys. J.,* **327**, .

Cropper, M., 1986. *Mon. Not. R. astr. Soc.,* **222**, 233.

Frank, J., King, A.R. & Lasota, J.P., 1988. *Astr. Astrophys.,* **193**, 113.

Frank, J. & Lasota, J.P., 1988. *Astrophys. J.,* , . in preparation

Garnavich, P., & Szkody, P., 1988. *Pub. Astron. Soc. Pac.,* , . in press

Hameury, J.M., King, A.R. & Lasota, J.P., 1986.(HKL1) *Mon. Not. R. astr. Soc.,* **218**, 695.

Hameury, J.M., King, A.R., Lasota, J.P. & Ritter, H., 1987. *Astrophys. J.,* **316**, 275.

Hameury, J.M., King, A.R., Lasota, J.P. & Ritter, H., 1988.(HKLR) *Mon. Not. R. astr. Soc.,* **231**, 535.

Hameury, J.M., King, A.R. & Lasota, J.P., 1988a.(HKL2) *Astr. Astrophys.,* **195**, L12.

Hameury, J.M., King, A.R. & Lasota, J.P., 1988b. *Mon. Not. R. astr. Soc.,* , . in press

Hameury, J.M., King, A.R. & Lasota, J.P., 1988c.(HKL3) *Mon. Not. R. astr. Soc.,* , . submitted

Hameury, J.M., King, A.R., Lasota, J.P. & Livio, M., 1988.(HKLL) *Mon. Not. R. astr. Soc.,* , . submitted

King. A.R. & Lasota J.P., 1979. *Mon. Not. R. astr. Soc.,* **181**, 653.

King, A.R. & Watson, M.G., 1987. *Mon. Not. R. astr. Soc.,* **227**, 205.

King, A.R., 1985. *Mon. Not. R. astr. Soc.,* **217**, 23p.

King, A.R., 1988. *Q. J. R. Astron. Soc.,* **29**, 1.

King, A.R., Frank, J. & Ritter, H., 1985. *Mon. Not. R. astr. Soc.,* **213**, 185.

Kuijpers, J. & Pringle, J.E., 1982. *Astr. Astrophys.,* **114**, L4.

Lamb, D.Q. & Melia, F., 1986, 'Evolution of Magnetic Cataclysmic Binaries' in K.O.Mason, M.G. Watson and N.E. White (eds.),*The Physics of Accretion onto Compact Objects*, Springer Verlag, Berlin, p. 113.

Osborne, J.P., 1988. *Mem. Soc. Astron. It.* , . in press

Penning, W.W., Schmidt, G.D. & Liebert J., 1986. *Astrophys. J.,* **301**, 881.

Ritter, H., 1988. *Astr. Astrophys.,* **202**, 93.

Schmidt, G., Stockman, H.S. & Grandi, S.A., 1986. *Astrophys. J.,* **300**, 804.

Taam, R.E. & McDermott, P.N. 1987. *Astrophys. J. Letters,* **319**, L83.

Warner, B., 1983. 'The Intermediate Polars' in M. Livio and G. Shaviv (eds), *Structure and evolution of close binary systems*, IAU symposium No 73, Reidel,Dordrecht, p. 155.

West, S.C., Berriman,G. & Schmidt, G.D., 1987. *Astrophys. J.,* **322**, L35.

Wickramasinghe, D.T., 1988. 'The Magnetic Cataclysmic Variables' in G.Coyle et al (eds.) *Polarized Radiation of Circumstellar Origin*, Vatican Press, Vatican.

RADIO EMISSION OF DWARF NOVAE

A.O.BENZ
Institute of Astronomy, ETH
8092 Zurich
Switzerland

ABSTRACT. With the recent detection of radio emission from dwarf novae a new channel of information on these semi-detached binaries has opened. The radiation is variable within days and highly polarized. Most likely it is emitted by some gyromagnetic process, which would allow to estimate the magnetic field. If it is synchrotron emission, the magnetic field strength in the source is of the order of 40 G; if interpreted by the cyclotron maser process the magnetic field is 800 - 1800 G depending on which harmonic of the local gyrofrequency is emitted. The radio emission seems to be limited to the later part of the optically high state of the accretion disk. This suggests an acceleration mechanism connected to the outburst.

1. Observations

A total of 37 different dwarf novae binary systems have been searched since 1983 for radio emission using the Effelsberg, Arecibo, and VLA telescopes (Benz et al., 1983; Cordova et al., 1983; Turner, 1985; Fürst et al., 1986; Echevarria, 1987; Benz and Güdel, 1989). Only three were detected as radio sources, one of them with the VLA (Benz and Güdel, 1989). All were found variable on the scale of days. Two sources have been compared with the visual light curve and found to be related to the high state of visual luminosity. No radio emission was ever found in low state, but of 18 single observations of dwarf nova systems known to be optically high only 4 have detected radio emission (two in SU UMa and two in EM Cyg). The phase of radio emission is short and may be limited to the later part of the optical outburst as indicated by Figure 1. The low detection rate even in optically high state suggests that there may be more restrictions on radio emissivity.

2. Theory of Radio Emission

Radio flux density is usually compared to the thermal emission of a black body of the size of the source. Using the Rayleigh-Jeans approximation an effective "brightness temperature" can be derived:

$$T_b a \approx \frac{c^2\ s\ D^2}{2K\nu^2\pi r^2} = 1.3\cdot 10^{10}\left(\frac{s}{0.3\ \mathrm{mJy}}\right)\left(\frac{10^{11}\ \mathrm{cm}}{r}\right)^2\left(\frac{D}{320\ rmpc}\right)^2\ \mathrm{K} \qquad (1)$$

G. Belvedere (ed.), Accretion Disks and Magnetic Fields in Astrophysics, 113–116.

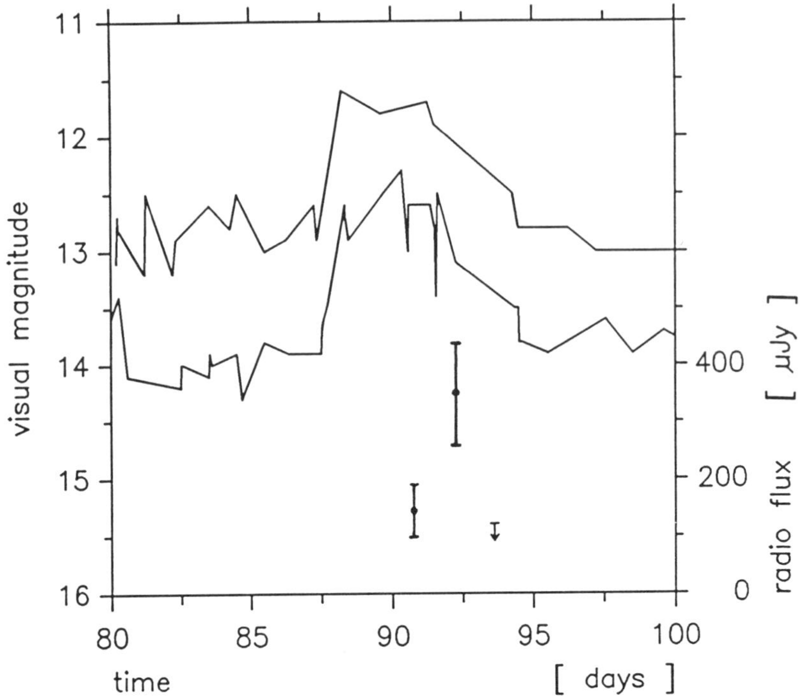

Figure 1. Visual light curves (top) of EM Cyg as observed by the VLA. The first curve shows time in Julian days - 2445592 (December 1983) and is shifted by + 1 magnitude. The 3 σ upper limit of Fürst et al. (1986) is indicated in the bottom part (scale at right). The second curve and the two radio detections refer to October 1987 with a time scale of Julian days - 2447000 (from Benz and Güdel, 1989).

where s is the observed flux density at the observing frequency ν, D is the distance to the source, and r the source radius. The chosen parameters refer to EM Cyg. Within its disk size of about $2 \cdot 10^{10}$ cm, and the binary separation of $1.5 \cdot 10^{11}$ cm (Jameson et al., 1981) no thermal source compatible with equation (1) is known. Similarly the thermal emission of the mass outflow generally believed to be associated with the optically high state is many orders of magnitude below the measured radio flux (Benz et al., 1983).
Synchrotron emission of nonthermal electrons can in principle produce high brightness temperatures up to an effective temperature, which is for typical values of the power-law index $\delta \approx 3$ of the electron energy and an aspect angle $\theta = 60°$ to the magnetic field (Dulk and Marsh, 1982):

$$T_{eff} = 2.8 \cdot 10^8 s^{0.755} \quad \mathrm{K} \tag{2}$$

where s is the observed frequency in units of the electron gyrofrequency. Equations (1) and (2) can be compatible only if either s takes a value well over 100 or if the radio source is considerably larger than the binary separation. In the first case the number of energetic electrons is large, since only the high energy tail contributes to the observed emission, and their pressure exceeds the pressure of the source magnetic field $B = 2\pi\nu mc/es \approx 6.4$ G (using the minimum energy hypothesis $T_b \approx T_{eff}$ and synchrotron approximations by Dulk and Marsh, 1982). The electrons cannot be confined and would adiabatically expand, thus quenching the synchrotron emission. In the case of a source twice as large as the binary dimension the harmonic number $s \approx 47$, thus $B \approx 40$ G. This magnetic field then easily dominates the particle pressure. We have, however, no other evidence for such high field strengths in the required volume.

As a possibly more attractive process we also consider coherent radiation of weakly relativistic electrons by the cyclotron instability (maser). The process directly converts about 0.1 % of the kinetic particle energy into radiation, if the electrons have a loss-cone velocity distribution. A likely place for such a condition is above the poles of the white dwarf to where collisionless particles are guided along the magnetic field, and magnetically reflected if their transverse velocity is large enough. The total energy requirement for superthermal electrons would be about 10^{29} erg/s, 7 orders of magnitude below optical luminosity during the associated outburst for EM Cyg and less for SU UMa (Benz et al., 1983). The emission is at the frequency

$$\omega = s\frac{eB}{mc} \tag{3}$$

where s now is a small integer depending on the ratio of plasma frequency to gyrofrequency. The density profile for an isotropic outflow is

$$n(r) = \frac{\dot{M}}{mpv4\pi r^2} \tag{4}$$

Taking a dipole magnetic field for the white dwarf with $B(r) = 10^5(r_*/r)^3$ Gauss, the source is at r = 35000 km for s = 2 and for a radius of the white dwarf of $r_* = 7000$ km. This altitude is insensitive to the assumed harmonic and to the surface magnetic field. Equation (4) with the mass loss rate and velocity observed in similar systems then yields a plasma

frequency $\nu_p = (e^2 n/\pi m_e)^{1/2} = 0.8$ GHz. Since the flow near a dipole is anisotropic, ν_p may exceed the local gyrofrequency, $\nu_B = 2.5$ GHz. In fact, the density could temporarily be high enough to quench the maser. Fürst et al. (1986) have proposed this as an explanation for the absence of radio emission at the maximum of the outburst. For $\nu_p \approx \nu_B$ the maser most efficiently emits at s = 1 or 2 (Sharma and Vlahos, 1984). This suggests a magnetic field strength in the radio source region of 1750 or 875 Gauss, respectively. If $0.5 \lesssim \nu_p/\nu_B \lesssim 1.0$ the s = 1 emission in ordinary mode dominates. For $1.0 \lesssim \nu_p/\nu_B \lesssim 1.5$ the radiation is purely extraordinary and at s = 2.

Other sites of strong fields, such as the late-type companion or the disk, are expected to have values of $\nu_p/\nu_B \gtrsim 10$. The electromagnetic maser instability is several orders of magnitude slower in growth under such conditions and may well be dominated by other processes (e.g. Sharma and Vlahos, 1984). The steep potential of a white dwarf combined with relatively high magnetic fields is therefore an ideal site for maser emission.

The variability and high polarization observed in the radio emission of EM Cyg supports the maser model. It is far less demanding in energy than any synchrotron jet model.

The electrons may be accelerated at the shock front of the outflow, from where electrons can stream back to either the white dwarf or the companion. The radiation then would be similar to stationary metric type IV radio bursts in the Sun, except that the smaller ratio of

ν_p/ν_B of a white dwarf allows more efficient masering. About 1% of the outflow energy (10^{37} erg) would have to be transferred to the returning electrons assuming maser radio emission. From the energy point of view a more attractive alternative is acceleration (or heating to high temperatures) near the accretion disk, from where electrons can easily propagate along converging field lines to the poles of the white dwarf.

3. Conclusions

Radio emission of dwarf novae is most likely caused by some form of gyromagnetic emission and thus carries information on the magnetic field in the source.
For synchrotron emission the magnetic field is of the order of 6 - 40 G. A synchrotron source most likely would be larger than the binary system. The magnetic field may be strong enough to contain it, but a slow expansion due to MHD instabilities is likely.
If this process is maser, which may be confirmed by spectral observation, the source magnetic field is 875 or 1750 G depending on which of the two most probable maser mode is most efficient. This question is related to the electron density in the source. A density $\lesssim 10^{11} cm^{-3}$ would lead to the higher B value and vice versa. Since electromagnetic cyclotron masers operate best in low density, strong magnetic field sources, the most likely site of maser emission, is near the white dwarf.
The radio emission for both types of emission is a considerable requirement for the number of energetic electrons. The particles therefore seem to be accelerated by or near the prime source of energy, the accretion disk. From there they may be guided to the converging fields over the poles of the white dwarf.
Spectral observations are needed to determine the emission mechanism. More data on the radio emitting phase of dwarf novae will give a better picture of the conditions for radio radiation and particle acceleration. Eclipses finally may determine the location of the radio source.

Acknowledgements

I would like to thank Manuel Güdel for discussions and comments on this work. I acknowledge the AAVSO observational material sent to us by the Director, Janet A. Mattei, which was necessary for this work. Most of all I thank Gaetano Belvedere for the superb organization of this workshop in a splendid part of the world.

References

Benz, A.O., Fürst, E., Kiplinger, A.L.: 1983, Nature **302**, 45.
Benz, A.O. and Güdel, M.: 1989, Astron.Astrophys., submitted.
Cordova, F.A., Mason, K.O., Hjellming, R.M.: 1983, Publ.Astron.Soc.Pacific **92**, 69.
Dulk, G.A. and Marsh, K.A.: 1982, Astrophys.J. **259**, 350.
Echevarria, J.: 1987, Astrophys.Space Sci. **130**, 103.
Fürst, E., Benz, A.O., Hirth, W., Kiplinger, A., and Geffert, M.: 1986, Astron.Astrophys. **154**, 377.
Jameson, R.F., King, A.R., and Sherrington, M.R.: 1981, M.N.R.A.S. **195**, 235.
Mattei,J.A.:1988, AAVSO observations, private communication.
Sharma, R.R. and Vlahos, L.: 1984, Ap.J. **280**, 405.
Turner, K.C.: 1985, Radio Stars (eds. R.M. Hjellming and D.M. Gibson), D. Reidel Publishing Company, Dordrecht, p. 283.

JETS AND MAGNETIC FIELDS

R. N. Henriksen
C.E. Saclay DPhg/SAp
91191 Gif sur Yvette
cedex France

ABSTRACT. In this paper we examine some key results to date on the properties of jets from relativistically compact objects. The approach is historical because in a confused theoretical situation it is useful to evaluate results on the basis of their frequency of recurrence. Some important results are isolated but a comprehensive review is not attempted. The various categories of theory associated with jets from non-relativistically compact objects are surveyed in the same way. It is claimed that one simple set of (nearly) dimensional relations goes a long way! There is much repetition in this area. Finally some new work on the magnetic two-point correlation functions is briefly reported. We believe that such a stochastic approach is preferable to the usual semi-deterministic dynamo theory.

1. Jets from Relativistically Compact Objects

Such objects as rotating black holes and neutron stars can in principle easily produce particles of relativistic energy. Moreover, there is distinct evidence from VLBI observations of the nuclei of active galaxies, and from such galactic objects as SS433, CygX-3, and perhaps ScoX-1 that relativistically fast beams of particles do occur. Such beams are unlikely to be thermally launched, particularly if they are baryons, and so a low entropy engine must be sought. Relativistically compact objects remain the most likely possibility.

The details of the mechanism(s) by which such objects produce steady or even unsteady relativistic beams *at infinity* remains unclear to say the least. The mechanism of all models known to me however is of the *Pulsar magnetosphere* type. That is they consist of relativistic, charge separated, electromagnetically force-free, plasma flows external to certain boundaries. Originally the only boundary considered was that provided by a central neutron star (Goldreich and Julian, 1969) , whose surface was shown to be unstable to the emission of charged particles in vacuo. This is the *pure* Pulsar model in which all of the energy of the radiated particles must come from the spin energy of the central object. Generally we may refer to it as the *Spinar* option.

Subsequently, with the growing recognition of black holes, accretion discs became an important boundary in the magnetosphere(e.g.Blandford and Payne,1982) , and they have been used also in Pulsar models (Michel and Dessler,1981). Such models have both the *spin* of the central object and the accretion power as sources of energy, and they may be either explosive or steady.There are many variations in which the mix is varied (e.g. Leblanc and Wilson,1970; Henriksen and Reinhardt and references therein,1977; Blandford and Payne,1982 ; Blandford

G. Belvedere (ed.), Accretion Disks and Magnetic Fields in Astrophysics, 117–128.

and Znajek,1977). When accretion power is predominant, the steady mechanism is closely related to the more recent proto-stellar jet models of Pudritz and Norman(e.g.1986).The most rigorous current studies of this sort are those due to Lovelace and colleagues(1987), and to Camenzind(e.g. these proceedings). For the *Spinar* problem , the most recent and self-consistent work is by Fitzpatrick and Mestel (1988 a,b).

Definitive, consensual, results have however been rare. By these I mean the kind of reliable *rule of thumb* results that may be taken up from the specialist by the working astronomer. Rather than review the whole area, I will try in this talk to isolate several reliable results by using the criterion of *Darwinian Survivability*. That is, accepting the concept of *memes* (R.Dawkins-The Selfish Gene), I will seek those results which have persisted, or which have been rediscovered or which have otherwise appeared, from more than one source over the short history of the subject, and which are most related to relativistic beams.

I begin with the notion of relativistic *winds*. *Winds*, as a manifestation of the tail of the particle energy spectrum in a central object , have certainly proven to be a good *meme* since their introduction by Parker. There are by now many elaborations, but the essentials for the production of relativistic winds are probably still given by the zero temperature, monopolar, electromagnetic theory(e.g.Michel,1969; Goldreich and Julian, 1969,1970; Henriksen and Rayburn 1971,1972). It is important to realize that such calculations fail utterly to deal with the global description of the *Spinar* mechanism,but they are likely to have some order of magnitude relevance in *piecewise* portions of the Spinar magnetosphere. I shall also not discuss the effects of any possible vacuum gaps in the magnetosphere as they do not produce ion beams.

The first result which I offer you is an answer to the question:"What is the characteristic energy of a particle on injection into the wave zone of the Spinar?" I suggest the following *meme*. The discussion unfolds normally in terms of a parameter

$$\sigma \equiv \frac{e\Delta\psi}{mc^2}, \tag{1}$$

where $\Delta\psi$ is the potential drop over the *working* part of the magnetosphere and m is the rest mass of the relevant charged particle (e). Of course in this form the parameter is just "wishful thinking", for one requires a magnetospheric model to give the value of $\Delta\psi$. The relativistic, monopolar, wind theory gives

$$\sigma = \frac{\Phi^2}{4\pi r_c^2 c}\frac{1}{f}, \tag{2}$$

where Φ is the magnetic flux sr^{-1}, r_c is the light cylinder radius and f is the rest mass flux sr^{-1}. The significance of this expression (Henriksen and Rayburn,1971) is that σ^{-1} gives the mass-energy flux circulating at r_c in the electro-magnetic field in units of the radial rest mass flux. Thus σ being very large is the limit of a *field dominated wind*, which is therefore the low entropy, relativistic limit that we are seeking as a central engine (Henriksen and Rayburn,1971,1972).

Now Michel (1969) and Goldreich-Julian (1970) have shown that the solutions which go to infinity give a particle Lorentz factor $\gamma \sim \sigma$. However Henriksen and Rayburn (1971, 1972) argued that such solutions should be matter dominated, and therefore only valid for small σ. They suggested that the correct relativistic solutions were the stellar *breeze* solutions . They showed also that these were the analogues of the Gold (1968,1969) *magnetic sling* model, and that one of them was uniquely selected by requiring it to touch the (relativistic) Dicke-Alfven singular point. Subsequently the solution gives a singularity at the light cylinder and so must be regarded only as a near zone or "injection" mode. It predicts however that at the Dicke -Alfven point (where a new intermediate-wave zone solution must begin)

$$\gamma \sim \sigma^{2/3}. \tag{3}$$

I claim that this result has shown good Darwinwian survivability, for it was rediscovered in 1975 by Asseo,Kenneland Pellat,and then by a late convert (Michel,1987) ,in the form

$$\gamma \sim \sigma^{2/3} ln(r/r_0). \tag{3'}$$

Here the formalism is simply that of a test particle in crossed electric and magnetic fields, so that *the model is again field dominated.* The logarithmic factor really describes a non-discontinuous injection (r_0 is the injection radius and r is the current radius) unlike the Dicke-Alfven injection above. It is most likely of order unity. We rediscovered the result ourselves (Henriksen and Norton,1975a) in the form $\gamma \sim \gamma_{\phi 0}^2$, where $\gamma_{\phi 0}$ is the rotational Lorentz factor at the Dicke-Alfven point; using a multi-particle approach. The Lorentz factor satisfies $\gamma_{\phi 0}^2 \sim \sigma^{2/3}$ in the relativistic wind theory. This work emphasized particularly clearly the distinction between *wave zone* solutions and *magnetic sling* solutions in terms of an analytic model.The arguments fit together reasonably well, but one should note one odd feature, namely that the magnetic sling solution is in fact more *rigid* than even a vacuum wave! For such waves would be sheared at the light cylinder. This is presumably due to the currents flowing in the magnetosphere. In fact in the time dependent case these currents prevent waves from propagating radially across the light cylinder(see below) and *are the basic self- collimating mechanism.*

This robust result predicts in concrete terms

$$\gamma \sim (\Phi_{24}^{4/3} P_{ms}^{-4/3} f_{14}^{-2/3}) 10^6, \tag{4}$$

where $\Phi_{24} \equiv 10^{-24}\Phi(gausscm^2)$,$P_{ms} \equiv 10^3 P(s)$ and$f_{14} \equiv 10^{-14} f(gms^{-1})$. For the Crab Pulsar this gives $10^4 \Phi_{24}^{4/3} f_{14}^{-2/3}$. Long ago (Henriksen and Rayburn 1972) it was pointed out that $\Phi_{24} \sim 10^{-1}$ and$f_{14} \sim 10^{\cdot 3}$ gave an explanation for the inter-pulse radiation cutting-off at$\geq 400 Mhz$ (electrons and ions obtain the same γ in these models which are not electrostatic). The idea of light cylinder emission was ably argued for many years by D.C.Ferguson(see e.g. IAU95). I wonder if this is still not tenable?

I note finally on this question that the simple theory also gives that the angular momentum and energy loss rates are $\propto \sigma^{-1/3}$ while the mass loss rate $\propto \sigma^{-1}$, so that $\gamma \sim \sigma^{2/3}$ .The mechanism is thus rather inefficient compared to the electrostatic ideal wherein $\gamma \sim \sigma$ and vanishly few particles of very high energy are emitted(this is due to $\mathbf{E} + \beta \wedge \mathbf{B} \approx \mathbf{0}$). Interestingly enough if density or mass flux perturbations from the Pulsar are all of equal duration, then such engines would produce a particle spectrum $dN/dE \propto E^{-3/2}$, which would become $\propto E^{-1}$ if the energy emitted were constant in each interval. However we must remember that the particles have yet to formally escape to infinity, and that magnetospheric *gaps* may also provide a different family of electrostatically accelerated particles.

I turn now to a second question : "can the relativistic wind be a collimated beam?" If so, then we have discovered one reasonably well understood mechanism for clean, relativistic jets. Even if there were some thermal contamination in the rest frame of the beam, relativistic streaming alleviates the confinement problem by the $\frac{1}{\gamma}$ time dilation factor in any case. Thus we can concentrate on providing an initial beam.

The spinar mechanism probably must be self-collimating because we expect as above the field energy to dominate any matter present.This leads naturally into another Darwinianally robust theory,the force-free theory.This has been rediscovered many times, but the early formulations were due to Mestel (1971, 1973), Cohen and Rosenblum (1972), Scharlemann (1974), Scharlemann and Wagoner (1973) and Henriksen and Norton (1975a) in the axially symmetric case. For the oblique Spinar progress has been slower, but notable results were obtained by Endean (1972), Henriksen and Rayburn (1974) and Henriksen and Norton (1975b).

Briefly the whole force-free theory can be summarized in terms of vector quantities *projected* onto rotating axes (there is no observer transformation). These are ($'$ denotes projected quantities–our presentation follows that of Henriksen and Norton, 1975b)

$$\mathbf{B}' \equiv \mathbf{B} - \beta_\phi \wedge \mathbf{E}, \tag{5a}$$
$$\mathbf{j}' \equiv \mathbf{j} - \rho_e c \beta_\phi \tag{5b}$$
$$\mathbf{E}' \equiv \mathbf{E} + \beta_\phi \wedge \mathbf{B} = -\nabla\psi \tag{5c}$$
$$\beta'_{\mathbf{a}} \equiv \beta_{\mathbf{a}} - \beta_\phi \tag{5d}$$

where $\beta_\phi \equiv \omega \wedge \mathbf{r}$, β_a is the total velocity of the ath species, and other quantities have their usual meanings. Equation 5(c) is a general integration of Faraday's law when the motion is steady in the rotating frame. When the charges are supposed to stream purely along the field lines in the rotating frame then Amperes law may be written in terms of our projected vectors as

$$\nabla \wedge \mathbf{B}' = \frac{4\pi}{c}\mathbf{j}', \tag{6a}$$
$$\mathbf{j}' = \kappa\mathbf{B}, \tag{6b}$$

where 6(b) follows from the *bead on a wire* constraint (Henriksen and Rayburn 1971) $\beta'_{\mathbf{a}} = \alpha_{\mathbf{a}}\mathbf{B}$. This constraint necessarily applies with axial symmetry (Mestel,1973). The scalar specific current function κ is evidentally constant on a magnetic field line, $\mathbf{B}$, because $\nabla.\mathbf{B} = \mathbf{0}$. This becomes the elegant and useful constraint (Henriksen and Norton, 1975b) in cylindrical coordinates (ϖ, ϕ, z)

$$\frac{\partial(\kappa, B'_z)}{\partial(\varpi, \phi)} + \frac{\partial(\kappa, B'_\varpi)}{\partial(\phi, z)} + \frac{\partial(\kappa, \varpi B_\phi)}{\partial(z, \varpi)} = 0. \tag{7}$$

We have also the last of Maxwell's equations in the form

$$\nabla.\mathbf{E} = \mathbf{4}\pi\rho_{\mathbf{e}}. \tag{8}$$

There are only two more equations(although unfortunately for each particle species!) which are required if corrections for the finite mass of the particles are to be made. One is the lovely Hamiltonian integral found by Endean (1972a,b) and known also to us (Henriksen and Norton, 1975a,b) and to Leon Mestel (e.g. Fitzpatrick and Mestel 1988) but apparently to few others. This gives for steady conditions in the corotating frame that

$$\epsilon_a - \beta_\phi . \frac{e_a \mathbf{A}}{m_a c^2} - \frac{l_a}{r_c} = \epsilon'_a(\kappa), \tag{9}$$

where κ we remember is a corotating stream-line (and field line when the bead on a wire constraint is used) constant. Here a is a species label, $\mathbf{A}$ is the usual vector potential, and (V is the usual scalar potential and $\psi = V - \beta_\phi.\mathbf{A} + constant$)

$$\epsilon_a \equiv \gamma_a + \frac{e_a V}{m_a c^2}, \tag{10a}$$
$$l_a \equiv r sin(\theta)\gamma_a \beta_{a\phi}. \tag{10b}$$

In the axially symmetric case (Henriksen and Norton 1975a,b) $\beta_\phi.\mathbf{A}$ is also constant on a field line (stream line), and the integral reduces to the Jacobi form

$$\epsilon_a - \frac{l_a}{r_c} = \epsilon'(\kappa). \tag{11}$$

Finally one has

$$\beta'_{\mathbf{a}}.\nabla\gamma_a \equiv d\gamma_a/dt = \frac{e_a\beta_a.\mathbf{E}}{m_a c^2}. \tag{12}$$

Well that is hard work, but it is a complete theory at the simplest level. Unfortunately it is already intractable, but *all* of the usual analytic results can be obtained as various limits. I will not trouble you with all of them. I note however that Leon Mestel has for a long time been a most articulate advocate of this approach. Of course he would argue that a dissipation term must be taken into account in equations (9) and (12) before a globally consistent model may be found.But I am less ambitious here and will ignore the radiation reaction.

But I promised you a result. The first of them is really an independent confirmation of the relativistic wind theory result. For equation (9) can be written as (Henriksen and Norton 1975b)

$$\gamma_a = \frac{\epsilon'_a(\kappa) - e_a\psi/m_a c^2}{1 - \beta_\phi^2 - \alpha_a\varpi B_\phi/r_c}, \tag{13}$$

and in the absence of electrostatic acceleration ($\psi = constant$) and in the limit of the unsheared mafnetic field ($B_\phi \to 0$) we recover $\gamma \propto \gamma_\phi^2$ as in the *breeze* relativistic wind. But we already had that.

The celebrated *Pulsar equation* has reappeared again (it has a history of recurrence going back to Cohen and Rosenblum,1972) in the work of Lovelace and collaborators and of Camenzind (this meeting) in the magnetospheres of their accretion discs. Thus it satisfies our criterion. One obtains it most simply as the zero current limit, constant potential (no electrostatic effects) limit. Then equations (6a),(5c) give respectively

$$\mathbf{B}' = \nabla\Omega,$$

$$\mathbf{B}' = \frac{\mathbf{B_P}}{\gamma_\phi^2} + \mathbf{B}_\phi,$$

whence $\nabla.\mathbf{B} = 0$ may be written as

$$\nabla.[\gamma_\phi^2(\nabla\Omega - \beta_\phi^2\nabla_\phi\Omega)] = 0. \tag{14}$$

That is it. It is written explicitly for example in Henriksen and Norton (1975b) . It cannot be used across the light cylinder! Therefore models which produce beams from accretion discs by way of such a magnetosphere are *necessarily dealing with beams that emerge initially along the rotation axis, within the light cylinder* (cf Lovelace et al. 1987). Since this is true I am able to offer another result which, if it hasn't already 'survived', may be at least visibly returning.

Henriksen and Norton (1975b) were able to generalize the *Pulsar equation* to allow for non-axial symmetry and non-zero current functions(more suitable for open structures perhaps). They found what they referred to as *helicon wave* solutions. I refer you to the original paper for the somewhat tedious detail, but the key idea is that the solutions are found in terms of the independent variables

$$\Psi = \phi + k_z z - \omega t, x = \varpi/r_c, \tag{15}$$

and the dependent variable

$$u \equiv B_z/\gamma_\phi^2 - k_z x B_\phi = u(\kappa), \tag{16}$$

with the complete solution given by

$$B_\varpi = (1/(\kappa x))\partial_\Psi u,$$
$$B_\phi = -[k_z x u + \frac{(1-x^2)}{\kappa}\partial_x u]/(1-k_x^2 x^2), \tag{17}$$
$$B_z = [u - \frac{k_z x}{\kappa}\partial_x u]/(1-k_x^2 x^2).$$

It is noteworthy that these solutions can formally be extended beyond $x = 1$ where they take the character of obliquely propagating plane waves. However , to our dismay at the time, Leon Mestel pointed out to us that the radial component of the Poynting flux vector satisfied

$$\langle S_\varpi \rangle_\Psi \propto \langle u \frac{du}{d\Psi} \rangle_\Psi \equiv 0. \tag{18}$$

Thus energy could only propagate in the z direction in these solutions. That seemed a catastrophe at the time because standing waves required a reflecting outer boundary. Purists as we were , we were determined to find solutions functioning in a vacuum. However now, with surrounding accretion discs and tori more widely available, it may be a very useful beam solution. The energy flux to infinity is simply (κ constant)

$$S_z(x) = \frac{c}{4\pi}\langle \beta_\phi B_z B_\phi \rangle_\Psi, \tag{19}$$

where the fields are given by equations (17). The function $u(x, \Psi)$ is given in the original paper. In general the distribution of the flux in x will be complicated, but it tends to be annular in general because of the β_ϕ factor. With an accretion disc (or bar for non-axial symmetry) , a superposition of modes must be taken. The usual order of magnitude estimates are however straightforward from equation(19).

Thus maybe these things do produce collimated beams of waves and particles. The significance of the presence of waves , as opposed to the steady models presented here by Lovelace and Camenzind, is that they can go on accelerating test particles by the *wave surfing* mechanism (Gunn and Ostriker,1969). Perhaps they can naturally dissipate their own energy in this way, gradually being absorbed by a beam of matter which continues to carry the momentum. Some such conversion mechanism appears to be needed (see e.g. summary by Henriksen). The helical magnetic structure seen in objects such as M87 (Hardee, 1988) can also be explained by these waves. For the short waves are absorbed first(indeed each mode should persist for only a few k_z^{-1}). Thus on any given scale we will tend to observe helical structure on that scale(shorter waves are damped and longer waves contribute only a uniform background).

2. Jets from not-so-compact objects

Here we confront the problem of protostellar jets such as HH34 (Buehrke et al. 1988), probably the bipolar molecular outflows, as well as non-relativistic outflows from AGN. This problem is not so *clean* as that of the previous section, but at least the various models fall into recurring categories.

Perhaps the earliest model involving a collapsed (or rather collapsing) object was due to Leblanc and Wilson (1970). Here the magnetic field was frozen into a collapsing star, where it was wound differentially and eventually produced a bipolar explosion along the axis. The calculation was numerical, and the outburst was a single rapid event, probably more relevant to a supernova such as 1987A.

Subsequently, in a long series of papers beginning in 1972 and ending in 1976 Henriksen and coworkers explored accretion onto a magnetic field using the *inverse* (not simply time-reversed because the boundary conditions are very different) magnetic wind theory of Weber and Davis (1967). See e.g. Henriksen (1976) and Henriksen and Reinhardt (1977) for references including those papers. Many Darwinian results were obtained, of which the most important for our purposes , was that the magnetic field provided a barrier near the Dicke-Alfven point (on open field lines) maintained by the Alfven wave transport of angular momentum (as opposed to the more frequently discussed but Rayleigh- Taylor unstable magneto-acoustic support over the closed field lines).In fact the problem is much more complicated than that, and under some conditions (such as when the barrier is broken) the field actually can provide an enhanced accret ion rate. But near equipartition this story is reasonably accurate.

We called the barrier an *Alfven gate* and showed that it could only be breached by a sufficiently large mass accumulation, which was attained at equipartition (between gravitational,magnetic and kinetic energy densities). When the accretion rate is low (so that it takes many rotational priods to accumulate this mass), Henriksen and Chia (1972) and Henriksen (1976) argued that the behaviour would be that of a relaxation oscillator, with long inactive times and an active period comparable to a rotation period. This was before the discovery of bursting X-ray objects! Unfortunately only a subset of those proved susceptible to this mechanism, but that is another story. The application to objects such as active galactic nuclei (AGN) and SS433 seemed obvious then, and has become more compelling since (see e.g. Shukre, Manchester, and Allen, 1983 for a rediscovery of magnetic accretion, but note that they believe that outbursts are radiatively driven and occur *along* the magnetic axis). The rather special geometry that we used seems however to have *blinded* others to the fact that many of the quantitative results have the force of dimensional analysis. No other geometries have been solved exactly, and much of the work involving accretion onto magnetic collapsed objects merely repeats this early work in more explicit contexts (this includes for example the possibility of *either* spin-up or spin-down during the accretion and oscillations at non-corotating frequency). But that is a slightly different story, and I will not write down again those results(best summarized in Henriksen and Chia,1972; Henriksen and Reinhardt, 1974; Henriksen,1976; Henriksen and Reinhardt,1977).

Our purpose is to make jets; as for example in SS433, AGN, proto-stellar objects and maybe supernovae. Essentially we believed that we could drive a flow perpendicular to the accretion disc plane by a mechanism similar to that of Leblanc and Wilson. The duration of the mechanism was not much different however, being only a rotation period P. This was alright (but never really calculated) for either an aligned or an perpendicular rotator , but what of the oblique case? Does this uncoiling spring emerge along the rotation axis, or perhaps perpendicular to the magnetic axis? The latter would give a precessing jet, at least at the base. This question remains unresolved. However if the mechanism works in general(and it should be calculated) the dimensional theory predicts:

1. The object should be spinning down and in the burst mode(a steady state is possible for the right parameters) the duration should be $\sim P^{-1}$.
2. The density at the Alfven point is

$$\rho_a = \frac{\pi\Phi^2}{P^2(GM)^2}, \tag{20}$$

where Φ is the magnetic flux sr^{-1} linked to the material and M is the mass of the central object.

3. The Alfven radius is

$$r_a = \left(\frac{GMP^2}{(2\pi)^2}\right)^{1/3}. \tag{21}$$

4. The Alfven speed is

$$c_a = \frac{2\pi r_a}{P}. \tag{22}$$

5. The specific angular momentum and the specific energy carried by the flow are respectively

$$| L | = \frac{2\pi r_a^2}{P},$$
$$| E | = c_a^2. \tag{23}$$

6. The accretion rate is (sr^{-1})

$$A = \frac{\Phi^2}{2GMP}. \tag{24}$$

7. The total luminosity (the process is inefficient to the extent that $r_a > r_*$,the radius of the central object) is

$$L \approx A c_a^2 \Omega \tag{25}$$

(Ω is the active solid angle), and the spin-down torque is

$$\tau \approx \frac{2\pi r_a^2}{P} A\Omega. \tag{26}$$

I claim that despite the lack of a global theory in both a social and a geometrical sense, these solutions are in fact rather Darwinian.

The mechanism is really a boundary layer effect (see e.g. Henriksen and Reinhardt, 1974) taking place near the Alfven surface, and another boundary effect has been suggested recently. Torbett (1986) has proposed a similar deposition of energy (non-magnetic) at the boundary between an accretion disc and the stellar surface. Here an oblique shock accompanies the resolution of the mismatch, and once again the energy is thought to emerge perpendicular to the disc. The amount of energy tapped is necessarily similar to that available above, except that it occurs right at the stellar surface. Clearly however this depends on the absence of any more gradual coupling between the disc material and the star.

The work of Pudritz, Pudritz and Norman (e.g.1986 for references) also involves accretion in magnetic fields, but the details are rather different in that the fields couple the disc to material in a *corona* above it rather than primarily to the central object. Otherwise the effect of the magnetic field is similar in general operation. However, the accretion disc boundary condition imposes additional constraints, and the authors have made cogent self-consistency argumments. These relate among other things the accretion rate through the disc to the angular momentum loss in the vertically propagating helical fields. This model is strikingly similar to the relativistic spinar with the disc boundary condition. It has moreover had some success in addressing the problem of proto stellar jets and bipolar molecular outflows, but there is some concern regarding the range of scales over which the mechanism must operate , and the solutions internal and external to the disc are not yet clearly related. Ralph is here to speak for himself, however and I will say little more about this model. I do observe that a steady state clearly requires dissipative processes in the disc. The actual origin of the physical forces can be most easily seen by writing the equations for the motion in the *corona* in what I call Field-line orthogonal or $FLOT$ coordinates. The disc may again be treated as a boundary condition.

Perhaps the most cogent work on protostellar jets to date is that of Arieh Koenigl, who is unfortunately unable to be with us. This is a thermal mechanism, wherein a vigorous stellar wind blows a bubble in the surrounding medium. This then distorts and expands according to the density variation in the environment (there are some elegant special solutions) and generally transforms itself from a *wind* to a collimated beam, while finding the path of least resistance.

There may even be evidence for secondary winds induced by the primary wind. Moreover the Rochester group has reported seeing bubbles around YSO's with an i/r array camera. If sufficiently vigorous winds are produced by proto-stars, there seems to be little objection to this mechanism in that context at the moment. The stars are certainly doing something extraordinary. One attractive idea is that the driving radiation for the wind is produced by a non spherically symmetric accretion shock, possibly of the kind suggested by Torbett.

Several variants of this last idea exist. One is due to Vincent Icke (1980) and is developed by one of the speakers at this meeting. A beam of particles is driven up the axis of a hot disc by the pressure of its radiation field. These models place an optically thin medium above the disc and rely on geometrical focussing to launch a well-defined beam. Such a model is very closely related to a radiation driven stellar wind. A sub-variant (Madej,Loken and Henriksen,1987) uses a thick accretion torus to focus the radiation field. However that calculation has not yet treated the effect of the radiation pressure on the dynamics of the disc properly.

One point to bear in mind in this discussion is the existence on a small scale of the high velocity H_α jets discovered by Mundt (L1551) and collaborators. These are really analogous to the VLBI for AGN . A wind mechanism will do for these jets (since they are not relativistic) if it is collimated sufficiently close to the star. Thereafter it presumably has to interact with and entrain its surroundings in various ways in order to initiate the large scale bi-polar outflows. Analytic calculations by Henriksen(1987) suggest that either symmetric or asymmetric flows can be produced in this way.

Another variant due to Choe and Henriksen (1986) is notable in that it purports to produce *pulsed jets*. The basic mechanism is a polar accretion shock which heats its surroundings by radiation until a wind or jet is eventually driven along the polar axis. But this halts the accretion flow and the cycle must begin again. Unfortunately the mechanism requires very low mass proto-stellar objects in order to produce low luminosity objects like L1551.

Of all these various categories of non-magnetic theories, those ideas with Darwinian survivability are probably the winds and the radiation driven winds. They satisfy the universality conditions needed to meet the observation that it is *easy*, indeed almost unavoidable, to make these outflows. However, theories that relate the properties of the central object in detail to the properties of the jets are required ultimately. In this respect the theories by Pudritz and Norman and by Choe and Henriksen have some advantages. The distinction between *pulsed jets* and interstellar *bullets* of an earlier suggestion by Norman and Silk, is otherwise moot for example.

I think that I have covered most of the categories of jets produced by compact objects. I have not been able to be very selective. I suspect that i/r arrays, detailed i/r spectra (Mitchell et al., 1988) and the increasing use of very high frequency receivers with the world's sub-millimetre telescopes, will change that. We can also anticipate sub- millimetre interferometry and sub-millimetre array cameras. It should be a fruitful area for research during the next $5-10$ years.

There is however one other model that has appeared very recently, by Shu et al.(1988). The major innovation is to regard the accretion disc as an extention of the protostar, thereby cutting the Gordian knot of the boundary layer problem. The other elements are really a combination of elements from the various theories discussed above. The key idea is that by having the equipartition point at the *junction* of the protostar and disc, material can be easily lifted through this *window* into a magnetically dominated region. Once there it is ejected in a centrifugally driven wind in the usual way. The mechanism is said to be self-regulating in that the accretion rate will be coupled to the mass loss rate in the wind The authors also seem to have rediscovered the theory of MHD in corotating reference frames and along field lines (!) which is a plus for the theory at least by our criteria. The model seems interesting but we shall have to await more details. The *letter* does not make the global field structure very clear, nor does it demonstrate the nature of the asymptotic jet, nor even estimate its aymptotic velocity. Moreover, not even the authors claim that it is the whole story, being prepared to have a *normal* stellar wind blowing open the closed field lines (but somehow managing not to disturb the equipartition region). Nevertheless, the mechanism is a kind of steady version of the *coiled spring* mechanism discussed above, with the spring region magnified for detail. As such, it is a step forward.

3. Magnetic Fields in the Interstellar Medium

The question of magnetic fields in star forming regions is vexed for two reasons, one observational and one theoretical. The observational reason is simply that there are very few direct measurements of the magnetic field (e.g. Zeeman splitting at 21cm) and no direct maps. The normal mapping technique (see e.g. Heyer et al., 1987) is by optical polarization of background bright stars, but such low levels of polarization are easily contaminated and are limited, in any case, to regions of low obscuration. A new technique (the i/r polarization method is similar) is needed.

But even when the field structure is reliably known, there remain problems of interpretation. Most observers, presented with a flattened or filamentary averaged cloud structure and a mean magnetic field aligned perpendicularly, conclude that the field has *controlled* the evolution of the cloud. However in a turbulent region driven ultimately by gravity , one may equally well argue that the field has been *controlled* by (through the action of some statistical dynamo) the large scale order (particularly vorticity) of the cloud! This raises the spectre of the status of dynamo theory in a region of compressible turbulence.

Indeed Hoyng (1987a,b 1988), in some perceptive and iconoclastic work, has already raised the issue of dynamo theory, and has criticized the usual turbulent treatment as being too deterministic. I tend to agree with him, and in fact I, together with a student B.Chaboyer, have been developing a statistical theory for magnetic 2-point correlations, following the work of Henriksen (1988). I have only time here to present some preliminary results.

The theory is developed in terms of two tensors

$$\begin{aligned} U^{i,j} &\equiv \langle B^i B'^j \rangle, \\ S^{ij,k} &\equiv \langle v^i B^j B'^k \rangle, \end{aligned} \tag{27}$$

where $\langle () \rangle$ denotes the ensemble average of () and quantities are evaluated at 2 points O, O'. These tensors are taken to be locally isotropic so that they have many symmetries. Then by taking moments of the MHD equation in the form

$$\frac{\partial \mathbf{B}}{\partial t} = \nabla \wedge (\mathbf{v} \wedge \mathbf{B}) - \nabla \wedge \left(\frac{\eta}{4\pi} \nabla \wedge \mathbf{B} \right), \tag{28}$$

and averaging with the use of the tensor symmetries (plus $\nabla'_k S^{ij,k} = 0$) gives the elegant result ($\mathbf{L}$ is $\mathbf{OO'}$)

$$\partial_t U = \left(\frac{\eta}{2\pi L^2} \right) \partial_L \left(L^2 \partial_L U \right) + \left(\frac{4}{L^2} \right) \partial_L \left(L^2 S \right), \tag{29}$$

where

$$U \equiv Tr(U^{i,j}) \qquad S \equiv S^{12,2} - S^{21,2}, \tag{30}$$

and η need not be constant.

The quantity S is a source term related to the lack of mirror symmetry. Moreover from $\nabla_i U^{i,j} = 0$ there follows

$$U = \left(\frac{1}{L^2} \right) \partial_L \left(L^3 U^{1,1} \right). \tag{31}$$

Now these relations do not form a closed set, and indeed they would not, even if we sought another averaged relation from the Navier-Stokes equations, because this would involve $R^{i,j}$, $R^{ij,k}$ (see Henriksen 1988) and $F^{ij,k} \equiv \langle B^i B^j v'^k \rangle$. The only general solution seems to be by way of *self-similar closure* (Henriksen 1988).

However in very simple cases, some conclusions may be based on these equations alone. Thus, restoring mirror symmetry so that $S = 0$ and treating only a steady state, we find the solution

$$\begin{aligned} U &= -C_1/L + C_2, \\ U^{1,1} &= C_2/3 - C_1/2L + C_3/L^3, \\ U^{2,2} &= -C_1/4L + C_2/3 - C_3/2L^3, \end{aligned} \tag{32}$$

where C_1,C_2,C_3 are arbitrary real constants. Thus the correlations all tend to be constant at large separations, and the trace increases inwardly as L^{-1}. Since $\rho \propto L^{-1}$ in these regions by observation, we see that the 2-point magnetic energy satisfies $U \propto \rho$ at small scales. At large scales both U and ρv^2 become constants, which suggests equipartition conditions. The field of course never *controls* the medium in these calculations. I suspect that we have good Darwinian survivability here , but much more work remains to be done. With that, I rest my case.

References

Asseo,E., Kennel, C.F., and Pellat, R. 1975, Astr.Ap. **44**,31.
Blandford, R., and Payne, D. 1982, M.N.R.A.S., **199**,883.
Blandford, R.D., and Znajek, R.L. 1977, M.N.R.A.S. **179**,433.
Choe, Seung-U., and Henriksen, R.N. 1986, Ap.J. **305**,131.
Cohen, J.M., and Rosenblum, A. 1972, Ap. and Sp.Sci. **6**,130.
Endean, V.G. 1972a, Nature Phys.Sci. **237**,72.
Endean, V.G. 1972b, M.N.R.A.S. **158**.
Fitzpatrick, R. and Mestel, L. 1988, M.N.R.A.S. **232**,277(I),303(II).
Gold, T. 1968, Nature **218**,731.
Gold, T. 1969, Nature **221**,25.
Goldreich, P.G., and Julian, W.H. 1969, Ap.J. **157**,869.
Goldreich, P.G., and Julian, W.H. 1970, Ap.J. **160**, 971.
Gunn, J.E., and Ostriker, J.P. 1969, Phys.Rev.Lett. **22**,728.
Hardee, P. 1988, These proceedings.
Henriksen,R.N., and Rayburn, D.R. 1971, M.N.R.A.S. **152**,323.
Henriksen,R.N., and Rayburn, D.R. 1972, Ap. Lett. **11**,107.
Henriksen, R.N., and Chia, T.T. 1972, Nature Phys.Sci. **240**,133.
Henriksen, R.N., and Reinhardt, M. 1974, Astr.Ap. **28**,47.
Henriksen, R.N., and Rayburn, D.R. 1974, M.N.R.A.S. **166**,409.
Henriksen, R.N., and Norton J.A. 1975, Ap.J. **201**,431(a),719(b).
Henriksen, R.N., and Reinhardt, M. 1977, Ap. and Sp.Sci. **49**,3.
Henriksen, R.N. 1976, Ap.J.(Lett.) **210**,L19.
Henriksen, R.N. 1988, Ap.J. **331**,359.
Heyer, M.H. et al. Ap.J. **321**,855.
Hoyng, P. 1987, Astr.Ap. **171**,348(a),357(b).
Hoyng, P. 1988, Ap.J. **332**, 857.
Icke, V. 1980, Astr.J. **85**, 329.
Leblanc, J.M., and Wilson,J.R. 1970, Ap.J. **161**,541.
Lovelace, R., Wang, J.C.L., and Sulkanen, M.E. 1987, Ap.J. **315**,504.

Madej, J., Loken, C., and Henriksen, R.N. 1987, Ap.J. **312**,652.
Mestel, L. 1971, Nature Phys.Sci. **233**,149.
Mestel, L. 1973, Ap. and Sp.Sci. **24**,289.
Michel, F.C. 1969, Ap.J.**158**, 727.
Michel, F.C., and Dessler, A.J. 1981, Ap.J. **251**,654.
Michel, F.C. 1987, Ap.J. **321**, 714.
Mitchell, G.M., Allen, M., and Maillard, J.-P. 1988, Ap.J.(Lett.) **333** ,L55.
Pudritz, R. and Norman, C. 1986, C.J.P. **64** , 407 and references therein.
Scharlemann, E.T. 1974, Ap.J. **193**,217.
Scharlemann, E.T., and Wagoner, R.V. 1973, Ap.J. **182**,951.
Shu, F.H., Lizano, S., Ruden, S.P., and Najita, J. 1988, Ap.J.(Lett.) **328**,L19.
Shukra, C., Manchester, R.N., and Allen, D.A. 1983, Nature **303**,501.
Torbett, M.V. 1986, C.J.P. **64**.
Weber, E.J., and Davis, L.Jr. 1967, Ap.J. **148**,217.

HYDROMAGNETIC FLOWS FROM RAPIDLY ROTATING COMPACT OBJECTS AND THE FORMATION OF RELATIVISTIC JETS

M. CAMENZIND
Landessternwarte Königstuhl
D-6900 Heidelberg 1
Federal Republic of Germany

ABSTRACT. Strong plasma outflows can be driven by plasma injected into the magnetosphere of rapidly rotating compact objects. The rapid rotation also builds up a strong toroidal component of the magnetic field, which efficiently pinches the outflowing plasma leading to the formation of collimated relativistic jets. The appearance of the light cylinder due to this rapid rotation requires a fully relativistic treatment of the entire problem. We present the most general theory of axisymmetric relativistic plasma flows including the effects of the gravitational background, the structure of the magnetic flux tube and the finite pressure of the plasma. With this reduction of the relativistic MHD we end up with a nonlinearly coupled system consisting of the hot wind equation and the relativistic Grad-Schlüter-Shafranov equation for the poloidal magnetic flux. We show self-consistent computer solutions of this system resulting in the formation of supermagnetosonic jets flowing along the rotational axis of the central object.

1. Introduction

Magnetized winds are produced by various types of rotating systems. Rapidly rotating young stars and disks in star forming regions can drive strongly magnetized winds (Pudritz, 1988). Winds and jets might also be driven by rapidly rotating compact objects and accretion disks in close binary systems. Finally, Radio Galaxies (such as Cyg A) and Quasars are well known jet sources indicating that plasma and magnetic fields must be driven away from a central region not bigger than a few light years into the large scales of a few hundreds of kiloparsecs. If superluminal motion as observed in some of these Quasars is due to the underlying bulk speed, these jets must have Lorentz factors between 2 and 20. Such high Lorentz factors can be achieved by means of the magnetic sling effect, which occurs when plasma is injected into a rapidly rotating magnetosphere of a supermassive disk in the center of a galaxy.

The description of such highly relativistic magnetized plasma flows must be based on relativistic magnetohydrodynamics (MHD). In the following, we present the most general axisymmetric theory for adiabatic relativistic plasma flows including the gravitational background of the rapidly rotating object (either for a

G. Belvedere (ed.), Accretion Disks and Magnetic Fields in Astrophysics, 129–143.

non-collapsed fast rotator or for a rotating black hole), the structure of the magnetic flux tube and finite pressure effects. By using the general conservation laws on a stationary and axisymmetric gravitational background, we arrive at an explicit equation for the poloidal velocity of the plasma flow. We also discuss the equation for the structure of the magnetosphere, which is known in the laboratory as the Grad-Schlüter-Shafranov equation. Both equations form a nonlinear set of equations which have to be solved simultaneously on a computer. In this way, self-consistent solutions can be constructed for various inner and outer boundary conditions.

We consider in particular flows initiated from a disk surrounding a supermassive rotator and flows directly injected into the magnetosphere of an accretion disk. We, however, do not discuss the origin of the magnetic fields in the disk, but assume their existence by giving a particular surface flux distribution. As a result, we find solutions with semi-relativistic speeds for supermassive objects and collimation of these outflows towards the rotational axis due to the strong pinch effect exerted by the toroidal component of the magnetic field. Finally, we discuss the *universality* of the magnetic sling effect, which can be applied to outflows from young stellar objects, outflows from rapidly rotating neutron stars and white dwarfs and to jets formed in the center of galaxies.

2. Magnetized Disk-Winds

Accretion processes occur in many types of astrophysical systems. There is great evidence for the existence of accretion disks around young stellar objects, and for collimated outflows driven by the magnetic field of the rotating star or the circumstellar disk (Pudritz and Norman, 1986). Accretion disks also exist in binary systems consisting of a star filling its Roche-lobe and a neutron star or a white dwarf. In particular, low-mass X-ray binary systems (LMXBs) consist of a rapidly rotating neutron star and a low-mass companion star, which is feeding an accretion disk via Roche-lobe overflow. In systems with binary periods less than about 10 hours, the accretion rate in the disk is about 10^{-10} $M_\odot$ yr^{-1}, systems with larger binary periods have accretion rates close to the Eddington accretion rate for neutron stars, $\dot{M} \simeq 2 \cdot 10^{-8}$ $M_\odot$ yr^{-1} (Verbunt, 1988). When the surface magnetic field of neutron stars settles to values around $B_* \simeq 10^9$ Gauss, the interaction between the inner part of those accretion disks and the magnetopshere leads to a spin-up of the neutron star until it reaches an equilibrium period P_{eq} given by

$$P_{eq} = 1.5\,\mathrm{ms} \left(\frac{\alpha I_{N+1}}{0.4}\right)^{\frac{3}{7}} \left(B_{*,9}\, R_{*,6}\right)^{\frac{6}{7}} \left(\frac{H}{\dot{M}_{18}\, R}\right)^{\frac{3}{7}} \left(\frac{R_*}{2.5 R_G}\right)^{\frac{3}{14}}. \tag{1}$$

α is the viscous parameter for standard accretion disks, $I_{N+1} \simeq 0.4$, R_* is the radius of the neutron star, R_G its Schwarzschild radius, and the relative height H/R of the disk is of order unity for these high accretion rates. As a result, we get a neutron star rotating close to the critical frequency and an inner edge of the disk located at a few neutron star radii

$$R_{\rm in} = 2.8\,R_* \left(\alpha\, I_{N+1}\right)^{2/7} \left(B_{*,9}\, R_{*,6}\right)^{4/7} \left(\frac{H}{\dot{M}_{18}\, R}\right)^{2/7} \left(\frac{R_*}{2.5 R_G}\right)^{1/7} \tag{2}$$

In the rotational equilibrium, this inner edge coincides with the corotation radius. The interaction of this rapidly rotating magnetosphere with the disk plasma outside of the inner edge could lead to mass-outflow driven by centrifugal forces (Anzer and Börner, 1983; Arons, 1987). Angular momentum lost by this outflow will then balance the angular momentum gained by the neutron star over the inner accretion process. We end up therefore in LMXBs with a magnetized rapid rotator and mass-outflow driven by this rotation.

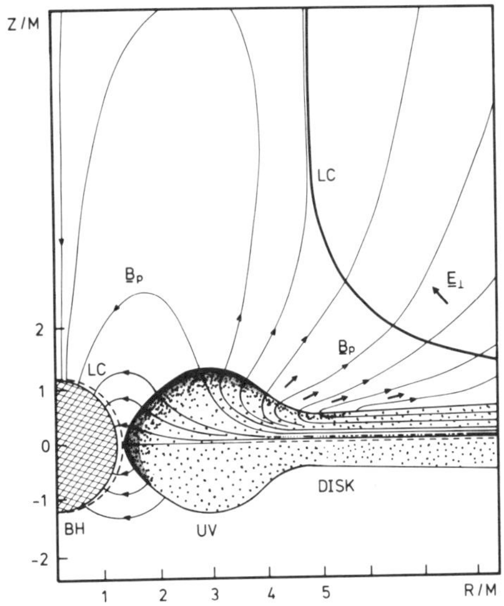

Fig. 1. The interaction of a strongly magnetized disk with a rapidly rotating black hole, a = 0.98. The angular momentum of the disk is carried away by a magnetized wind initiated from a corona above the disk. This wind flows through the light cylinder (LC), which has a complicated structure due to the differential rotation of the field lines.

Similar systems might occur in the center of active galaxies and Quasars (Camenzind, 1986, 1988). The formation of rapidly rotating compact objects is very probable in the center of a galaxy. Matter around the center of a galaxy has in general too much angular momentum for the onset of a dramatic collapse, the formation of a centrally condensed object (a rapid rotator or perhaps a rapidly rotating black hole) and an accretion disk around this object is a natural consequence of the excessive angular momentum. For those systems, mass loss is an essential ingredient in the evolution of supermassive rotators and accretion disks. It is interesting in this respect that the accretion disk will also have an inner edge at a few supermassive stellar radii provided the central object carries a magnetic field of at least a few kiloGauss (Camenzind, 1986)

$$R_{\rm in} = 2.5\, R_* \left(\alpha\, I_{N+1}\right)^{2/7} \left(B_{*,4}\, R_{*,15}\right)^{4/7} \left(\frac{H}{\dot{M}_{26}\, R}\right)^{2/7} \left(\frac{R_*}{40 R_G}\right)^{1/7} . \tag{3}$$

Due to the appropriate scaling of the various parameters, these systems are just a scaled up version of the neutron stars in LMXBs.

When the central object has collapsed to form a rapidly rotating black hole, the accretion disk itself can build up a strong magnetosphere (Blandford and Payne, 1982; Lovelace et al., 1987). Here, plasma can be injected into this magnetic structure from a corona above the disk and lead to supermagnetosonic outflows driven away centrifugally (see Fig. 1). This magnetized wind could extract the entire angular momentum of the disk.

3. Relativistic MHD for Axisymmetric Configurations

In this section we show how the MHD equations can be reduced to a tractable set of equations on a general stationary and axisymmetric gravitational background given by the metric (with signature $+---$)

$$ds^2 = g_{tt}\, dt^2 + 2g_{t\phi}\, dt\, d\phi + g_{\phi\phi}\, d\phi^2 + g_{AB}\, dx^A dx^B . \tag{4}$$

This metric represents the exterior gravitational field of either a rigidly rotating non-collapsed object (a neutron star or a supermassive rotator) or of a rotating Kerr black hole. The basic equations of MHD (Lichnérowicz, 1976) are then particle number conservation

$$N^\alpha_{;\alpha} = 0 \quad , \quad N^\alpha = n u^\alpha \, , \tag{5}$$

magnetic flux conservation

$$(u^\alpha\, B'^\beta - u^\beta\, B'^\alpha)_{;\alpha} = 0 \, , \tag{6}$$

and conservation of total energy-momentum

$$T^{\alpha\beta}_{;\beta} = 0 \, , \tag{7}$$

where $T^{\alpha\beta}$ is the total energy-momentum of the system

$$T^{\alpha\beta} = (\rho + P)\, u^\alpha\, u^\beta - (P + \frac{B'^2}{8\pi})\, g^{\alpha\beta} - \frac{1}{4\pi}\, B'^\alpha\, B'^\beta . \tag{8}$$

n is the proper particle density, ρ the total energy density, P the pressure, u^α the four-velocity of the plasma, and B' the magnetic field as seen in the plasma frame

$$B'_\alpha = u^\beta * F_{\alpha\beta} \quad . \tag{9}$$

The corresponding electric field E' seen by the comoving plasma vanishes due to the MHD assumption $u^\alpha F_{\alpha\beta} = 0$. We found in the last years a way to reduce this complicated system of equations for *axisymmetric* configurations.

3.1. THE QUESTION OF THE ALFVEN POINT

Axisymmetric MHD flows can be described in terms of several invariants. A full account for this procedure can be found in (Camenzind, 1986, 1987a,b). The essential idea is based on the existence of two Killing fields $k = \partial_t$ and $m = \partial_\phi$ on stationary and axisymmetric space-times. The basic conservation laws, which follow from these Killing fields, have been written down in a covariant way by Bekenstein and Oron (1978).

Stationarity and axisymmetry of the flow provide the existence of a stream-function $\Psi = \Psi(x^A)$, which is constant on poloidal flow lines. As a consequence of the MHD condition and of the symmetries of the Maxwell field F, the electromagnetic field is restricted to the form

$$F_{t\phi} = 0, \tag{10}$$

$$F_{tA} + \Omega^F(\Psi)\, F_{\phi A} = 0, \quad A = 1,2. \tag{11}$$

$\Omega^F(\Psi)$ is constant along flow lines and represents the angular frequency of the field lines. A second integration constant $\eta(\Psi)$ describes the particle-flux per unit flux tube

$$\sqrt{-g}\, n\, u^{1(2)} = -\eta(\Psi)\, F_{23(31)}, \tag{12}$$

$$\sqrt{-g}\, n\, u^t\, (\Omega - \Omega^F) = -\eta(\Psi)\, F_{12}. \tag{13}$$

$\Omega = u^\phi / u^t$ is the angular frequency of the plasma flow. Due to the existence of the two Killing fields, we find two conserved currents derived from the energy-momentum tensor, $P^\alpha_E = T^\alpha_\beta\, k^\beta = EN^\alpha$ and $P^\alpha_L = -T^\alpha_\beta\, m^\beta = LN^\alpha$. They involve the total energy E and the total angular momentum L along the flow and contain the redshift factor u_t and the specific angular momentum $l = -u_\phi / u_t$. With suitable expressions for the magnetic fields (see Camenzind, 1987a) we can solve these expressions for u_t and l

$$u_t = \frac{E}{\mu}\,\frac{1 - M_g^2 - \Omega^F L/E}{1 - M_g^2 - \Omega^F\, l(\Omega^F)}, \tag{14}$$

$$l = \frac{l(\Omega^F)(1 - \Omega^F L/E) - M_g^2\, L/E}{1 - M_g^2 - \Omega^F L/E}. \tag{15}$$

M_g is a rescaled Mach number defined as

$$M_g^2 = \frac{M^2}{g_{tt} + \Omega^F g_{t\phi}} \tag{16}$$

for the Alfven Mach number M, $M^2 = 4\pi\mu\eta^2/n$, $\mu = (\rho+P)/n$ being the relativistic specific enthalpy. The quantity $l(\Omega^F)$ is the angular momentum of a corotating system with respect to the zero angular momentum frame defined by $\omega = -g_{t\phi}/g_{\phi\phi}$

$$l(\Omega^F) = -\frac{g_{\phi\phi}\,(\Omega^F - \omega)}{g_{tt} + \Omega^F g_{t\phi}}\,. \tag{17}$$

It is important to note that the Mach number M is formed with the relativistic enthalpy $\mu = (\rho + P)/n$ and not with the rest mass. Secondly, there is a relativistic correction of the form $\Omega^F L/E$, which is important when magnetic fields carry an appreciable amount of energy. The third correction is of gravitational origin and involves the metric coefficients g_{tt} and $g_{t\phi}$. The redshift factor u_t contains as an additional correction in the denominator the term $\Omega^F\, l(\Omega^F)$, which is due to the existence of the light cylinder. In general, a "light cylinder" $\Omega^F\, l(\Omega^F) = 1$ means that the corotating field

$$\xi = k + \Omega^F m \tag{18}$$

spans a null surface defined by the condition $(\xi, k) = 0$, i.e.

$$g_{tt}(X_L) + 2\Omega^F g_{t\phi}(X_L) + (\Omega^F)^2 g_{\phi\phi}(X_L) = 0. \tag{19}$$

In the absence of gravity, this is the well known relation for the light cylinder in a flat space time, $R_L = c/\Omega^F$. In a Kerr space time e.g. we also find an inner light cylinder surface located close to the horizon of the black hole (Camenzind and Endler, 1988; see Fig. 1).

The position of the Alfven point now follows from the critical points of the expressions for the redshift factor u_t and the specific angular momentum l. The equation for the redshift factor u_t would become singular at the point $x = x_A$ defined by

$$M_{gA}^2 = 1 - \Omega^F l_A(\Omega^F), \tag{20}$$

unless we require at the same time

$$M_{gA}^2 = 1 - \Omega^F L/E. \tag{21}$$

From these two relations we get the position x_A in terms of

$$l_A(\Omega^F) = L/E. \tag{22}$$

It is easy to show that the expression for the specific angular momentum becomes singular at the same point, but does not produce any additional constraints. The above expression reduces to $R_A^2 \Omega^F = L/E$ in Minkowski space and is well known in the Newtonian case, $R_A^2\,\Omega^F = L/m$.

The expressions found above for the redshift factor u_t, the specific angular momentum l and the location of the light cylinder and Alfven surface are completely general and valid for any stationary and axisymmetric space time. They can be applied to any flow along a given magnetic flux tube anywhere in the meridional plane. This procedure generalizes the relations found by Okamoto (1978) for cold plasma flows on Minkowski space and the solutions found by Kennel et al. (1983) for the equatorial plane in a monopole geometry including pressure effects for a pair plasma. Since $\Omega^F L/E \leq 1$, relativity automatically requires the Alfven point to be inside the light cylinder.

3.2. THE HOT WIND EQUATION FOR ADIABATIC FLOWS

The poloidal velocity of the plasma streaming along a given flux tube in the rotating magnetosphere follows from the normalisation of the 4-velocity, $u^\alpha u_\alpha = 1$. Since the redshift factor u_t and the specific angular momentum l are determined by the conservation laws for the total energy and angular momentum, the solution of the poloidal velocity defined as $u_p^2 = -u^A u_A$, $A = 1,2$, follows from the expressions discussed in the last section

$$u_p^2 + 1 = \frac{u_t^2}{\Delta}(-g_{\phi\phi} - 2l\, g_{t\phi} - l^2\, g_{tt}). \tag{23}$$

$\Delta = g_{t\phi}^2 - g_{tt}\, g_{\phi\phi}$ is the square of the Weyl radius for axisymmetric metrics. The above equation for u_p is still implicit, since the Mach number M also contains the poloidal velocity. By using the definition of η, $\eta/n = (g_{tt} + \Omega^F g_{t\phi})u_p/B_p$, $B_p^2 = -B^A B_A$, $x = R/R_L$, we can, however, decompose this Mach number in a suitable way as

$$M^2 = 4\pi m\eta c^2 \,\frac{\mu}{mc^2}\,\frac{\eta}{n} = \frac{\mu}{mc^2}\,(g_{tt} + \Omega^F g_{t\phi})\, u_p\, x^2\, \frac{f(R)}{\sigma_*(\Psi)}, \tag{24}$$

with a similar decomposition for the specific enthalpy μ

$$\mu = mc^2 \left\{1 + \mu_1 \left(\frac{u_{\mathrm{p,in}}\, x_{\mathrm{in}}^2}{u_p^2 x^2}\right)^{\Gamma-1}\right\}, \tag{25}$$

$$\mu_1 = \frac{\Gamma}{\Gamma - 1}\,\frac{P_{\mathrm{in}}}{n_{\mathrm{in}}\, mc^2}\, f^{\Gamma-1}, \tag{26}$$

$$f(R) = \frac{B_{\mathrm{p,in}}\, R_{\mathrm{in}}^2}{B_p R^2}, \tag{27}$$

$$\sigma_*(\Psi) = \frac{B_{\mathrm{p,in}}\, R_{\mathrm{in}}^2}{4\pi\eta(\Psi) mc R_L^2}. \tag{28}$$

With all these parameters, the above implicit equation for u_p can be transformed to a polynomial of degree 16 for a polytropic index $\Gamma = 5/3$, when the quantity $z_p = u_p^{1/3}$ is used instead of u_p

$$\sum_{n=0}^{16} A_n\left(\frac{R}{R_L}, E, L, \frac{P_{\mathrm{in}}}{n_{\mathrm{in}} mc^2}; \frac{f}{\sigma_*}; g_{tt}, g_{t\phi}, g_{\phi\phi}\right) z_p^n = 0. \tag{29}$$

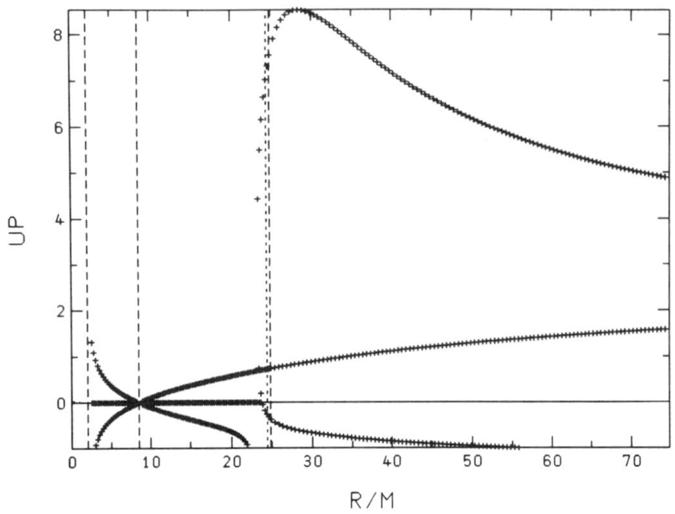

Fig. 2. The various solutions of the hot wind equation in the background of a Kerr black hole ($a = 0.1$) for the following parameters: $P_{\rm in}/n_{\rm in}mc^2 = 0.00025$, $u_{p,\rm in} = 0.03$, $\sigma_* = 20.0$, with a critical energy of $E_c = 24.164\,mc^2$, and $R_A = 0.983\,R_L$. The cylindrical radius R is in units of M.

The detailed form of the coefficients A_n for the hot wind equation can be found in Camenzind and Endler (1988). The above coefficients generalize the form of the hot wind equation derived for Minkowski space in Camenzind (1987a). The solutions of this hot wind equation depend on the initial parameters $R_{\rm in}$, R_L, $B_{p,\rm in}R_{\rm in}^2$, $u_{p,\rm in}$, $P_{\rm in}/n_{\rm in}mc^2$, and the particle injection law $\eta(\Psi) = d\dot{N}/d\Psi$. Of central importance for the acceleration of the plasma is the form of the flux tube function f, defined by equation (27). f is constant for a monopole geometry, but has a complicated form for an open flux tube in the rapidly rotating magnetosphere. The equation for the poloidal velocity (or for the Lorentz factor) has two additional critical points, the slow and fast magnetosonic points. The parameters L and E are now constrained by the requirement that the physical wind solution passes through the Alfven point and the fast magnetosonic point.

In Fig. 2 we show the various branches of solutions of the hot wind equation for a finite initial pressure in the background of a Kerr black hole with a constant flux tube, f = constant, $R_L = 25\,M_H$. The inner dashed line corresponds to the horizon of the hole. The injection of plasma occurs near the radius of maximum energy production in the disk. The physical wind solution starts with finite velocity near the slow magnetosonic point, crosses an unphysical branch at the Alfven point near the light cylinder and a second time at the fast magnetosonic point, and reaches finally a constant outflow velocity. Since the inertia of the plasma is included in our treatment, the light cylinder is no critical point of the wind equation. Without inclusion of gravity, the slow magnetosonic point would not appear in the solutions.

3.3. THE RELATIVISTIC GRAD-SCHLÜTER-SHAFRANOV EQUATION

One of the essential points of the computation of self-consistent MHD winds is the particular form of the toroidal current j_ϕ, which determines via Ampère's equation

$$\nabla \wedge \mathbf{B}_p = \frac{4\pi}{c} j_\phi \mathbf{e}_T , \quad \mathbf{B}_p = \frac{1}{R} (\nabla \Psi \wedge \mathbf{e}_T) , \tag{30}$$

the structure of the inner and outer magnetosphere. From here on we neglect the gravitational backround, since this gives an unnecessary complication of the problem. With the introduction of the magnetic flux function Ψ, which we now use as our stream function, this equation can be transformed to divergence equation of the form

$$R\nabla \cdot \left\{ \frac{1}{R^2} \nabla \Psi \right\} = -\frac{4\pi}{c} j_\phi . \tag{31}$$

It is well known that for non-rotating magnetic structures and static plasma equilibria, the current is simply given by two functions of Ψ (Semenzato et al., 1984)

$$j_\phi = cR\, \partial_\Psi P(\Psi) + cB_\phi\, \partial_\Psi (RB_\phi) . \tag{32}$$

Plasma confinement in the laboratory and in solar filaments is based on this equation, known as the Grad-Schlüter-Shafranov (GSS) equation. When plasma is moving along the magnetic flux tubes, the form of this current follows from the general force balance perpendicular to the flux tubes. For this purpose one considers the relativistic Lorentz force

$$n(\mathbf{u} \cdot \nabla)(\mu \mathbf{u}) + \nabla P = \rho_e \mathbf{E} + \frac{1}{c} (\mathbf{j} \wedge \mathbf{B}) . \tag{33}$$

The projection of this equation perpendicular to the flux tubes provides then the toroidal current j_ϕ (for a derivation, see Lovelace et al., 1986; Camenzind, 1987b). In the traditional approach to pulsar magnetospheres (Michel, 1973, 1982) and jet formation from disks (Lovelace et al., 1987), the inertial terms on the left hand side are neglected (massless limit). This must, however, lead to complications beyond the Alfven surface, where inertial effects of the plasma are important for the decoupling from the magnetic structure. All the troubles with the boundary conditions at the light cylinder (see e.g. Michel, 1982) come from this assumption, unless only sub-Alfvenic solutions are treated as in Lovelace et al. (1987).

The current density resulting from this general force-balance has various components, $j_\phi = j_\phi^{(n)} + j_\phi^{(0)} + j_\phi^{(GJ)}$, inertial currents, a current due to the toroidal field and a Goldreich-Julian current

$$j_\phi^{(GJ)} = c \frac{R}{R_L} \frac{\rho_{GJ}}{1 - M^2 - x^2} , \quad \rho_{GJ} = -\frac{\Omega^F B_z}{2\pi c} . \tag{34}$$

This latter current and part of the inertial currents can be absorbed by the left hand side of Ampère's equation. In this way we obtain the relativistic version of the *Grad-Schlüter-Shafranov* equation

$$R\nabla \cdot \left\{ \frac{1 - M^2 - x^2}{R^2} \nabla \Psi \right\} = -\frac{4\pi}{c} j . \tag{35}$$

For rigidly rotating magnetospheres, $\Omega^F(\Psi) = \Omega_*$, the remaining current j has still two different components

$$j = j^{(n)} + j^{(0)}, \tag{36}$$

the first one being of plasma origin, the second one due to the toroidal magnetic field and particle injection

$$j^{(n)} = \mu\gamma ncR\,\{\frac{1}{\mu}(\partial_\Psi E - \Omega\,\partial_\Psi L) + \frac{1}{\rho + P}\,\partial_\Psi P - \gamma\,(1 - \Omega l)\,\partial_\Psi ln(\mu)\}\,, \tag{37}$$

$$j^{(0)} = cR\,\frac{B_\phi^2 + M^2 B_p^2}{4\pi}\,\partial_\Psi ln(\eta)\,. \tag{38}$$

The above current and the toroidal magnetic field are regular at the Alfven surface, which, however, also appears as a critical surface on the left hand side of the GSS equ. (35). Since the Alfven surface is a critical surface of the GSS equation, the nominator in the current j_ϕ must also vanish there. According to equ. (34), this condition restricts the vertical component B_z of the magnetic field. The relativistic GSS equation is elliptic between the surface of the slow magnetosonic points and the surface of the fast magnetosonic points. Outside of this latter surface, the equation becomes hyperbolic, similar to its Newtonian counterpart (Sakurai, 1985).

The classical force-free pulsar equation (Michel, 1973; Scharleman and Wagoner, 1973) follows from this analysis in the low density limit $\eta \mapsto 0$, $E \mapsto \infty$, $L \mapsto \infty$, but $\eta E \mapsto$ finite and $\eta L \mapsto$ finite. In this limit, the Alfven surface moves globally towards the light cylinder, $x_A \mapsto 1$, and energy is only carried away by the Poynting flux. In general, however, $x_A < 1$, and inertia is important outside of the Alfven surface. For applications of the force-free limit to jet solutions, see also Bekenstein and Eichler (1985).

4. Supermagnetosonic Jets from Self-consistent Magnetospheres

We developed numerical schemes for the simultaneous solution of the hot wind equation and the GSS equation for given boundary conditions (Camenzind, 1987b). For this purpose, the GSS equation (35) is brought to a non-dimensional form by rescaling the coordinates in terms of the light cylinder R_L, $X = R/R_L$, $Z = z/R_L$, and the magnetic flux Ψ in terms of the total magnetic flux Ψ_G enclosed by the grid

$$X\,\nabla\cdot\left\{\frac{1 - M^2 - X^2}{X^2}\,\nabla\Psi\right\} = -\frac{1}{\sigma_0^2}\,j\,. \tag{39}$$

The quantity

$$\sigma_0 = \frac{\Psi_G}{4\pi\eta_0 mcR_L^2} \simeq \frac{\Psi_G^2}{2\dot{N}_{\rm inj} mcR_L^2} \tag{40}$$

acts as the inverse of the coupling constant for the toroidal current j, $\Psi_G \simeq 10^{34}$ Gauss cm^2 for supermassive objects, but $\Psi_G \simeq 10^{21}$ Gauss cm^2 for neutron stars in LMXBs.

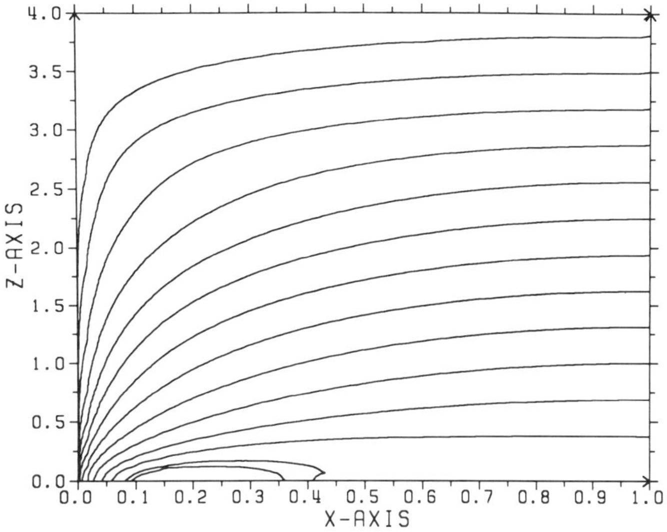

Fig. 3. The inner low-density magnetosphere interacting with an accretion disk sourrounding a rapid rotator. The isocontours show the poloidal magnetic field lines for a given dipole distribution at the stellar surface. The cylindrical coordinates R and Z are in units of light cylinder radii.

The effect of the Goldreich-Julian current, which has been absorbed by the left hand side of the GSS equation, is demonstrated in Fig. 3 for a magnetosphere filled with extremely low-density, $\sigma_0 \gg 1$, and η = constant. In this limit, the Alfven surface coincides with the light cylinder, and the boundary conditions at this Alfven surface, $B_z = 0$, completely deform the inner part of the magnetosphere. When plasma is filled into this structure, the Alfven surface will move away from the light cylinder towards the surface of the compact object. This is a quite dramatic effect when compared with Newtonian situations (Sakurai, 1985), where the Alfven surface is far away from the light cylinder and the inner dipolar structure is not drastically altered. The structure of the magnetic flux tubes near the light cylinder has a great influence on the acceleration of the outflowing plasma.

The GSS equation is solved subject to the outer boundary conditions that the escaping plasma and magnetic fields are in pressure equilibrium with a perfectly conducting ambient medium. The center of a galaxy is not a vacuum region, but is permeated by a plasma with a typical pressure given by the conditions in the broad line region, $P_{\rm ex} \simeq 10^{-2}$ dyn cm^{-2}. Figure 4 and 5 show a self-consistent solution for a black hole surrounded by a disk carrying a prescribed magnetic flux distribution. The outflowing plasma, injected at the foot points of each field line, consists of normal ions and electrons feeded into the magnetosphere from the disk. This outflow encounters a finite pressure at $R \simeq 10\,R_L$, $\sigma_0 \simeq 10$. This solution shows the general features expected to occur for open magnetospheres. The inner magnetosphere is largely deformed due to the sharp boundary conditions required at the Alfven surface (very near to the light cylinder). The magnetic field lines

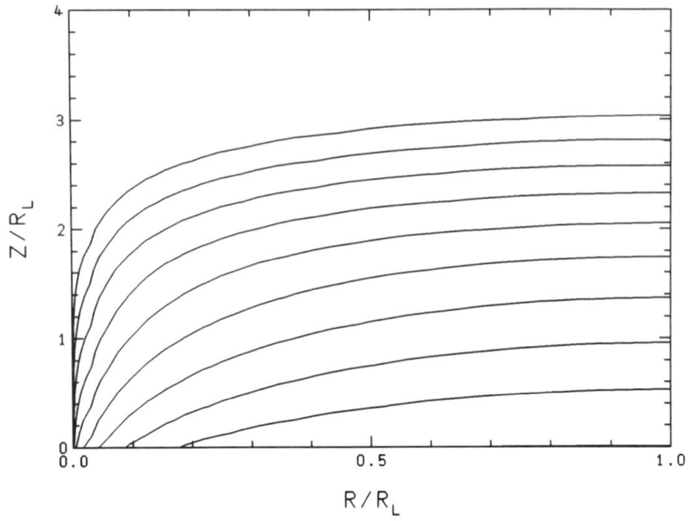

Fig. 4. The inner magnetosphere of an accretion disk around a black hole. The hole itself also carries a magnetic flux accreted over a long period. This part of the magnetosphere has no great influence on the structure of the magnetosphere from the surface of the disk.

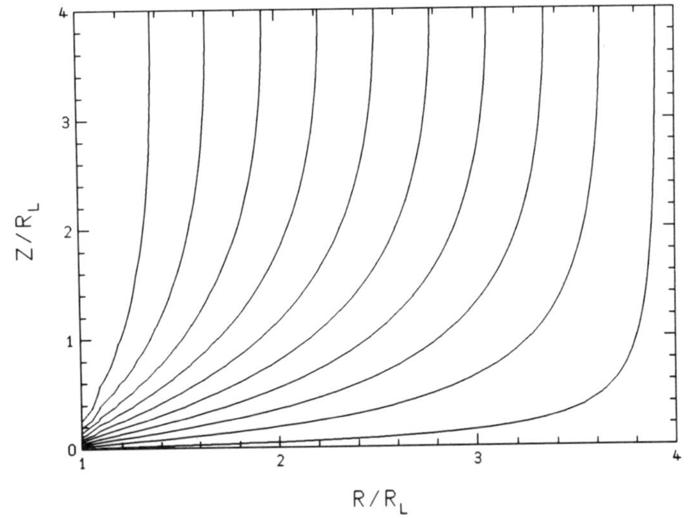

Fig. 5. The exterior magnetosphere corresponding to the interior solution shown in Fig. 4. The outflow is collimated due to the presence of a strong pressure at $R_j \simeq 10\,R_L$. These types of solutions are interesting for the problem of jet formation in AGNs (Camenzind, 1988). The plasma streams along the collimated flux tubes.

cross the Alfven surface without any cusp and the exterior solution depends on the global boundary conditions at infinity. In this example, we assumed that the black hole itself carries some remnant flux swamped in by the accretion process. In a more realistic solution, the field lines from the black hole would stay inside the light cylinder carrying only Poynting flux. This part of the magnetosphere has, however, no essential effects on the global structure of the field lines from the disk which cross the light cylinder.

The deformation of the magnetosphere due the strong boundary conditions at the Alfven surface has a great effect on the acceleration of the plasma. For this purpose, we first give an estimate for the overall scaling parameter σ_*, which essentially determines the asymptotic outflow velocities. For a monopolar magnetosphere, the Lorentz factor at infinity follows from the relation (Michel, 1969; Okamoto, 1978; Camenzind, 1987a)

$$\gamma_\infty = \sqrt{1 + \sigma_*^{2/3}}. \tag{41}$$

For supermassive accretion onto a black hole of 10^9 $M_\odot$, the parameter σ_* is of order 10,

$$\sigma_* \simeq \frac{\Phi_*^2}{2\dot{M}_{\rm jet} c R_L^2} \simeq 10\, \Phi_{34}^2\, \dot{M}_{-1\,\rm jet}^{-1}\, R_{L,16}^{-2} \tag{42}$$

for a mass flux $\dot{M}_{\rm jet} \simeq 0.1$ $\mathrm{M}_\odot$ yr^{-1}, a total magnetic flux $\Phi_* \simeq 10^{34}$ Gauss cm^2 and a light cylinder $R_L \simeq 10^{16}$ cm. These are parameters appropriate for strong Quasars with a central mass $M_H \simeq 10^9$ $\mathrm{M}_\odot$, such as 3C 273 (Camenzind and Courvoisier, 1984).

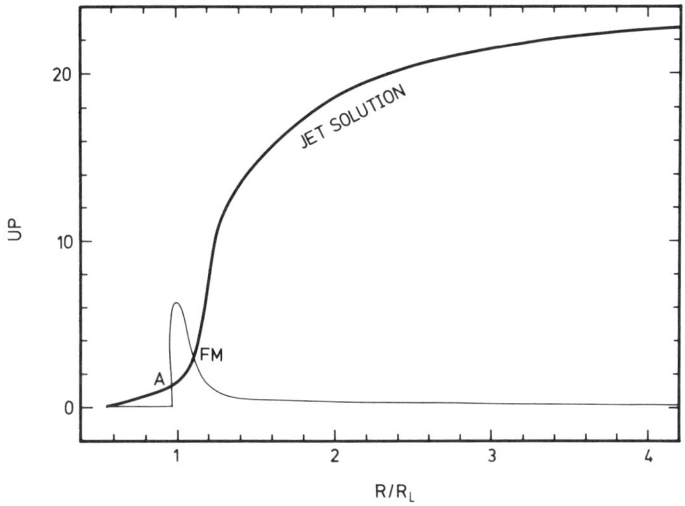

Fig. 6. The solution of the hot wind equation for a realistic flux tube function as determined by the global solutions shown in Fig. 4, 5. The cylindrical radius is given in units of GM_H/c^2, the poloidal velocity in units of c, $u_p = \gamma v_p/c$.

The strong deformation of the magnetic flux tube near the light cylinder has the effect of a *magnetic nozzle* (Camenzind and Endler, 1988). In Fig. 6 we show the solution of the hot wind equation along a magnetic flux tube starting from the inner edge of the accretion disk. We observe a strong acceleration beyond the light cylinder due to this nozzle effect, the final Lorentz factor is now roughly given by the value of σ_* (from numerical experience),

$$\gamma_\infty \leq \sigma_* \quad , \quad \text{for a nozzle.} \tag{43}$$

This nicely conforms with the values of the Lorentz factors required by superluminal motion for Quasars (Porcas, 1987). The magnetic sling effect is the only mechanism which can accelerate a heavy plasma to such high Lorentz factors.

The question of the mass outflow is an interesting point for this type of theory. In a rotational equilibrium, the angular momentum carried inwards by accretion must be balanced by the angular momentum loss due to the outflow, i.e.

$$\dot{M}_{\rm acc}\, j_{\rm in} = \dot{M}_{\rm w}\, R_A^2\, \Omega_*, \tag{44}$$

with $j_{\rm in} = R_{\rm in}^2\, \Omega_{\rm in}$ as the specific angular momentum at the injection point. Since for $\sigma_* > 1$ the Alfven radius is practically the light cylinder radius, we arrive at the following expression for the mass loss

$$\dot{M}_{\rm w} = \dot{M}_{\rm acc}\, (R_{\rm in}/R_L)^2. \tag{45}$$

For all these systems, mass loss is only a small fraction of the accretion rate.

5. The Universality of the GSS Equation

The solutions of the GSS equation depend on three characteristic quantities

a. the value of the coupling constant $\sigma_0 \simeq \sigma_*$,

b. the particle injection function $\eta(\Psi)$,

c. the inner and outer boundary conditions for Ψ.

The GSS equation is completely scale-free in the sense that the absolute scale of a problem is always given in terms of the light cylinder radius. The solutions for a rapidly rotating neutron star with accretion disk have the same form as those for a supermassive rotator surrounded by an accretion disk, or as the solutions for a disk around a supermassive black hole. When $\sigma_0 > 1$, the outflowing plasma can only be collimated on a scale which is much greater than the light cylinder, i.e. $R_{\rm jet} \geq 10\, R_L$. For supermassive objects in the center of a galaxy, the light cylinder is therefore also the fundamental radius which determines the extensions of the jets formed in the central part of the galaxy. For plasma escaping from the innermost part of the accretion disk, we have $R_L \simeq 4\, R_{\rm inj}$, and therefore with $R_{\rm inj} \simeq 10\, M_H$ we then find $R_L \simeq 40\, M_H \simeq 6 \cdot 10^{15}\, M_{H,9}$ cm.

The research reported in this paper is partly supported by the Sonderforschungsbereich 328 in Heidelberg.

6. References

Anzer, U., Börner, G.: 1983, Astron. Astrophys. **122**, 73

Arons, J.: 1987, in IAU Symposium **125** (D.J. Helfand and J.H. Huang eds.) Reidel, Dordrecht, p. 207

Bekenstein, J.D., Oron, E.: 1978, Phys. Rev. **D18**, 1809

Bekenstein, J.D., Eichler, D.: 1985, Ap. J. **298**, 493

Blandford, R.D., Payne, D.G.: 1982, Mon. Not. Roy. Astron. Soc. **199**, 883

Camenzind, M.: 1986, Astron. Astrophys. **156**, 137

Camenzind, M.: 1987a, Astron. Astrophys. **162**, 32

Camenzind, M.: 1987b, Astron. Astrophys. **184**, 341

Camenzind, M.: 1988, in "High Energy Astrophysics", (G. Börner ed.) Springer Proceedings (Springer-Verlag, Berlin), p. 118

Camenzind, M., Courvoisier, T.J.-L.: 1984, Astron. Astrophys. **140**, 341

Camenzind, M., Endler, M.: 1988, in preperation

Kennel, C.F., Fujimura, F.S., Okamoto, I.: 1983, J. Ap. Geophys. Fluid Dyn. **26**, 147

Lichnérowicz, A.: 1976, "Relativistic MHD", Benjamin, N.Y.

Lovelace, R.V.E., Mehanian, C., Mobarry, C.M., Sulkanen, M.E.: 1986, Ap. J. Suppl. **62**, 1

Lovelace, R.V.E., Wang, J.C.L., Sulkanen, M.E.: 1987, Ap. J. **315**, 504

Michel, F.C.: 1969, Ap. J. **157**, 1183

Michel, F.C.: 1973, Ap. J. Letters **180**, L133

Michel, F.C.: 1982, Rev. Mod. Phys. **54**, 1

Okamoto, I.: 1978, Mon. Not. Roy. Astron. Soc. **185**, 69

Porcas, R.W.: 1987, in "Superluminal Radio Sources" (J.A. Zensus and T.J. Pearson) Cambridge University Press, Cambridge, p. 12

Pudritz, R.E.: 1988, in "Galactic and Extragalactic Star Formation" (R.E. Pudritz and M. Fich eds.) Kluwer, Dordrecht, p. 135

Pudritz, R.E., Norman, C.: 1986, Ap. J. **301**, 571

Sakurai, T.: 1985, Astron. Astrophys. **152**, 121

Scharleman, E.T., Wagoner, R.V.: 1973, Ap. J. **182**, 951

Semenzato, S., Gruber, R., Zehrfeld, H.R.: 1984, Comp. Phys. Report **1**, 389

Verbunt, F.: 1988, in "The Physics of Neutron Stars and Black Holes" (Y. Tanaka ed.) Kyoto, in press

THREE DIMENSIONAL SIMULATION OF ACCRETION DISKS WITH SMOOTHED PARTICLE HYDRODYNAMICS

D. A. P. MOLTENI and G. GIANNONE
Istituto di Fisica dell'Università, via Archirafi 36
90123 Palermo
Italy

ABSTRACT. We performed three dimensional simulations of the mass transfer and the accretion disk formation in a close binary system. We used the Smoothed Particle Hydrodynamic interpolation method to solve the fluid dynamic equations. This method seems quite promising for its simplicity and efficient computing profile. Our results are compared to standard theoretical and numerical models. We find that the stream of cold gas falling on the compact star forms the disk in the conventional way, a strong shock develops and penetrates deep into the disk. However, since our numerical viscosity is still high we cannot establish if spiral shocks are forming.

1. Introduction

The process of mass transfer from a normal star onto a compact one, forming a close binary system, is clear, at least in its general features. The physics of the phenomena occurring in the accretion disk around the collapsed component is also known, but the actual disk structure is far from being clear.

The usual models are essentially based on the hypothesis of keplerian motion of the gas and on "ad hoc" prescriptions for the viscosity (see Verbunt, 1982 for a review). Recently numerical simulations by Sawada et al. (1987) found that spiral shocks develops within the accretion disk around the compact star in a close binary system. This could be an important alternative to the conventional turbulent viscosity regime.

In this context numerical simulations are very useful. An important point is the reliability of the numerical procedure used to integrate the differential equations of the fluid dynamics. Comparing different methods results profitable to establish beyond doubt the physical properties of the revealed structures. Since this method is very convenient for 3D problems we started to study three dimensional phenomena in accretion disks, here we refer to some preliminary runs.

2. The method

The Smoothed Particles Hydrodynamic method (SPH) is essentially a lagrangean scheme where points represent parcel of fluid moving according pressure and body forces. A full account of the method can be found in Monaghan (1985). Summarizing we can say that every bounded function can be approximated by the interpolating formula:

G. Belvedere (ed.), Accretion Disks and Magnetic Fields in Astrophysics, 145–150.

$$\rho(\mathbf{r}) = \int W(\mathbf{r}-\mathbf{r}', h)\rho(\mathbf{r}')d\mathbf{r}'$$

where W is the interpolating Kernel, a continuous function whose width is h and that in the limit for h->0 it is the Delta distribution function. Applying this transform to the fluid equation in the rotating frame we obtain the following set of equations for the motion and the energy of i-th "pseudoparticle":

$$\frac{D\mathbf{v}_i}{Dt} = -m\sum_j\left(\frac{p_j}{\rho^2_j} + \frac{p_i}{\rho^2_i}\right)\nabla_i W_{ij} + \mathbf{g}_i$$

$$\frac{D\varepsilon_i}{Dt} = \frac{m}{2}\sum_j\left(\frac{p_j}{\rho^2_j} + \frac{p_i}{\rho^2_i}\right)\mathbf{v}_{ij}\cdot\nabla_i W_{ij}$$

where

$$\mathbf{g}_i = -2\omega\,\mathbf{k}\wedge\mathbf{v}_i + \omega^2\mathbf{r}_i - \nabla\phi_{grav}$$

D_/Dt stands for the usual substantial derivative. The continuity equation is identically satisfied in this scheme. The method is excellent in conserving energy and angular momentum.

An artificial viscosity is however needed to handle shocks. We used the artificial viscosity by Lattanzio et al. (1985) with an artificial thermal diffusion term that reduces shock smearing, they contribute to the momentum and energy equations with terms like:

$$(\alpha\mu_{ij} + \beta\mu_{ij}^2)$$

where

$$\mu_{ij} = \frac{2h(\mathbf{v}_i - \mathbf{v}_j)\cdot(\mathbf{r}_i - \mathbf{r}_j)}{(c_i + c_j)(r_{ij}^2 + \eta^2)}$$

c_i is the sound speed at the i-particle, $\mathbf{v}_{ij} = \mathbf{v}_i - \mathbf{v}_j$, r_{ij} is the distance between the i and j particles, η must be a small value less then η^2, α and β parameters of the order of unity. This numerical artificial viscosity produces a viscosity force of order $f_{num}=\alpha C_{sound}\rho hF$, where F is a function of the derivatives of the velocity field (Monaghan, 1985).

3. Parameters

The simulation we here discuss has the following physical and numerical parameters:

3.1. GAS

Ideal, no cooling, politropic index γ=1.1, no viscosity;

3.2. BINARY SYSTEM

Equal star masses = 1 M_o ; distance of the star centres $A=2.4x10^{11}$cm , implying an orbital period of 12.6 hours ; star radius: primary 0.02A, secondary 0.5A: temperature of the secondary atmosphere 10^4 K;

3.3. NORMALIZATION

Normalization values: distance A for lengths, $V_o=(G(M_1+M_2)/A)^{1/2}$ for the velocity, so the orbital period is normalized to 2π ; reference density $\rho_o=10^{-9}$ gr cm^{-3}; pressure is normalized to $P=\rho_o V_o^2$ and the energy per unit mass to V_o^2; the temperature is scaled to $T_o=(\gamma-1)mV_o^2/K_b$;
resolution length (the Kernel width) H=0.0075; integration domain: a sphere of radius 0.55 around the primary star;

3.4. BOUNDARY CONDITIONS

Near the compact star we eliminate particles within a 4h distance, corresponding to a free inflow condition, similar to other simulations, this alters the structure of the disk few h values near the inner edge.

Injection from the Lagrangean point: new pseudoparticles are generated in fixed positions within a stratum of "injectors", representing the star atmosphere, whenever the particles leave an injector free; This procedure leads to a realistic overflow process.

4. RESULTS

Our aim is to perform realistic 3D simulations. Up to now we have preliminary results, but some seem worth mention. A stationary status was reached after about 3 orbits, the total number of particles is about 6540 , that is a very convenient number of points compared to the mesh points for finite difference methods with similar spatial resolution. The disk size is about 0.2A, in agreement with theoretical predictions (Paczinski, 1977). The disk height and temperature are in agreement with an αP thin disk model, i.e. $Z_{disk}=r(3\gamma-3)^{1/2}$ and T=GM/r, independent from the assumptions on the stress tensor $t_{r\phi}$, provided that the dissipated energy goes entirely into thermal energy. The only quantity affected by the stress is the density and its value differs from the theoretical one. If we assume that the stress is due to the numerical viscosity we can determine its value fitting the data. We found that a $t_{r\phi} = KC_{sound}\rho\Omega_k h$, where Ω_k is the keplerian angular velocity, fits quite well the density behaviour. Since the basic "molecular" expression of the stress for a keplerian disk is $t_{r\phi} =3 \eta\ \Omega_k/2$ we deduce that the numerical viscosity coefficient, within the disk, is about $\eta_{num}=\alpha C_{sound}\rho h$.

Fig. 1 shows respectively the distribution of the disk height Z, Temperature (adimensional), and particle number, within rings large dr=0.02 versus the distance r from the primary star. Fig.2 shows the same quantities computed within 24 sectors of rings of amplitude dr=0.05 up to r=0.3, and shows that the disk is perturbed in its symmetry around M1 down to r=0.1. Fig.3 shows the Mach field of a quarter of the disk (i.e. the ratio of the particle speed to its sound speed). Fig. 4 shows the “yz” view of the particles between x_2=–.08 and x_1=–.06, that exhibits the asymmetrical pattern of the disk due to the transverse impinging of the gas with the penetrating cold stream.

The accretion rate is 10^{15} gr sec^{-1} , the ratio of the accreted matter to that injected is 0.7, however we cannot say if the remaining .3 escapes the binary system due to our limited integration domain.

Further simulations are in progress, with different accretion rates and politropic indices of the gas and successively taking account the radiation contribution to energy and pressure.

We are grateful indebted to prof. J.J. Monaghan for its kind supplying the core of its SPH code and to prof. L Scarsi director of the CNR IFCAI Institute whose computer facilities we are using.

References

Monaghan J.J. , *Comp. Phys. Repts.*,**3**, 71, 1985

Lattanzio J. C., Monaghan J. J., Pongracic H., Schwarz M. P., *S.I.A.M., Sci. Stat. Comp.*, 1985

Paczinski B., *Astrophys. J.*, **216**, 822, 1979

Sawada K., Matsuda T., Inoue M. and Hachisu I., *M.N.R.A.S.*, **224**, 307, 1987

Verbunt F., *Space Sci. Rev.*, **32**, 379, 1982

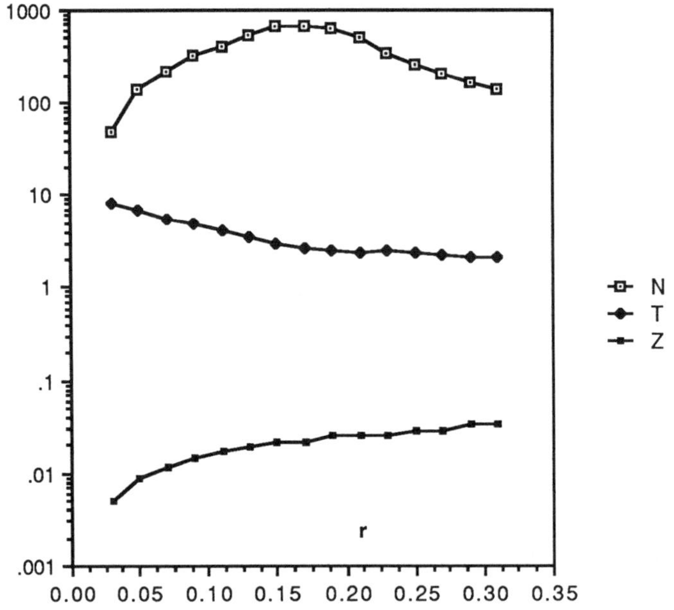

Fig.1 Number of particles, mean adimensional Temperature, disk heigth Z within rings large dr=0.02 versus distance r from the primary star. The Z and T value are in agreement with theory.

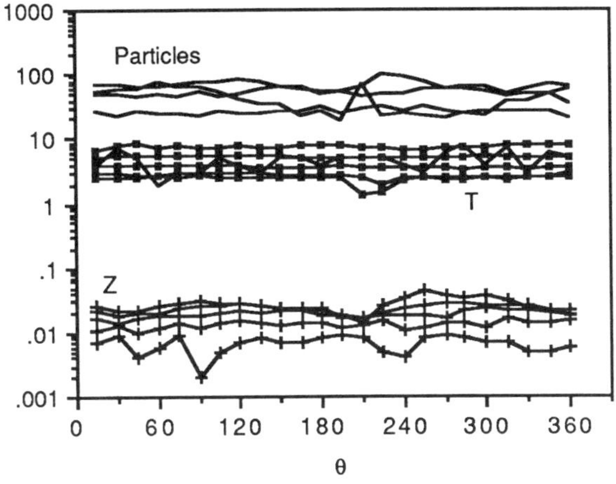

Fig.2 Number of particles, disk heigth Z and Temperature versus angle around the primary. $\theta = 0$ is along the x axis , x axis connects the stars centres. It is apparent the asymmetry in the disk at least down to r=0.15.

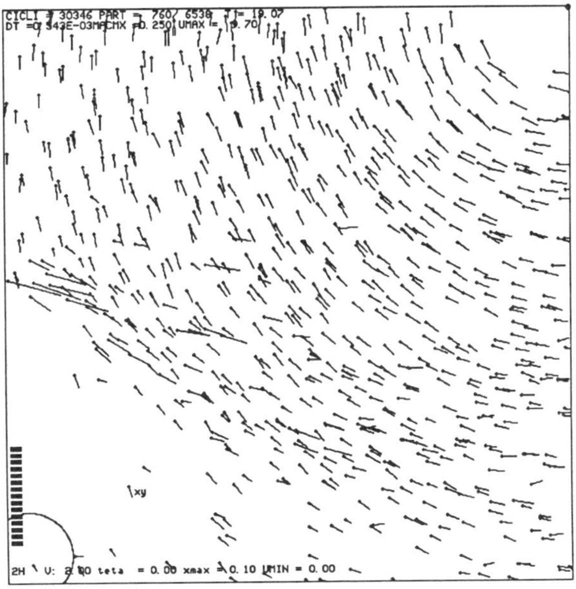

Fig.3 The "xy" view of Mach field of a quarter of the disk: the ratio of the velocity of each particle to the speed of sound computed at the same particle position; velocity vectors start from the small crosses; the reference velocity is 0.25.

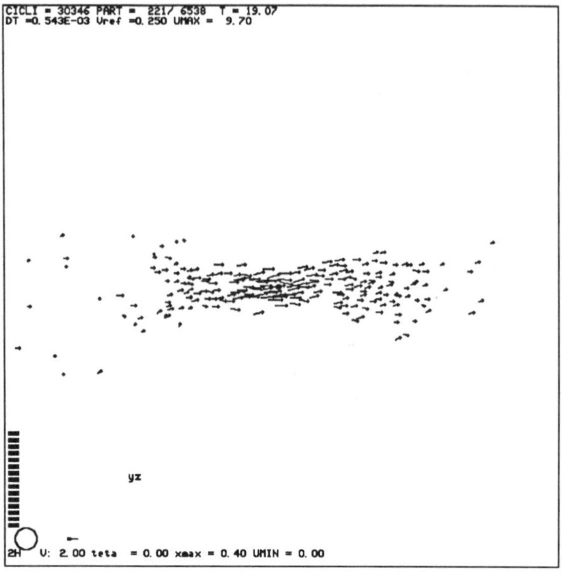

Fig.4 The "yz" view of the particles between x_2=-.08 and x_1=-.06, the reference velocity is 0.25

THE EARLY STAGES OF STAR FORMATION

L Mestel
Astronomy Centre, University of Sussex
Falmer, Brighton BN1 9QH
England

ABSTRACT. The presence of flux from the galactic magnetic field alters radically the problems of equilibrium, collapse and fragmentation of self-gravitating gas clouds. An adequate theory must incorporate the Alfvénic turbulent motions that can yield support along the magnetic field even in a cool cloud; magnetic transport of angular momentum; flux leakage, primarily through ambipolar diffusion; and the effect of the field and the associated turbulence on the masses of the first optically thick fragments.

1. MAGNETIC SUPPORT OF MOLECULAR CLOUDS

A realistic study of the structure of interstellar gas clouds and so also of the dynamics of star formation must take account of the galactic magnetic field. In the simplest picture, a uniform field B_o is supposed immersed in a uniform background medium of density ρ_o. Compression eg by a spiral density wave or an expanding supernova shell followed by radiative cooling leads to accumulation of dense gas down the field lines to form a cloud of mass M, threaded by magnetic flux F. The self-gravitation of the cloud will cause trans-field motions which distort the frozen-in field and so generate magnetic forces that oppose gravity. Simple application of the virial theorems (Chandrasekhar & Fermi 1953; Mestel 1965; Strittmatter 1966; Mestel & Paris 1984) to spherical or oblate clouds yields the appropriate non-dimensional parameter

$$f = F/\pi^2 G^{½} M. \qquad (1)$$

Provided f is less than a critical value f_c, so that M exceeds the critical mass $M_c(F) = F/\pi^2 G^{½} f_c$, then the magnetic forces alone are never strong enough to balance gravity, and a cool slowly rotating cloud with weak turbulent motions will spontaneously collapse in all three dimensions.

The assumption of strict flux-freezing during cloud formation enables definition of an accumulation length L_o by

G. Belvedere (ed.), Accretion Disks and Magnetic Fields in Astrophysics, 151–163.

$$M/F = \rho_o L_o/B_o. \tag{2}$$

The collapse condition $f < f_c$ then implies a minimum value for L_o:

$$L_o > (L_o)_c = \left\{ \frac{B_o}{G^{1/2}\rho_o} \right\} \frac{1}{\pi^2 f_c} \simeq 700 \frac{(B_o/3\times10^{-6})}{(n_o/1)} \text{ parsecs.} \tag{3}$$

As this is of the same order as the minimum Rayleigh-Taylor-Parker instability wavelength (Parker 1969, Mouschovias 1974), it may be argued that accumulation lengths of the order of a kiloparsec are not necessarily an embarrassment. Nevertheless, one is led to wonder whether during violent cloud formation, strict flux freezing does hold, or whether local tangling of the field leads to an effective diffusion of gas across the field — essentially the same "turbulent resistivity" needed by theories of both stellar and galactic dynamos (eg Weiss, this volume). If so, then the accumulation length becomes notional: the cloud mass M may still formally be written as $(2\pi\rho_o R_o^2/3)L_o$, with R_o defined by $F = \pi B_o R_o^2$, but the condition (3) on L_o no longer implies cloud formation through highly anisotropic flow of the background gas ρ_o.

At this point the discussion has a natural bifurcation.

(1) Super-critical clouds, with $f < f_c$ and $M > M_c(F)$. Not only is the field acting alone unable to prevent gravitational collapse, but it is also not an insuperable barrier to fragmentation into smaller gravitationally-bound masses, after further spontaneous anisotropic flow down the frozen-in field (Mestel 1965). The essential reason is that the collapse condition $f < f_c$ imposes a constraint on the accumulation length L along the field, but not on the lengths across the field. The condition that a roughly spherical cloud be super-critical in M/F is equivalent to a lower limit on $\rho L/B$, where ρ and B are mean values and L is the diameter: a cylinder of gas of radius R', with axis parallel to B_o and passing through the cloud centre, then automatically contains enough mass $(\rho L)(\pi R'^2)$ to form a super-critical sub-sphere by flow down the field to a density $(\rho L/2R')$. We return to this later.

(2) Sub-critical clouds, with $f > f_c$. Even if the cloud is cool, weakly turbulent and slowly rotating, indefinite gravitational collapse in the two trans-field dimensions is prevented by frozen-in flux; after dissipation of macroscopic kinetic energy the cloud must settle into a state of approximate magneto-gravitational equilibrium, with the scale along the field determined by the thermal and turbulent energies (Parker 1973, 1974; Mouschovias 1976a,b; Mestel & Ray 1985; Tomisaka et al 1988). The value of f_c is not very sensitive to the details of the flux/mass distribution within the cloud. When $f > f_c$, so that $M < M_c(F)$, then by (1) the

mean field B within a cloud of trans-field radius R is $\simeq kG^{1/2}M/R^2$, where the factor k decreases from $\simeq 2$ for a flattened cloud to $\simeq 1$ in a more nearly spherical cloud. If $f >> f_c$, so that $M << M_c(F)$, then equilibrium is achieved with a very slight distortion of the background field, and B stays close to B_o (cf. Mestel & Paris 1984, Mestel 1985b).

If the effective pressure in a molecular cloud were just the thermal pressure of gas at 10-20K, then the clouds should form highly flattened disk-like structures of semi-thickness z given by the one-dimensional virial theorem (Spitzer 1942; Ledoux 1951)

$$2\pi G\rho z^2 \simeq a^2 \tag{4}$$

where a is the sound speed; whence with $M \simeq 4\pi\rho zR^2/3$,

$$z/R \simeq (a^2/GM)^{3/4}(M/2\pi\rho)^{1/4} \simeq 4\times10^{-3}/(M/10^4M_o)^{1/2}(n/10^6)^{1/4} \tag{5}$$

when n is the number density of hydrogen molecules. There is evidence for flattened structures in some clouds (McCutcheon et al 1985;Monetti et al 1984; Kaifu et al 1984; Sargent et al 1988), but it is doubtful if it is as severe as suggested by (5) (Shu et al 1987). In fact, the measured line-widths Δv in typical molecular clouds indicate highly super-thermal (turbulent) motions (see Myers & Goodman 1988 for collected references to measurements of Δv and B). If the cloud is still disk-like, then the semi-thickness can be estimated from (4) as $z \simeq v_t/(2\pi G\rho)^{1/2}$, where v_t is the mean turbulent speed replacing the sound speed a, while if the turbulence is strong enough for the cloud to be better approximated as a sphere of radius R, then $v_t \simeq (3GM/5R)^{1/2} \simeq (4\pi G\rho/5)^{1/2}R$. Since factors like $(2/5)^{1/2} \simeq .63$ are in fact not significant in this preliminary discussion, for simplicity we shall normally adopt the factors appropriate to a spherical cloud, while still keeping distinct the length scales z and R respectively along and perpendicular to the overall direction of B. A strong roughly isotropic turbulence will contribute to support against gravity perpendicular to as well as along B; however, we shall continue to use the terminology of a "critical" mass/flux ratio appropriate to a cold, non-turbulent cloud, as in the cases of interest the expected turbulent contribution will at the most be of the same order as the magnetic. Thus when $f > f_c$,

$$B \triangleq G^{1/2}M/R^2 \triangleq (4\pi/3)G^{1/2}\rho z \tag{6}$$

$$\triangleq (5^{1/2}/3)(4\pi\rho)^{1/2}v_t \tag{7}$$

independent of the cloud mass: in a cloud with f close to the critical value, the equilibrium achieved by the lateral adjustment of B and ρ and the longitudinal adjustment of ρ yields inevitably an Alfvén speed $v_A = B/(4\pi\rho)^{1/2}$ close to the mean turbulent speed v_t, whether the cloud is quasi-spherical or significantly flattened along $\underset{\sim}{B}$. Equivalently, (6) and (7) yield

$$B \simeq (5v_t^2/3G^{1/2}z) \tag{8}$$

$$\simeq \left[\frac{20\pi}{9}\right]^{2/3} G^{1/2}\rho^{2/3}M^{1/3} \left[v_t^2/(GM/R)\right]^{2/3}. \tag{9}$$

Thus if v_t were constant, then from (7), comparison of clouds with different masses but all with $f \simeq f_c$ should yield $B \propto \rho^{1/2}$, whereas if v_t always adjusts itself to be the same fraction of the free-fall speed $(GM/R)^{1/2}$, then from (9), $B \propto \rho^{2/3}M^{1/3}$.

This theoretical picture of combined magnetic and turbulent support of molecular clouds (Henriksen & Turner 1984; Shu et al 1987; Myers 1987) has been applied by Myers & Goodman (1988) to 14 clouds where Zeeman splitting and line-widths can be accurately measured. As the observed field strengths are greater than the standard value of 3×10^{-6} for the background galactic field B_o, the clouds almost certainly have $f \sim f_c$, so that the theoretical estimate $B \simeq G^{1/2}M/R^2$ is appropriate. The clouds are assumed to be kept roughly spherical by kinetic support along the field, so that the observed radius R can be identified with the semi-thickness z. Since both R and the line width Δv (yielding v_t) are easier to measure than the column density ρz, the theoretically predicted field is estimated from (8) rather than from (6). The comparison with the observed B values is satisfactory (to a factor 2) over a range from 10^{-5}G to 10^{-2}G, both for different clouds and for three sub-regions within the Orion cloud. The kinetic support condition $v_t^2 \simeq GM/R \propto G\rho R^2$ is used to estimate ρ; comparison with (7) shows clearly the dependence on ρ, but the effect of v_t is within the observational scatter.

As noted above, in a near-critical cloud the turbulent velocity v_t required to yield support along the field is necessarily close to the Alfvén speed v_A corresponding to the mean density, suggesting that the motions are in fact due to a field of torsional Alfvén waves (Arons & Max 1975; Zweibel & Josafatsson 1983; and others). The simplest example illustrating the effect (Dewar 1970; Shu et al 1987) has a mean field Bz and waves with amplitudes $\delta\underset{\sim}{B}$, $\delta\underset{\sim}{v}$ that decline in the direction of propagation:

$$\delta\underset{\sim}{B} = \underset{\sim}{b} \sin (kz - \omega t) \exp(-k_1 z) \tag{10}$$

with $b.z = 0$, $b \simeq (\delta v/v_A)\bar{B}$. The exponential factor yields a non-zero time-average for the fluctuating Lorentz force density $(\nabla\times\delta B)\times\delta B/4\pi = -\nabla(\delta B^2/8\pi) = |b\exp(-k_1 z)|^2\ (k_1/8\pi)z$ along Bz. A non-zero damping constant k_1 is essential for a net input of momentum into the cloud, but the associated energy input need not be as heat, but may be as macroscopic kinetic energy, (eg rotational energy). The force will be able to keep the cloud moderately oblate if it is comparable with the lateral force density $\bar{B}^2/4\pi R$, requiring that the mean wave amplitude $b \simeq \bar{B}$, with $\delta v \simeq v_A$, and the damping length $1/k_1$ be of the order of the trans-field radius R.

The attraction _prima facie_ of this "Alfvénic turbulence" is that in a uniform medium the associated density fluctuations $\delta\rho$ are zero, so that even though the motions with $\delta v \simeq v_A$ will be strongly supersonic there will not be rapid dissipation through shock formation. The source of the motions could be e.g. sub-condensations within the cloud that are being magnetically braked (_cf_. below), or winds from proto-stars, or expanding HII regions from hot stars newly arrived on the main sequence; the essential point is that super-Alfvénic motions will dissipate rapidly through compressive shock formation, leaving just the longer-lived Alfvén spectrum to support the cloud. A difficulty is that the above analysis is directly applicable only if the semi-wavelength πk^{-1} is small compared with the scale of the cloud, but it is for small wavelengths that damping by ion-neutral collisions and second-order steepening is most effective. If there do not exist long-wavelength, purely torsional modes, with $\delta \mathbf{B}.\bar{\mathbf{B}} = 0$ and $\delta\rho = 0$ to first order, then waves travelling parallel to $\bar{B}$ will also rapidly steepen into shocks, especially if δv is close to v_A. An observational constraint, noted by Zweibel & Josafatsson (1983) is that the heat input from wave dissipation must not raise the cloud temperature above the observed maximum of $\simeq$ 20K. Their tentative conclusion is that in near-critical clouds with $\bar{B} > 10^{-4}$, highly supersonic Alfvén waves can yield cloud support for a few $\times 10^6$ years. When $f \gg f_c$, so that $B \simeq B_0$, and the density in a roughly spherical cloud ensures that $v_A \gg (GM/R)^{1/2}$, the Alfvén wave pressure along $\bar{B}$ could prevent the condensation of cooled gas that is to be expected in a non-turbulent cloud (Mestel 1965; Mouschovias 1976a,b; Mestel 1985a,b).

The anticipated rapid decay of super-Alfvénic turbulence ensures that the turbulent energy density $\frac{1}{2}\rho v_t^2$ never exceeds the magnetic energy density $B^2/8\pi$. A mildly _super_-critical cloud — with $M > M_c(F)$ — can be supported against its self-gravitation by the joint action of the forces exerted by the mean magnetic field and the turbulent δB and δv fields with $\delta v < v_A$. A decay in the turbulence following a cut-off in energy supply would precipitate collapse and possible fragmentation — following preferential flow down the mean field lines — even under strict flux-freezing. Sub-critical clouds however require some deviation from flux-freezing in order that the whole or part of the cloud can become super-critical and so ultimately form stars. Flux leakage is called "slow" if its time-scale, though short enough to be astronomically of interest, is long compared with the instantaneous free-fall time, so that the magnetic forces remain dynamically important. By contrast, even the strongest observed stellar magnetic fields when plausibly extrapolated inwards from the photosphere have an energy much below the thermal and gravitational energies of the star (in the present terminology, stars are enormously super-critical). At some epoch the bulk of the primeval flux must be lost, but on present evidence this does not occur until the later proto-stellar stages.

2. MAGNETIC BRAKING; FLUX LEAKAGE

A great advantage of flux leakage being slow even in a lightly ionized molecular cloud is that the strong field can transport away efficiently much of the embarrassingly high angular momentum expected from both the galactic rotation and the interstellar turbulence. (For brevity we speak of rotation and contraction of a "cloud", but the discussion applies equally to regions within a cloud, such as a dense core). The process is a simple application of familiar hydromagnetic principles. A contracting cloud initially in corotation with its surroundings will tend to spin up, so generating a shear along a line of a poloidal field B_p linking the cloud with the environs; the consequent toroidal field B_t propagates into and away from the cloud as a torsional Alfvén wave which tries to restore corotation. A rough estimate for the characteristic time of braking τ_b is given by the time of travel of the waves through a region of the low-density surroundings with the same moment of inertia as the cloud. The results are to some extent model-dependent (Mouschovias & Paleologou 1979, 1980; Mestel & Paris 1984), but for a near critical or super-critical cloud with roughly radial field lines linking the cloud and the background field $\underset{\sim}{B}_o$

$$\tau_b \triangleq (M/\rho_o^{1/2}F)(R/R_o)^{6/5} \tag{11}$$

where again $R_o = (F/\pi B_o)^{1/2}$ is the radius at which the external field reaches the value B_o. This should be compared with the instantaneous free-fall time τ_f:

$$\tau_b/\tau_f \triangleq (M/M_c(F))(R_o/R)^{3/10} \triangleq M/M_c, \tag{12}$$

the weak dependence on R_o/R being insignificant. Thus the ratio M/M_c plays a double role in the theory: it not only discriminates between sub-critical and super-critical clouds, but it determines whether the braking time is less than or greater than the free-fall time. If $M < M_c$, then corotation with the surroundings is achieved rapidly, but the degree of contraction is limited roughly to $R/R_o \simeq (1 - (M/M_c)^2)$ (Strittmatter 1966; Mestel & Paris 1984). Under slow flux leakage M_c decreases and the cloud contracts steadily, with corotation being maintained, in a time fixed by the leakage rate. The ratio of centrifugal force to gravity steadily declines, corresponding to efficient transport of angular momentum to the surroundings. By contrast, when $M > M_c$ a freely falling cloud or fragment will effectively conserve its angular momentum and so spin up, until centrifugal forces become comparable with gravity; the subsequent contraction will be in quasi-equilibrium and at the rate determined by magnetic braking (as long as flux leakage is slow). Detailed analytical and numerical work (Gillis, Mestel & Paris 1974, 1979; Mestel & Paris 1979, 1984) confirms these qualitative conclusions. There is also some tentative observational support (see references in Shu et al 1987). A correlation is found between higher rotation rates (40-50 times the usual values) in cloud cores with higher masses: molecular lines from NH_3, H_2CO yield $\Omega \simeq 25$-50 km sec^{-1} pc^{-1} in regions with diameters ~.01-.05 pc and masses 1-10$M_\odot$, corresponding to fragments or cloud cores in approximate centrifugo-gravitational balance. Similar conclusions hold for much larger masses (10^2-$10^3 M_\odot$) of molecular gas in the neighbourhood of three HII regions.

In a lightly ionized HI or molecular cloud, the relevant process of flux leakage over most of the parameter domain is "plasma drift", often called "ambipolar diffusion": the charged component (the "plasma") together with the inductively coupled $\underset{\sim}{B}$-field drifts relative to the neutral bulk of the gas with the velocity $\underset{\sim}{v}_d$ fixed by a balance between the Lorentz force (felt directly only by the plasma) and the plasma-neutral friction (Mestel & Spitzer 1956, and many papers since). Thus

$$\underset{\sim}{v}_d = (\underset{\sim}{j} \times \underset{\sim}{B}/c)/\alpha(n_i/n)\rho^2 \qquad (13)$$

with n_i and n the respective number densities of ions and neutral particles, and the frictional coefficient $\alpha = \langle\sigma_{in}(v_T)_n\rangle/m_n$, where m_n is a mean neutral particle mass, σ_{in} the momentum transfer collision cross-section, and $\langle\sigma_{in}(v_T)_n\rangle$ the collision rate averaged over the thermal velocity distribution of the neutral particles, whence the number of collisions per second felt by an ion (more massive than the neutral hydrogen and helium particles) is $1/\tau_{in} \equiv n\langle\sigma_{in}(v_T)_n\rangle$. For a molecular cloud with standard cosmical H_2 and H_e abundances, $\tau_{in} \triangleq .67 \times 10^8/n$ sec and $\alpha \triangleq 4 \times 10^{14} cm^3/sec$ (Osterbrock 1961; Nakano 1984). The ion-neutral friction is the means by which the magnetic force is transferred to act on the neutral bulk. There is therefore always a steady leakage of flux through the cloud as the field lines try to straighten themselves in the characteristic time

$$R/v_d = \alpha(n_i/n)\rho^2 R/|\underset{\sim}{j}\times\underset{\sim}{B}/c| \triangleq 3\alpha(n_i/n)/4\pi G$$

$$\triangleq 5 \times 10^{13}(n_i/n) \text{ yrs}, \qquad (14)$$

where $|\underset{\sim}{j} \times \underset{\sim}{B}/c|$ has been put equal to the mean gravitational force density $4\pi G\rho^2 R/3$ in a spherical cloud. In a highly flattened cloud there is an extra factor $\triangleq 4R/3\pi z$ in (14).

Applied to an HI cloud, where $n_i/n \triangleq 5 \times 10^{-4}$ typically and α is larger by a factor 2, the time (14) exceeds the Hubble time, so leakage is ignorable. Direct measurement of

n_i/n for molecular clouds is difficult and one has to rely on theoretical estimates. A balance of cosmic ray ionization against two-body recombination yields the law $n_i = Cn^{1/2}$ with C depending only weakly on the temperature but proportional to the square root of the metal depletion (Elmegreen 1979), whence $n_i/n \triangleq 10^{-7}$ at a density of $10^4/cm^3$. At densities much above $10^8/cm^3$ natural radioactivity dominates over cosmic rays as the ionizing agent, grains begin to carry a major fraction of the total charge, and n_i approaches a constant value. For recent detailed discussions of the problem, see Nakano & Umebayashi 1986a,b, Nakano 1988 and Norman & Heyvaerts 1985. It appears that up to $n \triangleq 10^{11}$-10^{12}, the time (14) is well below the Hubble time but significantly longer than the free-fall time, corresponding to "slow" flux leakage.

As f declines towards f_c, a sub-critical cloud adjusts steadily to a new state of equilibrium at higher ρ and also with higher B, in spite of the decrease in the total flux (analogous to the speeding up of a satellite as it loses angular momentum through friction). As pointed out by Nakano (1976, 1983), the sensitivity of flux leakage to the value of n_i/n can lead to a new fragmentational instability. In the dense core of a cloud the same formula (14) yields a local leakage time that is shorter than that from the cloud as a whole; the consequent more rapid contraction of the core increases the central density and so reduces the ratio n_i/n still further, etc., so that the core goes over into collapse ahead of the cloud as a whole.

Lizano & Shu (1988) model the evolution of turbulent magnetic clouds. The turbulent pressure P_t is taken as $K\log(\rho/\rho_o)$, so yielding $v_t^2 = dP_t/d\rho = K/\rho$, which is a fair approximation when v_t is supposed kept close to v_A but B does not vary much. Thus $P_t/(GM\rho/R) \propto \log(\rho/\rho_o)/\rho^{4/3}$ ultimately decreases as ρ increases through flux leakage. The models studied are all super-critical according to the definition based on (1), but initially are held in equilibrium by the turbulent pressure; flux leakage and the accompanying reduction in $P_t/(GM\rho/R)$ lead again to instability and collapse of cloud cores. At the onset of collapse B has in general not increased very much, so justifying a posteriori the adoption of $v_t \propto \rho^{-1/2}$, a law with some observational support (cf. Solomon & Sanders 1985), possibly because B has much the same value in all the clouds observed.

Since flux leakage is slow during the pre-opaque phases, the collapsing fragment will still have f near the critical value f_c, and f will be nearly constant during the free-fall (as long as no macroscopic process of flux destruction occurs — cf. below). The angular momentum of the fragment may have been reduced by large factors during the pre-collapse phase; but even if the bulk of the "angular momentum problem" has thus been magnetically resolved in these early phases, the subsequent collapse will again be with near angular momentum conservation. The centrifugal forces may catch up with gravity even in the pre-opaque phases, and will almost certainly do so well before main sequence or pre-main sequence densities are reached (Mestel 1965; Mouschovias 1977; Mestel & Paris 1979), so that dense rotating disk-like bodies are likely to form following dissipation of kinetic energy of collapse. The evolution of such bodies under the action of either primeval or dynamo-built magnetic fields is a major concern of the later stages of star formation.

3. THE MASSES OF PROTO-STARS

The simplest discussions of cloud collapse and star formation involve just the mutual interaction of self-gravitation and thermal pressure. The maximum spherical mass for which the internal pressure can prevent gravitational collapse is then the modified Jeans mass (Ebert 1955; McCrea 1957):

$$M_J \hat{=} 4\left[\frac{T}{\mu G}\right]^{3/2}\left[\frac{3}{4\pi\rho}\right]^{1/2} \hat{=} 0.005\left[\left[\frac{T}{15}\right]^{3/2}\left[\frac{\mu}{2}\right]^{-3/2}\left[\frac{n}{10^{10}}\right]^{-1/2}\right]M_\odot \qquad (15)$$

where T is the temperature, μ the mean molecular weight and n the number density. As long as $T/n^{1/3}$ decreases with increasing n, M_J decreases during collapse, and a cloud with M just greater than the initial value of M_J may subsequently fragment into a hierarchy of sub-condensations (Hoyle 1953). Analysis by Hunter (1962, 1964), Mestel (1965) and Lynden-Bell (1973) has elucidated how density perturbations of mass greater than the instantaneous value of M_J are indeed gravitationally amplified more rapidly than the mean density of the collapsing cloud. As long as $n < 10^{10}$ the optical depth τ through a mass M is small, the compressional heat generated during collapse is easily radiated and T stays contant or decreases, so M_J does decrease; but when $n > 10^{10}$ the gas becomes opaque, T increases roughly adiabatically with $T/n^{1/3} \propto n^{1/3}$, and M_J increases with n. A plausible estimate for the mass of this opacity-limited fragment is the value of M_J when $\tau = 1$ at $n \hat{=} 10^{10}$, yielding from (15) the very small value $\hat{=} 5 \times 10^{-3} M_\odot$ (cf. the detailed treatment by Low & Lynden-Bell (1976)).

Some of the persisting doubts as to whether Hoyle-type fragmentation will occur derive from the difficulty in following convincingly the dynamical evolution of a perturbed collapsing cloud into the non-linear domain, in which randomization of the kinetic energy of infall is shown to occur simultaneously with transformation of the cloud into a "gas" whose "molecules" are self-gravitating fragments (cf. Layzer 1963). Nevertheless, the Jeans mass and the opacity limit are basic concepts which should have analogues in a more realistic scenario. Thus consider a collapsing magnetic fragment, eg the mildly oblate core of a magnetic cloud that has become super-critical (with $f < f_c$) through slow flux-leakage, as discussed above. If the Alfvénic turbulence persists at the level with $v_t^2 \triangleq GM/R$ then the contraction will be isotropic without preferential flow down the field and so there will be no fragmentation. A significant drop in v_t will allow flattening and fragmentation, with the minimum fragment mass now $\triangleq (15/2)(1-f^2/f_c^2)^{-1}(2v_t^2R/3GM)^2M$ (Mestel 1965; Campbell & Mestel 1987), and unless f is very close to f_c this will not differ much from (15) with $T/\mu \equiv a^2$ replaced by v_t^2 and the density ρ given by (4) and (5). If the turbulence decays to subsonic levels, v_t must be replaced by a, which again remains low until τ reaches unity at $n \sim 10^{10}$, so yielding essentially the same very small opacity limit to the fragment masses. Thus the problem of the break-up of a collapsing super-critical fragment is linked intimately with the maintenance or decay of the Alfvénic turbulence.

As noted above, the remnant angular momentum is conserved during free-fall. If magnetic braking in the pre-collapse phase has been modest, spin-up during free-fall will yield centrifugal forces that become comparable with gravity in the pre-opaque phases. If the magnetic and angular momentum vectors are roughly parallel, the body will approach quasi-equilibrium with centrifugal force making up the difference between gravity and magnetic force in two dimensions, and again the turbulent-plus-thermal pressure balancing gravity in the third. Magnetic transport of angular momentum to the surroundings will cause overall contraction, but as long as flux leakage remains slow, further fragmentation again requires decay of the turbulence and associated flattening. In the other extreme case, with the magnetic and angular momentum vectors nearly orthogonal, <u>prima facie</u> there exist grossly stable cloud models in which it is the centrifugal force which prevents flattening along the field (Mestel 1965). Magnetic removal of angular momentum now leads to <u>isotropic</u> contraction even in the absence of turbulence; flattening with subsequent fragmentation now depends on flux leakage becoming sufficiently fast. In this model, the masses of the fragments that reach the opaque phase are fixed by the competition between magnetic braking and flux leakage. If leakage is due just to slow microscopic processes such as ambipolar diffusion, the predicted masses may now be embarrassingly large (Campbell & Mestel 1987). However, the actual flux-loss rate could sometimes be significantly faster. The gravitationally distorted magnetic field of a fragment may suffer some reconnection across pinched equatorial zones (Mestel & Strittmatter 1967; Mestel & Ray 1985); the regions with closed field lines will then be subject to spontaneous hydromagnetic instabilities analogous to those observed in the laboratory and discussed in the stellar context by Wright 1973 and Markey & Tayler 1973 (but now without the limitations imposed on the unstable displacements by adiabaticity). The non-linear development of these instabilities may yield a faster, dynamically-driven flux leakage. Since flux loss will simultaneously reduce the turbulent energy to the new Alfvénic level, an accelerated rate will make fragmentation easier in both the aligned and orthogonal geometries; though one may still hazard the guess that the final fragment masses will tend to be higher in the orthogonal case.

The fragmentation problem in some form is relevant simply because there is no obvious reason why the bodies brought into gravitational collapse — eg the cores of molecular clouds evolving via ambipolar diffusion — should have masses of stellar order. It would in fact have been really embarrassing if the opacity criterion had yielded a minimum mass well above rather than well below $M_\odot$, for as already discussed, the combination of two or more of magnetic flux, Alfvénic turbulence and angular momentum can inhibit spontaneous break-up during the pre-opaque phases. It is however not clear that the ultimate mass spectrum is fixed essentially by the dynamics of the early phases, if only because opaque fragments are likely to retain enough angular momentum to form disk-like bodies which may often be the precursors of binary systems; and there will also be substantial mass loss through the observed bipolar flows.

These later stages of star formation will clearly be affected by any remnant primeval flux. Even if the slow ambipolar diffusion has been augmented by dynamically-driven flux loss, it is likely that at densities $n \triangleq 10^{10} - 10^{11}$ the optically thick fragments are still magnetically strong, with $f < f_c$. During subsequent nearly adiabatic contraction the temperature rises, and beyond $T \triangleq 10^3$ K thermal ionization increases n_i/n to the level where flux freezing is again a very good approximation. In the interval between $10 < T < 10^3$ ρ increases by 10^4 or more. According to Nakano & Umebayashi (1986b), in this interval the field no longer moves with the plasma. The electrons are largely attached to grains, and the current is carried by ions which suffer a large Ohmic dissipation due to ion-neutral collisions. The estimated remnant flux at $T = 10^3$K is less than that in the strongest magnetic Ap stars. However, the decline in the plasma density may sometimes be off-set from ionization by sub-cosmic rays generated locally in magnetic reconnection processes, or by X-rays emitted by newly formed stars in the same cloud complex (Silk & Norman 1983); if so, then both Ohmic dissipation and ambipolar diffusion will remain slow. Further, if Population II and (possibly) Population III stars formed from magnetized gas at $T \triangleq 10^4$K, collisional ionization will have kept flux loss through large-scale ambipolar diffusion at a very slow rate at all epochs. But as already noted, a proto-star entering the opaque phase with $f \sim f_c$ must certainly have lost the bulk of its flux before reaching the main sequence. This may in fact occur during pre-main sequence contraction through magnetic buoyancy, so supplying a possible energy source to power the T Tauri phenomenon.

REFERENCES

Arons, J. & Max, C.E., 1975. Astrophys. J.,**196**, L77.

Campbell, C.G. & Mestel, L., 1987. Mon. Not. R. astr. Soc., **229**, 549.
Chandrasekhar, S. & Fermi, E., 1953. Astrophys. J., **118**,116.
Dewar, R.L., 1970. Phys. Fluids, **13**, 2710.
Ebert, R., 1955. Zeits. f. Astrophys., **37**, 217.
Elmegreen, B.G., 1979. Astrophys. J., **232**, 729.
Gillis, J., Mestel, L. & Paris, R.B., 1974. Astrophys. Space Science, **27**, 167.
Gillis, J., Mestel, L. & Paris, R.B., 1979. Mon. Not. R. astr. Soc., **187**, 311.
Henriksen, R. & Turner, B., 1984. Astrophys. J., **287**,200.
Hoyle, F., 1953. Astrophys. J., **118**, 513.
Hunter, C., 1962. Astrophys. J., **135**, 594.
Hunter, C., 1964. Astrophys. J., **139**, 570.
Kaifu, N. et al., 1984. Astron. Astrophys., **134**, 7.
Layzer, D., 1963. Astrophys. J., **137**, 351.
Ledoux, P., 1951. Ann. Astrophys., **14**, 438.
Lizano, S. & Shu, F.H., 1988. Astrophys. J., (in press).
Low, C. & Lynden-Bell, D., 1976. Mon. Not. R. astr. Soc., **176**, 367.
Lynden-Bell, D., 1973. In Dynamical Structure and Evolution of Stellar Systems, Saas-Fee Lectures, Geneva Observatory.
Markey, P. & Tayler,R.J., 1973. Mon. Not. R. astr. Soc., **163**, 77.
McCrea, W.H., 1957. Mon. Not. R. astr. Soc., **117**, 562.
McCutcheon, W.H., Vrba, F.J., Dickman, R.L. & Clemens, D.P., 1985. Astrophys. J., **309**, 619.
Mestel, L., 1965. Q. Jl. R. astr. Soc., **6**, 161 and 265.
Mestel, L., 1985a. In Protostars and Planets II, eds. Black, D.C. & Matthews, M.S., p. 320, Tucson, Univ. Ariz.Press.
Mestel, L., 1985b. Phys. Scripta, **T11**, 53.
Mestel, L. & Paris, R.B., 1979. Mon. Not. R. astr. Soc., **187**, 337.
Mestel, L. & Paris, R.B., 1984. Astron. Astrophys., **136**, 98.
Mestel, L. & Ray, T.P., 1985. Mon. Not. R. astr. Soc., **212**, 275.
Mestel, L. & Spitzer Jr., L., 1956. Mon. Not. R. astr. Soc. **116**, 583.
Mestel, L. & Strittmatter, P.A., 1967. Mon. Not. R. astr. Soc., **137**, 95.
Monetti, A., Pipher, J.L., Helfer, H.L., McMillan, R.S. & Perry, M.L., 1984. Astrophys. J., **282**, 508.
Mouschovias, T.C., 1974. Astrophys. J., **192**, 37.
Mouschovias, T.C., 1976a. Astrophys. J., **206**, 753.
Mouschovias, T.C., 1976b. Astrophys J., **207**, 141.
Mouschovias, T.C., 1977. Astrophys. J., **211**, 147.
Mouschovias, T.C. & Paleologou, E.V., 1979. Astrophys. J., **230**, 204.
Mouschovias, T.C. & Paleologou, E.V., 1980. Astrophys. J., **237**, 877.
Myers, P.C. & Goodman, A.A., 1988. Astrophys. J.,**326**, L27.
Nakano, T., 1976. Publs. astr. Soc. Japan, **28**, 355.
Nakano, T., 1983. Publs. astr. Soc. Japan, **35**, 209.

Nakano, T., 1984. Fund. Cosmic Phys., **9**, 139.
Nakano, T., 1988. In Galactic and Extragalactic Star Formation, eds. Pudritz, R. & Fich, M., Reidel, Dordrecht.
Nakano, T. & Umebayashi, T., 1986a. Mon. Not. R. astr. Soc., **218**, 663.
Nakano, T. & Umebayashi, T., 1986b. Mon. Not. R. astr. Soc., **221**, 339.
Norman, C. & Heyvaerts, J., 1985. Astron. Astrophys., **147**, 247.
Osterbrock, D.E., 1961. Astrophys. J., **134**, 270.
Parker, D.A., 1973. Mon. Not. R. astr. Soc., **163**, 41.
Parker, D.A., 1974. Mon. Not. R. astr. Soc., **168**, 331.
Parker, E.N., 1969. Space Sci. Rev., **9**, 651.
Sargent, A.I., Beckwith, S., Keene, J. & Masson, C., 1988 (preprint).
Shu, F.H., 1987. In Star Formation in Galaxies, ed. Persson, C., p.743, Washington, NASA.
Shu, F.H., Adams, F.C. & Lizano, S., 1987. Ann. Rev. Astron. Astrophys., **25**, 23.
Silk, J. & Norman, C.A., 1983. Astrophys. J., **272**, L49.
Solomon, P.M. & Sanders, D.B., 1985. Protostars and Planets II, eds. Black, D.C. & Matthews, M.S., p.59, Tucson, Univ. Ariz. Press.
Spitzer Jr., L., 1942. Astrophys. J., **95**, 329.
Strittmatter, P.A., 1966. Mon. Not. R. astr. Soc., **132**, 359.
Tomisaka, K., Ikeuchi, S. & Nakamura, T., 1988. Astrophys. J., (in press).
Wright, G.A.E., 1973. Mon. Not. R. astr. Soc., **162**, 359.
Zweibel, E.G. & Josafatsson, K., 1983. Astrophys. J., **270**, 511.

ACCRETION DISKS AND JETS IN PROTOSTELLAR SYSTEMS

ARIEH KÖNIGL
Department of Astronomy and Astrophysics
The University of Chicago
5640 South Ellis Avenue
Chicago, IL 60637, USA

Abstract. There is growing observational evidence for the presence of Keplerian circumstellar disks around newly formed stars. There is also mounting evidence for the association of such disks with energetic bipolar outflows from these objects. In addition, the projected magnetic field vecotrs around these sources are generally found to be well aligned with the direction of the outflow and the axis of the associated disk, pointing to the possible dynamical role of ordered magnetic fields in the accretion and outflow processes. A general scenario which attempts to tie these various pieces of evidence together is presented. In this picture, the apparent association between disks and jets is attributed to the fact that the outflows are powered by accretion and actually are the main sink for the excess angular momentum and the liberated gravitational energy of the accreted matter. According to this interpretation, the coupling between the inflow and the outflow is achieved through the ordered magnetic field that threads the disk and gives rise to a centrifugally driven wind from the disk surface. This interpretation is illustrated with the help of a self-consistent accretion-outflow model of magnetized accretion disks wherein magnetic stresses transport all the excess angular momentum. In this steady-state model, the shearing and advection of the magnetic field lines inside the disk are balanced either by ambipolar diffusion or by Ohmic resistivity. Radially self-similar solutions for the disk magnetic field configuration are presented, and it is demonstrated that they can be matched continuously onto centrifugally driven wind solutions which become super-Alfvénic and magnetically collimated at a finite height above the disk.

1. Introduction

Important observational clues have been obtained in recent years concerning the nature of circumstellar disks and jets in young stellar objects (YSOs). These observations provide evidence for the presence of accretion disks and energetic outflows in these sources and point to a probable relationship between them as well as to the possible role of magnetic fields in the accretion and outflow processes. The relevant findings may be summarized as follows:

(1) There is growing evidence that low-mass ($M \lesssim 2M_\odot$) YSOs, including both deeply embedded infrared sources (DEIS) and classical T Tauri stars (CTTS), are surrounded by circumstellar disks with estimated masses $\sim 0.01 - 1M_\odot$ on scales $\sim 10^2 - 10^3$ AU (see Strom, Strom, and Edwards 1988 for a review). The most direct evidence has come from

G. Belvedere (ed.), Accretion Disks and Magnetic Fields in Astrophysics, 165–178.

molecular line and millimeter continuum interferometric measurements as well as from high-resolution infrared observations. For example, the presence of a rotating, Keplerian disk of size ~ 4000 AU and mass $\sim 0.1 M_\odot$ around a $1 M_\odot$ star has been inferred in this way in HL Tau (Sargent and Beckwith 1987). Additional evidence has been provided by an analysis of infrared and UV spectra (e.g., Adams, Lada, and Shu 1987; Myers *et al.* 1987; Hartmann and Kenyon 1987; Kenyon and Hartmann 1987; Bertout, Basri, and Bouvier 1988) and of optical emission line profiles (e.g., Appenzeller, Jankovics, and Ostreicher 1984; Edwards *et al.* 1987). Furthermore, there are strong indications that dense, Keplerian disks are also associated with high-mass YSOs (e.g., Vogel *et al.* 1985; Persson, McGregor and Campbell 1988).

(2) YSOs are frequently found to drive energetic, bipolar outflows which sweep up the ambient gas into an expanding shell that is detectable through molecular line observations (e.g., Lada 1985). These sources often give rise also to collimated high-velocity ($\sim 200-400$ km s^{-1}) ionized jets (e.g., Mundt 1985). In the case of DEIS, the inferred momentum discharges ($\sim 10^{-7} - 10^{-3}$ $M_\odot$ km s^{-1} yr^{-1}, corresponding to mass loss rates $\sim 10^{-9} - 10^{-5}$ $M_\odot$ yr^{-1}) are usually significantly larger than those associated with the optically visible jets, which suggests that more massive, neutral winds are present in these sources. Direct evidence for such winds has, in fact, been obtained in recent atomic (e.g., Bally and Stark 1983; Lizano *et al.* 1988) and molecular (Mitchell *et al.* 1988; Koo 1989) spectroscopic measurements. The inferred wind kinetic luminosity L_w in both low-mass and high-mass YSOs is correlated with the total bolometric luminosity L, but generally $L_w \lesssim 0.1\ L$. However, L/c is typically much smaller than the estimated momentum discharge in the wind, indicating that the outflow is not driven by radiation pressure.

(3) There is now mounting evidence for a strong link between circumstellar disks and energetic outflows in YSOs. One of the more compelling arguments in favor of such an association in low-mass YSOs derives from a comparison of DEIS and CTTS with the so-called "naked" T Tauri stars (NTTS; Walter *et al.* 1988). NTTS are similar to CTTS in all respects except that they do not exhibit infrared excesses (a signature of circumstellar material) and ultraviolet excesses (commonly attributed to emission from the boundary layer of an accretion disk) and that they lack the strong low-ionization and forbidden line emission as well as the P Cygni line profiles which are the hallmark of energetic outflows in CTTS. The simultaneous absence of evidence for substantial disks and strong winds in NTTS points to a likely connection between the accretion and outflow processes in YSOs. The properties of DEIS and CTTS suggest that the basic relationship lies in *the outflows being powered by accretion.* In particular, the observed correlation between L_w and L is then understood from the fact that the luminosity excesses in YSOs are probably also due to accretion (Strom *et al.* 1988). In a similar vein, the fact that DEIS and CTTS with comparable bolometric luminosities also have similar outflow parameters indicates that the wind driving mechanism is the same in both types of objects and is tied to the presence of an active accretion disk. Additional support for the disk-wind connection comes from observations of CTTS of the YY Ori type, which show evidence for concurrent mass outflow and inflow (manifested, respectively, in P Cygni and inverse P Cygni line profiles) as well as for the presence of an opaque disk (indicated by the absence of redshifted forbidden line emission). The evidence for the association between accretion disks and energetic outflows is discussed in detail by Strom, Strom, and Edwards (1988), who also point out that the correlation between disk and wind signatures in T Tauri stars apparently extends to higher-mass YSOs as well.

(4) Finally, evidence continues to accumulate for the possible dynamical role of ordered magnetic fields in the accretion and outflow processes. Polarization measurements have revealed that the axes of the collimated optical jets as well as of the swept-up molecular lobes are aligned within 30° of the ambient field direction in over 70 percent of the bipolar outflow sources that have been examined (e.g., Cohen, Rowland, and Blair 1984; Strom *et al.* 1986). In those cases where the circumstellar disks are resolved, their minor axes are found to be parallel to the outflow axes and are thus also aligned with the local magnetic field direction. Although dense molecular cores are apparently not magnetic pressure-dominated (e.g., Myers and Goodman 1988), so that their contraction is not directly controlled by magnetic stresses (see Heyer 1988), the above results suggest that the magnetic field nevertheless influences the orientation of the embedded circumstellar disk axes, probably through magnetic braking effects (e.g., Mouschovias and Paleologou 1980). Further corroboration of the importance of magnetic fields in the bipolar outflow phenomenon comes from the fact that a local alignment between stellar jet axes and the ambient magnetic field is detected even in giant molecular clouds where the field appears to play only a minor role in the global dynamics (e.g., Vrba, Strom, and Strom 1988).

In this contribution I outline a general scenario for accretion-driven outflows in protostellar systems that attempts to account for these different observational findings in a unified manner. The basic idea of this scheme is that the observed outflows represent centrifugally driven winds that carry most of the excess angular momentum and liberated gravitational energy out of the associated accretion disks. A simple model of such a disk-wind system is presented in Section 2, and its application to YSOs is outlined in Section 3. It is argued that this model is compatible with the observational data and, moreover, that it provides a natural framework for interpreting some of the basic radiative and evolutionary properties of newly formed stars.

2. Self-Similar Disk-Wind Configurations

One of the most promising mechanisms for powering astrophysical jets is the tapping of rotational kinetic energy in their central "engines" by means of magnetic fields. The original discussion of this mechanism in the context of MHD disks was carried out by Blandford and Payne (1982; hereafter BP), who focused on the application to active galactic nuclei. The application to bipolar outflows in YSOs was first considered by Pudritz and Norman (1983). In particular, it was shown that, unlike the previously proposed radiative schemes, centrifugally driven winds originating in a protostellar disk could readily account for the large momentum discharges inferred in these outflows (Pudritz 1985). The operation of this mechanism in the context of star formation was also studied with the help of numerical simulations by Uchida and Shibata (1985).

Although the question of whether the observed bipolar outflows originate in the YSO itself or in the surrounding disk is still open, it is of particular interest to consider rotationally driven winds that are associated with accretion disks. This is because such outflows provide a natural means of transporting angular momentum away from the accreted matter and, in fact, can be the main sink for the excess angular momentum in the disk. An explicit demonstration of this fact was given by BP, who considered a self-similar model of a cold, axisymmetric, centrifugally driven wind from a Keplerian disk that is threaded by a "frozen-in," open magnetic field. They demonstrated that, if the magnetic field is sufficiently strong (so that the effective lever arm used in calculating the torque exerted by the wind is significantly larger than the radius of the disk at the base of the outflow), then only a small fraction of the inflowing mass must be expelled from the disk surface in order to carry away all the excess angular momentum. BP further showed how the outflow, which is initially magnetic pressure-dominated and directed at a relatively large ($\geq 30°$) angle to the rotation axis, could become super-Alfvénic and highly collimated (by magnetic hoop stresses) farther out from the disk surface. Although the solutions considered by BP are special in that the wind passes only through a single critical point, the general properties of their model continue to hold even when the assumptions of self-similarity and negligible thermal pressure are dropped (Sakurai 1985, 1987).

As discussed by BP, the initial bending of the disk magnetic field away from the rotation axis is required in order for the centrifugal force at the base of the wind to exceed the gravitational pull of the central object. This bending of the poloidal field component must, however, be associated with nonmagnetic stresses since the balance between magnetic tension and the magnetic pressure gradient in a magnetically dominated region (such as the wind zone immediately above the disk) would, in fact, lead to a "concave in" field geometry. BP concentrated on the wind properties and did not attempt to also derive a solution for the steady-state field configuration in the disk. Because of the differential rotation in a Keplerian disk (which winds up the field lines), any such steady-state model necessarily must incorporate the effect of field diffusion. A unified disk-wind model which takes this effect into account has recently been constructed by Königl (1989). This model incorporates two alternative flux removal mechanisms which may operate in a weakly ionized molecular gas: ambipolar diffusion – the drift of ions and electrons (to which the magnetic field is frozen) relative to the "sea" of neutrals, and Ohmic resistivity. Both of these mechanisms could be relevant in protostellar environments, with ambipolar diffusion being the more important below $\sim 10^{11}$ cm^{-3}, and Ohmic diffusivity dominating at higher densities (e.g., Nakano and Umebayashi 1986). The magnetic field structure in the disk is obtained by solving the induction equation together with the angular momentum conservation equation (assuming axisymmetry and an even z-symmetry), and the solutions are required to match continuously onto the radially self-similar wind configurations of BP. The general formulation of the problem in the case of an ambipolar diffusion-dominated disk is described in this Section. The solutions obtained for this case should be applicable on the scales that are currently resolved by molecular line observations. The corresponding solutions for a resistive disk are presented in Königl (1989), where possible applications to other cosmic jet sources are also discussed. An analogous disk-jet model, involving a viscous, resistive disk with an odd field symmetry and an electromagnetic (rather than hydromagnetic) outflow had previously been considered by Lovelace, Wang, and Sulkanen (1987) in the context of active galactic nuclei.

2.1 MHD EQUATIONS

Assuming a steady-state ($\partial/\partial t = 0$) and axisymmetric ($\partial/\partial\phi = 0$) disk and using cylindrical coordinates (r, ϕ, z), the angular momentum conservation equation for the dominant (neutral) component takes the form

$$\frac{\rho v_r}{r}\frac{\partial}{\partial r}(r v_\phi) = \frac{B_r}{4\pi r}\frac{\partial}{\partial r}(r B_\phi) + \frac{B_z}{4\pi}\frac{\partial B_\phi}{\partial z}, \tag{1}$$

where ρ is the density, $\mathbf{v}$ and $\mathbf{B}$ are, respectively, the velocity and magnetic field vectors, and where we set $v_z = 0$ (which should hold for a geometrically thin disk). Equation (1) expresses the condition that the transport of angular momentum anywhere within the disk is due entirely to magnetic stresses (acting in both the $\hat{\mathbf{r}}$ and $\hat{\mathbf{z}}$ directions). In lieu of the radial momentum equation, we substitute the assumption that the disk is very nearly Keplerian,

$$v_\phi = (GM/r)^{\frac{1}{2}} \tag{2}$$

(where M is the mass of the central object and G is the gravitational constant), which requires that the inflow speed $u \equiv -v_r$ satisfy $u \ll v_\phi$. Provided that the disk is sufficiently thin, the z-variation of v_ϕ can be neglected, and we similarly consider ρ and v_r to be functions of r only.

As was discussed above, the dynamics in the disk interior must be dominated by nonmagnetic stresses. In this case, the vertical structure of the disk is determined by the balance between gravity and the nonmagnetic pressure gradient force (which may include both thermal and turbulent contributions). For a thermal pressure-dominated inflow, this equation can be combined with the energy conservation equation to give the variation of the temperature in the disk. However, we do not consider these relations explicitly and simply assume that they are consistent with the similarity scalings adopted below (see eqs. [9] and [10]). We further assume that the magnetic field configuration obtained by solving the transport equations for angular momentum and magnetic flux is maintained in a steady state by nonmagnetic forces.

The field evolution equations can be derived by considering the momentum conservation equation for the ionized disk component,

$$\frac{1}{4\pi}\,(\nabla \times \mathbf{B}) \times \mathbf{B} = \gamma \rho_i \rho (\mathbf{v}_i - \mathbf{v}), \tag{3}$$

and the flux-freezing condition

$$\nabla \times (\mathbf{v}_i \times \mathbf{B}) = 0. \tag{4}$$

Equation (3) (in which ρ_i and $\mathbf{v}_i$ are the ion density and velocity, respectively) expresses the balance between the $\mathbf{j} \times \mathbf{B}$ force and the drag force exerted on the ions by collisions with the neutrals, with the characteristic value of the collisional coupling coefficient taken to be $\gamma \approx 3.5 \times 10^{13}\ \mathrm{cm^3\,g^{-1}\,s^{-1}}$ (valid for $|\mathbf{v}_i - \mathbf{v}| \lesssim 2 \times 10^6\ \mathrm{cm\ s^{-1}}$; Draine, Roberge, and Dalgarno 1983). This equation, which was already assumed implicitly in writing down equation (1), is a good approximation to the full ion equation of motion in the limit $\rho_i \ll \rho$. The induction equation (4) is just a statement of the fact that the magnetic field lines move with the ions. Imposing axisym[illegible] the condition $\nabla \cdot \mathbf{B} = 0$, these relations can be combined to give

$$uB_z - \frac{1}{4\pi\gamma\rho_i\rho}\left[(B_z^2 + B_r^2)\left(\frac{\partial B_r}{\partial z} - \frac{\partial B_z}{\partial r}\right) + B_r B_\phi \frac{\partial B_\phi}{\partial z} - \frac{B_\phi B_z}{r}\frac{\partial}{\partial r}(rB_\phi)\right] = 0 \tag{5}$$

and

$$\frac{v_\phi}{r} + \frac{1}{4\pi\gamma\rho_i\rho r}\left[\left(B_z + \frac{B_\phi^2}{B_z}\right)\frac{\partial B_\phi}{\partial z} + \frac{B_r}{r}\frac{\partial}{\partial r}(rB_\phi) + \frac{B_r B_\phi}{B_z}\left(\frac{\partial B_r}{\partial z} - \frac{\partial B_z}{\partial r}\right)\right] = \omega, \tag{6}$$

where we made use of the assumption $v_z = 0$. Equation (5) expresses the steady-state balance between ambipolar diffusion and the radial advection of magnetic flux by the inflowing gas. The function $\omega(r,z)$ described by equation (6) satisfies $\mathbf{B}\cdot\nabla\omega = 0$, which implies that it is constant on the surfaces formed by rotating the poloidal projections of the field lines about the z-axis. As we verify in Section 2.2, ω maintains the same value for a given flux surface both inside and outside the disk.

It is instructive to evaluate equations (5) and (6) in the equatorial plane $z = 0$ (denoted by a subscript 0). For the adopted even field symmetry

$$B_r(r,z) = -B_r(r,-z),\ B_\phi(r,z) = -B_\phi(r,-z),\ B_z(r,z) = +B_z(r,-z), \tag{7}$$

we have $B_{r0} = B_{\phi 0} = 0$. It is then straightforward to show that ω is just the equatorial ion angular velocity for the given field line and can be expressed as

$$\omega = \Omega_{i0} = \Omega_0 \left[1 - \frac{(u_0/2r_0)}{\gamma\rho_{i0}}\right], \tag{8}$$

where $\Omega_i \equiv v_{i\phi}/r$ and $\Omega \equiv v_\phi/r$. This result can be interpreted as follows: $1/\gamma\rho_i$ is the collisional engagement time between the neutrals and ions, whereas, by the angular momentum conservation condition, $r/2u$ is the effective magnetic braking time of the neutral disk component. When the ratio of the engagement time to the braking time is small, the ions are well coupled to the neutrals and $\Omega_i \approx \Omega$, but the coupling weakens (and Ω_i/Ω decreases) as this ratio starts to increase. This is a manifestation of a general property of weakly ionized media, namely, that the neutrals cannot be magnetically braked on a time scale shorter than the neutral-ion momentum exchange time (see Königl 1987). In our model, this condition will be satisfied if u/v_ϕ is sufficiently small.

Motivated by the intention to match our disk solutions onto the centrifugally driven wind solutions of BP, we adopt the same similarity scalings that were employed by these authors. Specifically, we look for solutions where all the physical variables scale as power laws of the *spherical* radius along a given direction. Thus, we describe the field line that intersects the equatorial plane at a radius r_0 by the vector

$$\mathbf{r} = [\chi(\xi)r_0,\ \phi,\ \xi r_0], \tag{9}$$

so that $\chi(0) = 1$ (note that the roles of ξ and χ as defined here are reversed in the BP notation). The position of the disk surface will be denoted by the subscript 1 (with $\chi_1 \equiv \chi[\xi_1]$); clearly, under the similarity hypothesis, the disk scale height $h(r_0)$ scales linearly with r_0. The assumption of Keplerian rotation implies $u_0 \propto v_{\phi 0} \propto r_0^{-\frac{1}{2}}$, so that, by the mass conservation condition, $\rho_0 \propto r_0^{-\frac{3}{2}}$. Equation (1) then shows that $B_0 \propto r_0^{-\frac{5}{4}}$, as was already pointed out by BP. Finally, equation (3) indicates that ρ_{i0} must scale as $r_0^{-\frac{3}{2}}$.

In accordance with the ansatz (9), we write

$$\mathbf{B} = B_0(r_0)\mathbf{b}(\xi), \quad \rho_i = \rho_{i0}(r_0)\Lambda(\xi), \tag{10}$$

where, by the adopted symmetry (eq. [7]), $b_r(0) = b_\phi(0) = 0$, $b_z(0) = 1$, and where, of course, $\Lambda(0) = 1$. The field components b_r and b_z are related by $b_r/b_z = \chi'$ (where the prime denotes a derivative with respect to ξ), so $\chi'(0) = 0$. There are thus four independent functions of ξ: b_z, b_ϕ, χ, and Λ, for which we have four independent equations: (1), (5),

(6), and $\nabla \cdot \mathbf{B} = 0$. These equations can be expressed in terms of the variable ξ by using the similarity scalings of $v_{\phi 0}$, u_0, ρ_0, B_0, and ρ_{i0} and following the same procedure that was employed for deriving the wind equation in BP. It is also useful to cast the above four equations in a dimensionless form. One then identifies the following free parameters:

$$\alpha \equiv \frac{2\pi\rho_0 u_0 (GM/r_0)^{-\frac{1}{2}}}{B_0^2}, \quad \beta \equiv \frac{4\pi\gamma\rho_0\rho_{i0}r_0 u_0}{B_0^2}, \quad \epsilon \equiv \frac{u_0}{(GM/r_0)^{1/2}}. \tag{11}$$

The transformed equations are not reproduced here due to space limitations.

2.2 NUMERICAL RESULTS

We have obtained specific disk solutions by numerically integrating the transformed angular momentum conservation and induction equations from the equatorial plane at $\xi = 0$ (where the values of b_ϕ, χ', and χ are fixed at 0, 0, and 1, respectively) to the disk surface at $\xi = \xi_1$ for given choices of the free parameters α, β, ϵ, and ξ_1. A solution was deemed unphysical if, at any point within the integration interval, the variable Λ became negative. As a fiducial set of parameters, we have chosen $\alpha = \beta = 9, \epsilon = \xi_1 = 0.1$. This combination could represent the conditions inside a circumstellar disk of total mass 0.05 $M_\odot$ (on a scale of 10^{17} cm) and mass accretion rate 2×10^{-7} $M_\odot$ yr^{-1} at a distance $r_0 = 10^{15}$ cm from a solar-mass star, where the field amplitude B_0 is 1 mG and the ion density satisfies $\rho_{i0} = C\rho_0^{\frac{1}{2}}$ (where $C \approx 3 \times 10^{-16}$ cm$^{-\frac{3}{2}}$ g$^{\frac{1}{2}}$. Although any given circumstellar disk may be described by a somewhat different set of parameters, the general behavior of the solutions is not sensitive to the particular choice of fiducial values.

The solution obtained by using our fiducial set of parameters is depicted in Figures 1 and 2. Figure 1 shows the poloidal projection of the magnetic field lines in the disk. Owing to the self-similarity, all the field lines intersect the disk surface at the same angle ($\chi'_1 = 1.50$ in this case). This figure demonstrates that the bending of the field lines away from the rotation axis (corresponding to $b_r < 0$), which is necessary for a smooth matching onto the wind solution, is consistent with the requirement of a steady-state balance between flux advection and diffusion (eq. [5]). This figure should be contrasted with Figure 3 in BP, which depicts the "concave in" field geometry in the wind. The effect of ambipolar diffusion is illustrated in Figure 1 by means of the arrows, which represent the drift velocities of the ions relative to the inflowing neutral gas. Figure 2 shows how the three field components and the ion density vary with ξ. It is seen that, while b_z remains nearly constant, both b_r and $|b_\phi|$ increase rapidly with height along a field line and, in this example, become comparable to b_z near the disk surface. As discussed below, the relatively large values that can be attained by b_{r1} and $|b_{\phi 1}|$ have important consequences for the acceleration and collimation properties of the jets. The negative sign of b_ϕ is mandated by the magnetic braking requirement (see eq. [1]) and is consistent with the shearing of the radial field component as a result of the differential rotation in the disk. This figure also demonstrates the increase of Λ with ξ that is required, according to equation (6), for maintaining the value of the flux-surface constant ω as b_r and $|b_\phi|$ increase along the field line. The implied increase in the degree of ionization with distance from the midplane is consistent with the expected decline of ρ with z in a real disk as well as with our treatment of the wind above the disk surface as a single-component, ionized gas.

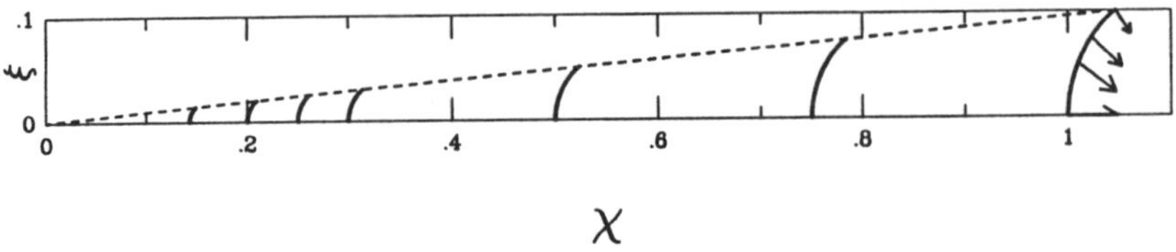

Fig. 1 – Poloidal magnetic field geometry in an ambipolar diffusion-dominated disk specified by the fiducial parameters $\alpha = \beta = 9$ and $\epsilon = \xi_1 = 0.1$. The heavy solid lines depict a few of the self-similar field lines, whereas the dashed line delineates the surface of the disk. The arrows represent the drift velocities of the ions relative to the neutral disk material, with the equatorial drift velocity satisfying $(v_{ir} - v_r)_o = u_o$. Only nonnegative values of χ and ξ are shown owing to the assumed reflection symmetry about the coordinate axes.

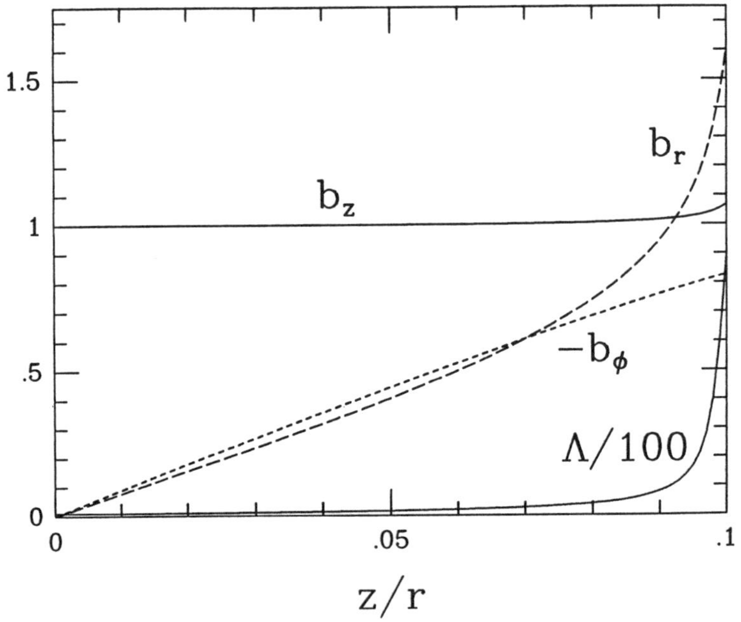

Fig. 2 – Variation of the normalized magnetic field components b_r, b_ϕ, b_z and ion density Λ along a field line in a self-similar, ambipolar diffusion-dominated disk with the same parameters as in Fig. 1.

Since the basic aim of our calculation is to construct continuous disk-jet solutions, we need to consider the matching conditions at the disk surface between the interior solutions derived in this section and the self-similar wind solutions derived by BP. From the relations $\nabla \cdot \mathbf{B} = 0$ and $\nabla \times \mathbf{E} = 0$, we infer the continuity of the normal component of the magnetic field and the tangential component of the electric field, respectively, across the disk-wind interface. These conditions as well as the continuity of the angular momentum and energy fluxes normal to the disk surface are satisfied provided that the magnetic field components and the flux-surface constant ω are continuous across the disk-wind interface.

In the formulation of BP, the self-similar, cold wind solutions are determined by three independent parameters: κ, the dimensionless ratio of the mass flux to the magnetic flux (which, in the single component model, is a field-line constant); λ, the dimensionless specific angular momentum (which includes both kinetic and magnetic contributions and is also a field-line constant in a highly ionized gas); and χ_1'. In the application to steady-state disk models there is a fourth independent parameter that must be considered, namely, the ratio $\nu \equiv v_{w\phi 1}/(GM/r_1)^{\frac{1}{2}}$ of the wind (subscript w) rotational speed to the Keplerian speed at the disk surface. BP set this ratio equal to unity, but in our case, using the matching condition for ω and the cold wind approximation, we find that it actually depends on the disk parameters and is given by $\nu = (1 - \epsilon\alpha/\beta)\chi_1^{\frac{3}{2}}$. For the fiducial solution depicted in Figures 1 and 2, $\chi_1 = 1.05$ and hence $\nu = 0.96$. Since the main focus of our present discussion is the disk solution, we simplify the matching procedure in this case by adopting the $\nu = 1$ wind solution of BP. A discussion of the more general wind solution which must be used when $(1 - \nu)$ is not much smaller than 1, will be given elsewhere.

The solutions that are of interest to us correspond to outflows that become super-Alfvénic at a finite height above the disk. The requirement that the solution pass through the Alfvén critical point fixes the value of χ_1' for a given choice of κ and λ. The solution curves $\chi_1' = const$ that are obtained in this way in the $\kappa - \lambda$ plane are displayed in Figure 3, which is patterned on Figure 2 in BP. Only solutions with $\kappa < 1$, which represent winds that are magnetically dominated in the vicinity of the disk, are considered. The region in the $\kappa - \lambda$ plane where the solutions remain everywhere sub-Alfvénic is bounded by the line $\lambda_{\rm min}(\kappa)$, given by $\kappa\lambda_{\rm min}(2\lambda_{\rm min} - 3)^{\frac{1}{2}} = 1$. Also plotted in Figure 3 are contours of constant $\mu(\kappa, \lambda) \equiv \kappa(2\lambda - 3)^{\frac{3}{2}}$, the parameter that determines the far-field behavior of the solutions (see BP). In particular, magnetic collimation of the jets occurs farther out from the disk surface the larger the value of μ.

The relation between the toroidal and the poloidal (subscript p) field components just above the disk surface is given, according to the BP solution, by $B_\phi/B_p = -\kappa(\lambda - \nu)/[1 + (\chi_1')^2]^{\frac{1}{2}}$. Using the matching conditions for B_p and B_ϕ as well as the condition $\nabla \cdot \mathbf{B} = 0$, we obtain the following relation between the disk and the jet solutions:

$$\Pi(\kappa, \lambda, \nu) \equiv \kappa(\lambda - \nu) = -\chi_1(\chi_1 - \xi_1 \chi_1') b_{\phi 1}. \tag{12}$$

For our fiducial solution, the right-hand side of equation (12) is equal to 0.78. Using the approximation $\nu = 1$, we find that the hyperbola $\Pi(\kappa, \lambda)$ corresponding to this value intersects the solution curve $\chi_1' = 1.50$ in the $\kappa - \lambda$ plane at the point $\kappa = 0.06, \lambda = 13.59$ (see Fig. 3). This demonstrates that, for the fiducial set of disk parameters α, β, ϵ, and ξ_1, there exists a unique self-similar jet solution, corresponding to a centrifugally driven wind that becomes super-Alfvénic at a finite height above the disk, which joins smoothly onto the steady-state disk solution and which can be determined with the help of the derived disk parameters χ_1, χ_1', and $b_{\phi 1}$. Figure 4 shows the behavior of the disk solution curves in the $\chi' - \Pi$ plane as the parameters α, β, ϵ, and ξ_1 are varied about their fiducial values. It is seen that physical disk solutions exist for a fairly wide range of parameters and that many of these solutions have comparatively large values of χ_1' and Π that would allow their matching to magnetically driven, asymptotically super-Alfvénic wind solutions. These conclusions are important for the interpretation of stellar outflow sources presented in the next Section.

Another result that is relevant to the discussion in Section 3 concerns the energy flux carried by the jets. In the cold wind approximation, this is just the Poynting flux at the base of the wind. Noting that this is equal to ω times the angular momentum flux $-r_1 B_{z1} B_{\phi 1}/2\pi$ and using the thin disk approximation ($h \ll r$, implying $r_1 \approx r_0$) as well as equations (1) and (2), we infer that the energy flux is approximately equal to $\omega h(r_0)\rho_0 u_0 v_{\phi 0}$ (neglecting the radial magnetic angular momentum flux, which is $\sim h/r$ times the vertical flux). By comparison, the divergence of the z-integrated radial energy flux density in the disk is approximately $-\Omega_0 h(r_0)\rho_0 u_0 v_{\phi 0}$, where we have neglected the magnetic contribution in comparison with the kinetic term (since it is, again, a factor $\sim h/r$ smaller). Using equation (8), we conclude that the jets carry a fraction $\sim (1 - \epsilon\alpha/\beta)$ $(= 0.9$ in the fiducial solution) of the gravitational energy liberated in the disk (with the remainder dissipated by ion-neutral collisions). The fact that this fraction can be O(1) in our model may be attributed to the adopted even symmetry of the field and to the assumption that the excess angular momentum of the accreted matter is transported exclusively by magnetic stresses.

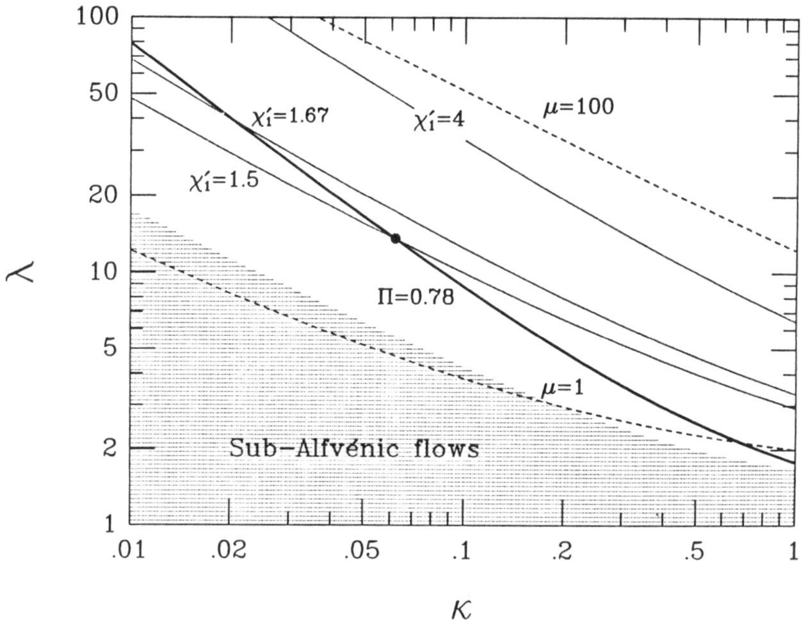

Fig. 3 – Solution plane for the cold, self-similar wind models of Blandford and Payne (1982). The solid hyperbola represents the constraint imposed by eq. (12) in the text (with ν set equal to 1) for the disk solution shown in Figs. 1 and 2 (which corresponds to $\Pi = 0.78$). The intersection point (*filled circle*) of this hyperbola with the $\chi_1' = 1.5$ curve yields the unique centrifugally driven wind solution that joins smoothly onto the given disk configuration.

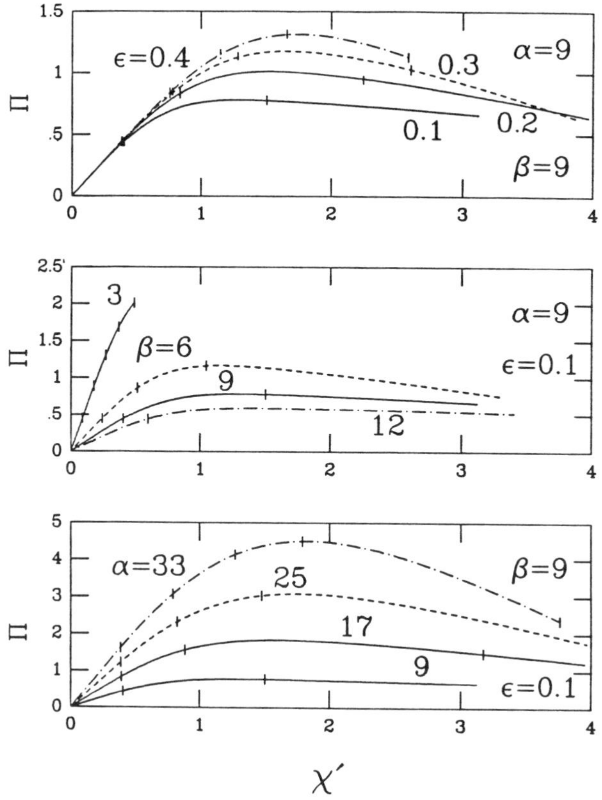

Fig. 4 – Solution space for the self-similar models of an ambipolar diffusion-dominated disk, illustrating the dependence on the free parameters α, β, ϵ, and χ_1. The tick marks denote successive values (0, 0.05, 0.1, ...) of χ_1. The curves terminate when ξ_1 reaches 0.25 or if the solution becomes unphysical ($\Lambda < 0$). The solid curve in each frame corresponds to the fiducial values of α, β, and ϵ that were used in Figs. 1 and 2.

3. Application To Young Stellar Objects

The disk-wind model described in the preceding Section is consistent with the observational evidence for circumstellar disks driving energetic bipolar outflows in YSOs (see Section 1). In particular, using the BP solution for a centrifugally driven wind in a Keplerian disk together with the assumption that the wind is the main means of angular momentum removal from the disk, it can be readily demonstrated that the wind kinetic luminosity is smaller than or of the order of the accretion luminosity (Königl 1986), which directly accounts for the correlation $L_w \propto L$ that characterized YSOs of all masses. However, as interpreted in the context of this model, the connection between accretion and outflows is not merely a reflection of the fact that the winds are powered by accretion but is of a more fundamental nature. Rather than being an incidental byproduct of the accretion process, the outflows could be a *necessary* ingredient in this process in that *they regulate the transport of angular momentum in the disk.* In this picture, the observed association between accretion disks and energetic outflows is a manifestation of the fact that the winds carry away most of the angular momentum that needs to be removed in order for the accretion to proceed. The coupling between the inflow and the outflow is achieved, in this scenario, through the ordered magnetic fields that thread the disk. This explains the commonly observed alignment of the outflow axis with the ambient magnetic field direction and justifies the even field symmetry adopted in our solutions. The model also incorporates the fact that YSO disks are evidently not magnetic pressure-dominated, although it stipulates that the more tenuous wind zone above the disk surface be magnetically controlled. In this connection it may be noted that our simplifying assumption of a single-fluid wind in which the magnetic field is well coupled to the bulk of the matter should be adequate even in cases where the wind is largely neutral if the neutral-ion collision time $1/\gamma\rho_{iw}$ is sufficiently short (see Hartmann and MacGregor 1982; Pudritz and Norman 1983), and that it is even better justified if the disk possesses a corona.

The interpretation of bipolar outflows as being an integral component of disk accretion has interesting implications for the evolution of YSOs. The similarity of the spectral properties (Adams, Lada, and Shu 1987), the luminosity distributions (Myers *et al.* 1987), and the wind characteristics (Strom, Strom, and Edwards 1988) in DEIS and CTTS has led the respective authors to suggest that the former are just the veiled predecessors of the latter. The transition between the two classes of objects is thought to occur through the action of the bipolar outflows, which disperse the obscuring molecular material on a time scale of $\lesssim 10^5$ yr (cf. Lada 1985). The similarity between CTTS and NTTS has motivated an analogous proposal for these objects, namely, that the former evolve into the latter (on time scales ranging from $< 10^6$ yr to $\sim 10^7$ yr) as a result of the dissipation of the associated accretion disks (e.g., Walter *et al.* 1988). The disappearance of the disks is conceivably also linked to the disruption of the parent molecular cores (which are the source of the accreted mass) by the energetic outflows that accompany the accretion process. The entire evolution from DEIS through CTTS to NTTS may thus be controlled by the feedback effect of the accreted material on the surrounding cloud, which is transmitted by the energetic outflows. The winds, therefore, may not just influence the instantaneous structure of the circumstellar disks by regulating the transfer of angular momentum, but they could also affect the long-range evolution of the disks by modifying the rate of mass accretion from the ambient medium. Furthermore, the combined accretion and outflow processes may also play a central role in the evolution of the stellar object itself. It has often been argued that the termination of accretion following the onset of an energetic outflow is the dominant factor that determines the final mass of a YSO (see Shu, Adams, and Lizano 1987). It was previously suggested that physical processes inside the YSO (specifically, the ignition of deuterium) might be responsible for triggering the wind. However, there are several apparent difficulties with this suggestion (e.g., Stahler 1988). According to the present interpretation, the end of the stellar accretion phase is not tied to any particular event in the YSO but rather is a consequence of a more or less continuous outflow which commences as soon as disk accretion begins. For a solar-mass star that forms out of a rotating molecular core, most of the stellar buildup may occur by disk accretion rather than by direct infall (e.g, Cassen, Shu, and Terebey 1985), so the evolution of the star is naturally linked to that of the disk (see Mercer-Smith, Cameron, and Epstein 1986). This interpretation does not preclude the possibility that the highly collimated optical jets (whose effect on the ambient cloud is often small) are associated with stellar winds (instead of simply representing the ionized inner regions of the disk-driven outflows), nor is it necessarily at variance with the suggestion (Shu *et al.* 1988) that the YSO could also drive a centrifugal outflow if it develops a sufficiently strong surface magnetic field. In addition, it is conceivable that the physical conditions along the disk vary in such a way that a strictly steady-state solution cannot be constructed for the entire accretion flow. In this case, mass would temporarily accumulate at certain locations in the disk, which could have significant implications to FU Orionis outbursts (Pudritz and Norman 1986; Hartmann and Kenyon 1987) as well as to the formation of binaries and planets. However, if the mass inflow is effectively continuous on a time scale of $\gtrsim 10^5$ yr, then the basic aspects of the proposed evolutionary scenario would remain unchanged.

The magnetized disk-wind model is also consistent with some of the peculiar radiative properties of YSOs that show strong evidence for the presence of disks and outflows. One such characteristic is the approximately flat near- and mid-infrared specific-luminosity spectrum (i.e., $\nu F_\nu \approx const$) in T Tauri stars with large IR and UV excesses and strong emission lines (Adams, Lada, and Shu 1987, 1988). A plausible interpretation of this spectral component is that it represents intrinsic emission from a circumstellar disk. However, the inferred radial distribution of the surface temperature, $T \propto r^{-1/2}$ (derived assuming a steady-state, optically thick, Keplerian disk), is inconsistent with the variation $T \propto r^{-\frac{3}{4}}$ predicted by viscous disk models. This prediction is a consequence of the fact that, in these models, all the energy that is liberated in the course of the accretion process is radiated locally from the disk surface. By contrast, in the magnetized disk model considered here, most of the liberated energy leaves the disk in the form of a Poynting flux that powers the centrifugally driven wind (see Section 2.2). Hence it is, in principle, possible to have a relatively shallow surface temperature distribution in this model without violating the angular momentum conservation relation. Another characteristic feature of T Tauri spectra is the double peaked nature of many of the measured forbidden line profiles (Appenzeller, Jankovics, and Ostricher 1984; Edwards *et al.* 1987). These lines have been interpreted as arising in a wind with either a hollow cone geometry or a latitude-dependent velocity that decreases on going from the stellar pole to the equator. The hollow cone geometry is clearly consistent with a disk origin for the wind, whereas the inferred latitude dependence could, in principle, be mimicked by a disk outflow with a radially decreasing distribution of final velocities, as expected for a centrifugally driven wind in a Keplerian disk (see BP). The compatibility with the observed radiative properties of YSOs lends further support to the proposed model. A more quantitative discussion of these points is, however, beyond the scope of this contribution and will be presented elsewhere.

References

Adams, F. C., Lada, C. J., and Shu, F. H. 1987, *Ap. J.*, **312**, 788.

__________. 1988, *Ap. J.*, **326**, 865.

Appenzeller, I., Jankovics, I., and Ostricher, R. 1984, *Astr. Ap.*, **141**, 108.

Bally, J., and Stark, A. 1983, *Ap. J. (Letters)*, **266**, L61.

Bertout, C., Basri, G., and Bouvier, J. 1988, *Ap. J.*, **330**, 350.

Blandford, R. D., and Payne, D. G. 1982, *M.N.R.A.S.*, **199**, 883 (BP).

Cassen, P., Shu, F. H., and Terebey, S. 1985, in *Protostars and Planets II*, ed. D. C. Black and M. S. Matthews (Tucson: University of Arizona Press), p. 448.

Cohen, R. J., Rowland, P. R., and Blair, M. M. 1984, *M.N.R.A.S.*, **210**, 425.
Draine, B. T., Roberge, W. G., and Dalgarno, A. 1983, *Ap. J.*, **264**, 485.
Edwards, S., Cabrit, S., Strom, S. E., Heyer, I., Strom, K. M., and Anderson, E. 1987, *Ap. J.*, **321**, 473.
Hartmann, L., and MacGregor, K. B. 1982, *Ap. J.*, **259**, 180.
Hartmann, L., and Kenyon, S. J. 1987, *Ap. J.*, **322**, 393.
Heyer, M. H. 1988, *Ap. J.*, **324**, 311.
Kenyon, S. J., and Hartmann, L. 1987, *Ap. J.*, **323**, 714.
Königl, A. 1986, *Can. J. Phys.*, **64**, 362.
_________. 1987, *Ap. J.*, **320**, 726.
_________. 1989, *Ap. J.*, in press.
Koo, B-C. 1989, *Ap. J.*, in press.
Lada, C. J. 1985, *Ann. Rev. Astr. Ap.*, **23**, 267.
Lizano, S., *et al.* 1988, *Ap. J.*, **328**, 763.
Lovelace, R. V. E., Wang, J. C. L., and Sulkanen, M. E. 1987, *Ap. J.*, **315**, 504.
Mercer-Smith, J. A., Cameron, A. G. W., and Epstein, R. I. 1984, *Ap. J.*, **279**, 363.
Mestel, L. 1961, *M.N.R.A.S.*, **122**, 473.
Mitchell, G. F., Allen, M., Beer, R., Dekany, R., Huntress, W., and Maillard, J-P. 1988, *Ap. J. (Letters)*, **327**, L17.
Mouschovias, T. Ch., and Paleologou, E. V. 1980, *Ap. J.*, **237**, 877.
Mundt, R. 1985, in *Protostars and Planets II*, ed. D. C. Black and M. S. Matthews (Tucson: University of Arizona Press), p. 414.
Myers, P. C., Fuller, G. A., Mathieu, R. D., Beichman, C. A., Benson, P. J., Schild, R. E., and Emerson, J. P. 1987, *Ap. J.*, **319**, 340.
Myers, P. C., and Goodman, A. A. 1988, *Ap. J.*, **329**, 392.
Nakano, T., and Umebayashi, T. 1986, *M.N.R.A.S.*, **218**, 663.
Persson, S. E., McGregor, P. J., and Campbell, B. 1988, *Ap. J.*, **326**, 339.
Pudritz, R. E. 1985, *Ap. J.*, **293**, 216.
Pudritz, R. E., and Norman, C. A. 1983, *Ap. J.*, **274**, 677.
_________. 1986, *Ap. J.*, **301**, 571.
Sakurai, T. 1985, *Astr. Ap.*, **152**, 121.
_________. 1987, *Pub. Astr. Soc. Japan*, **39**, 821.
Sargent, A. I., and Beckwith, S. 1987, *Ap. J.*, **323**, 294.
Shu, F. H., Adams, F. C., and Lizano, S. 1987, *Ann. Rev. Astr. Ap.*, **25**, 23.
Shu, F. H., Lizano, S., Ruden, S. P., and Najita, J. 1988, *Ap. J. (Letters)*, **328**, L19.
Stahler, S. W. 1988, *Ap. J.*, **332**, 804.
Strom, K. M., Strom, S. E., Kenyon, S. J., and Hartmann, L. 1988, *A. J.*, **95**, 534.
Strom, K. M., Strom, S. E., Wolff, S. C., Morgan, J., and Wenz, M. 1986, *Ap. J. Suppl.*, **62**, 39.
Strom, S. E., Strom, K. M., and Edwards, S. 1988, in *Galactic and Extragalactic Star Formation*, ed. R. Pudritz and M. Fich (Dordrecht: Reidel), in press.
Uchida, T., and Shibata, K. 1985, *Pub. Astr. Soc. Japan*, **37**, 515.
Vogel, S. N., Bieging, J. H., Plambeck, R. L., Welch, W. J., and Wright, M. C. H. 1985, *Ap. J.*, **296**, 600.
Vrba, F. J., Strom, S. E., and Strom, K. M. 1988, *A. J.*, **96**, 680.
Walter, F. M., Brown, A., Mathieu, R. D., Myers, P. C., and Vrba, F. J. 1988, *A. J.*, **96**, 297.

MASSIVE CLOUD OF ACTIVE STAR FORMATION AND THE SPINNING STREAMER AS ANGULAR MOMENTUM DRAIN

Yutaka UCHIDA[1], Akira MIZUNO[2], and Yasuo FUKUI[2]
[1]*Department of Astronomy, and Institute of Astronomy, University of Tokyo*
[2]*Department of Astrophysics, Nagoya University*

ABSTRACT. It was found that the ^{13}CO line-emitting regions with blue- and redshifts are clearly separated on north-west and south-east sides of the curved axis in the ρ Oph streamer (north), respectively (Uchida *et al.*1989a). A most likely interpretation is that the streamer is spinning around its curved axis. We discuss here the implication of this in connection to the evolution of the massive cloud of active star formation located at the south-west end of it by introducing a magnetodynamic picture which is related to the sweeping-magnetic-twist mechanism proposed by Uchida and Shibata (1985) for the production of molecular bipolar flows from forming stars.

1. Introduction

One of the conspicuous features seen in the CO line emission in the regions around the active star-forming clouds is the elongated structure called "streamer" extending out from these clouds. A typical streamer has a length of a few tens pc and a width of a few pc varying from a place to another with an enhanced density of $n \sim 10^{2.5 \sim 3.5}$ cm^{-3} (eg., Fukui *et al.* 1986).

Examples of the streamers are those in Ophiuchi or in Orion (eg., Sugitani *et al.* 1986). With the knowledge that here again the observed direction of the interstellar magnetic field tends to be aligned along the elongated structure as in the case of the molecular bipolar flows (Vrba *et al.* 1976), an intuitive consideration suggests that these streamers might also be "angular momentum drains" from the massive cloud of enhanced star formation just like molecular bipolar flows serving for the rotating disks around forming stars (Uchida and Shibata 1985). In this analogy, the massive cloud of enhanced star formation may correspond to the *disk* or *ring* (cf, Kaifu *et al.* 1984) in the case of single star formation in which the accretion to the central star is enhanced by the extraction of angular momentum (or, magnetic braking) from the disk material through the production of the magnetic twist and the bipolar flows in the Uchida-Shibata model. The entire sequence of the star formation may take place in a deeper part of the gravitational potential in such a cloud having smaller angular momentum, and more energetic event may accompany them.

In the present paper, we report, based on the re-examination of the data obtained by using the Nagoya 4m telescope, that a streamer in Oph (ρ Oph streamer (north)) shows a systematic separation of the blue- and redshifted regions on each side of the curved axis, indicating that the streamer may be spinning. A streamer, then, may not be an ad-hoc feature, but highly likely to be a structure playing an intrinsic role in the evolution of the cloud of active star formation (Uchida *et al.* 1989b), just like the bipolar flows are playing an intrinsic role in the process of star-formation by extracting angular momentum from the disk to allow the final infall of the material to the star (Uchida and Shibata 1985).

G. Belvedere (ed.), Accretion Disks and Magnetic Fields in Astrophysics, 179–183.

2. Observational Results

Uchida *et al.* (1989a) re-analyzed the Nagoya ^{13}CO data of ρ Oph streamer (north), and found a systematic splitting of the blue- and redshifted regions, with blue always on the north-western side and red on the south-eastern side of the curved axis of the streamer. This is in the same sense with the rotation velocity of the blob (Myers *et al.* 1978) sitting at the south-west end of the streamer, which is referred to as the "base of the streamer" in the following. Figures 1a and 1b (Sugitani *et al.* 1986) show overviews of the region observed with the ^{12}CO and ^{13}CO lines integrated over the wavelength with three boxes for representative parts for which the blue- and redshifted contours in ^{13}CO line are given in Figures 2a, b, and c. The position-velocity diagrams in these are shown in Figures 2d, e, and f, respectively. The general trend is that the blue- and redshifted contours tend to be separated on each side of the axis of the structure (blue on the north-western and red on the south-eastern side) as seen in Figure 2c with a correspondingly tilted PV-diagram as seen in Figure 2f (Uchida *et al.* 1989a).

Figures 2a and 2b, with the corresponding PV-diagrams, Figures 2d and 2e, respectively, however, show two peculiar parts which have a bit more complicated features, although they still have the same general trend. Figures 2b and 2e include IRAS 16285 – 2355, an infrared source measured by IRAS satellite, where the structure of the streamer in ^{13}CO shows a peculiarity. The width of the ^{13}CO-emitting belt, which we call a "dense streak" since it does not seem to be simply a high part of a smooth density profile of the streamer, is narrowed there, and IRAS 16285 – 2355 is located just on the narrow ridge of the contour, as if the streak is bunched in the formation of the infrared source. This locality is examined with the Nobeyama 45m telescope, and it was revealed that the general trend of splitting of the velocity persists down to the small scale of the narrowed streak with some complicate winding shape near the IR source, and that the IR source has a molecular bipolar flow (Mizuno *et al.* 1989). Figures 2a and 2d, on the other hand, have some contamination from the "base of the streamer" which may well have somewhat different kinematical characteristics from that of the streamer.

More detailed observational results and their interpretation will be discussed in Uchida *et al.*(1989a), and in Mizuno *et al.*(1989).

3. Discussion

If we attribute the systematic splitting of the blue- and redshifted regions to the spinning of the streamer around its axis, the amount of angular momentum is larger than the amount carried into it with the mass in contraction, consistent with the view that the excesss angular momentum is provided from the "base of the streamer" which contracted with much larger mass than that of the streamer. This means that the angular momentum was extracted from the "base of the streamer" through the magnetic tension, and this may explain the enhanced rate of the star formation in the "base of the streamer" as described in the introduction (Uchida *et al.* 1989a,b).

Another remarkable point is that the newly formed stars are seen to align along the sharp "streaks" as in the case of IRAS 16285 – 2355, and in some cases like the Orion streamer, in the form of an IR star chain (cf, Mouschovias and Morton 1985). According to our view, those "streaks", which have sharp contrast and a winding shape inside the streamer, may be due to the propagated effect of the active star formation in the "base of the streamer" along the pre-existed helical magnetic structure produced by the rotation of the "base of the streamer" before the star formation. A question may arise as to why the effect of a star formation can extend such a large distance as 10 ~ 20 pc. Our answer (Uchida *et al.* 1989b) will be that this may be due to the successive triggering of the second and the third star formation by the propagated effect of them. Then a related question may be whether the time scale for the influence to reach the entire length of a few tens of pc is short enough, even if such a mechanism of "chain reacting star formation" operates. Our answer to this is that it seems possible. Suppose that the magnetic twist with the molecular flows driven out from a system consisted of a forming star + accretion disk near the edge of the "base of the streamer" encounters a dense clump on

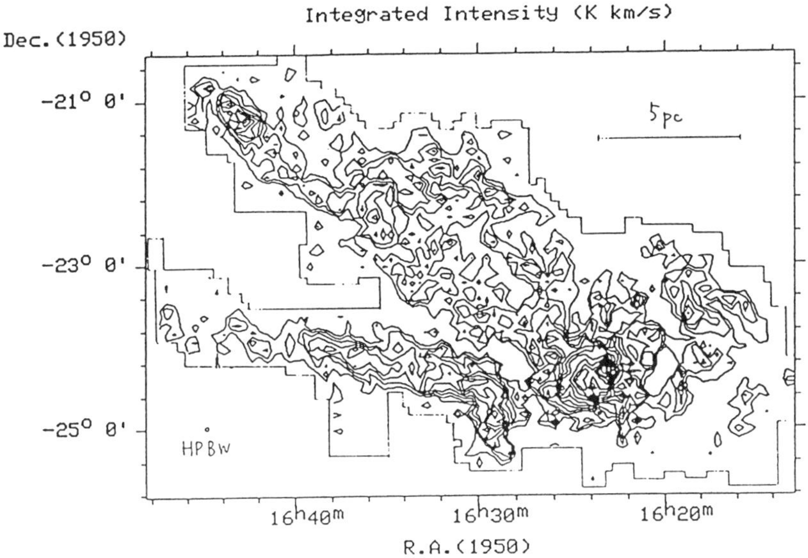

1(a)

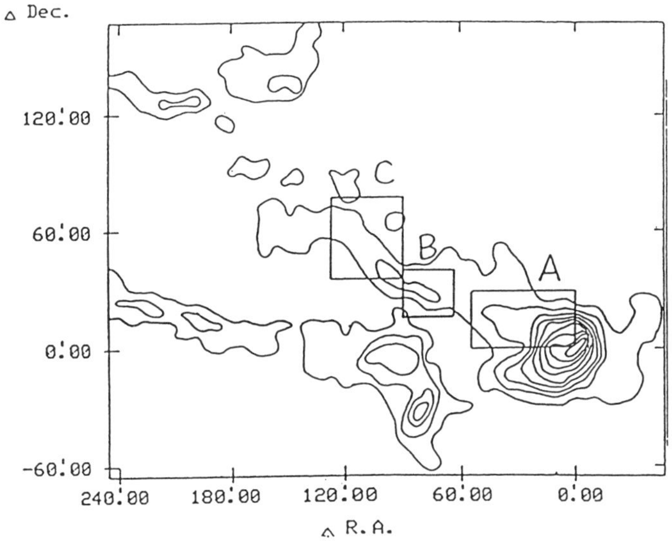

1(b)

Figure 1(a): Mapping of the ρ Oph region in the ^{12}CO 115 GHz line. The streamer we deal with is the elongated structure roughly diagonal in the figure with the "base of the streamer" at around ($16^h 23^m$, $-24°3$).

1(b): The same in the ^{13}CO 110 GHz line.

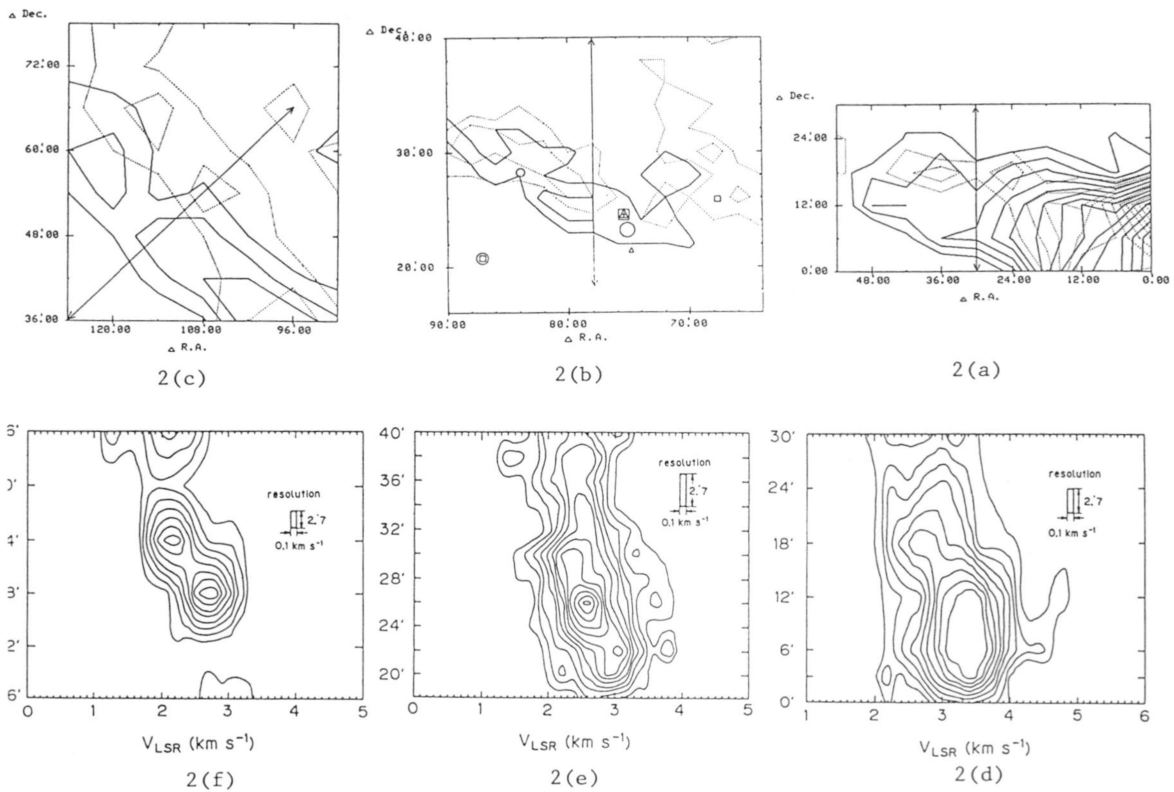

Figure 2: (a), (b), and (c) are the contours in the regions in the boxes A, B, and C in Figure 1(b) at blue- and red-wings. Solid lines and dotted lines in (a) are, $3.1 \pm 0.3 kms^{-1}$ and $2.5 \pm 0.3 kms^{-1}$, respectively. The same in (b) are, $2.8 \pm 0.1 kms^{-1}$ and $2.4 \pm 0.1 kms^{-1}$, respectively. The same in (c) are, $2.6 \pm 0.2 kms^{-1}$ and $2.2 \pm 0.2 kms^{-1}$, respectively. ΔRA and ΔDec are measured from the core of the "base of the streamer".
(d), (e), and (f) are P-V diagrams with the baselines shown in Figures 2 (a), (b), and (c), respectively. Abscissa is ΔDec measured from $-24°19'44''$.

its way of propagation along the large scale magnetic field which was made into a form of a large scale helix due to the rotation of the footpoint of the magnetic flux in the "base of the streamer". In the interaction of the disturbance with the clump, the magnetic twist accumulates in front of the clump due to the lower Alfven velocity in it, and exerts pinching effect on the clump (Shibata and Uchida 1989). This may trigger the contraction of the clump depending upon the condition. In case the contraction is triggered by the pinching on the incidence and accumulation of the B_ϕ packet, the state near the central part of the clump may be brought into the gravitationally unstable domain, and the formation of the system of the central star + accretion disk will take place in a much shorter time than otherwise (Habe *et al.* 1989). Such a triggering may occur for a cloud which is near the critical condition for the Jeans' criterion, and especially effective in a clump in which the magnetic pressure is effective in supporting the gravitational contraction, since the direction of the resultant magnetic force changes sign in the process of the magnetic pinching.

As soon as the central condensation with accretion disk is formed in the streak, it begins to emit the magnetic twist and flows into bipolar directions along it just by the same process. The backward-propagating packet unwinds the accumulated twist, and the forward-propagating one carries the same effect further forward along the flux tube to make the streak longer. The effect is amplified by the liberation of the gravitational energy of the triggered object, which plays the role of the "relay station", enabling the propagation of the effect to a distance much longer than the lobe length of a single bipolar flow. The propagation of the effect in between the "relay station" takes place in the Alfven velocity which can be higher than 10 km/s because of the stronger magnetic field bunched by the source and the clump on both sides, and of the lower density in between after the evacuation of the gas condensed into these objects on both sides, explaining the time scale for the formation of such an elongated structure.

More detailed discussion of the model and its theoretical significance on the dynamics of the star-forming regions will be given elsewhere (Uchida *et al.* 1989b).

Reference

Fukui, Y., Sugitani, K., Takaba, H., Iwata, T., Mizuno, A., Ogawa, H., and Kawabata, K., 1986, *Astrophys. J. (Letters),* **311**, L85

Habe, A., Uchida, Y., Ikeuchi, S., and Pudritz, R., 1989, in preparation

Kaifu, N., Suzuki, S., Hasegawa, T., Morimoto, M., Inatani, J., Nagane, K., Miyazawa, K., Chikada, Y., Kanzawa, T., and Akabane, K., 1984, *Astron. Astrophys.*, **134**, 7

Mizuno, A., Uchida, Y., Tohdoh, Y., Yamaoka, H., Fukui, Y., and Kaifu, N., 1989, in preparation.

Mouschovias, T. and Morton, S.A., 1985, *Astrophys. J.*, **298**, 205

Myers, P.C., Ho, P.T.P., Schneps, M.H., Chin, G., Pankonin, V., and Winnberg, A., 1978, *Astrophys. J.*, **220**, 864

Shibata, K. and Uchida, Y., 1986, *Publ. Astron. Soc. Japan*, **38**, 631

Shibata, K. and Uchida, Y., 1989, *Publ. Astron. Soc. Japan*, **41**, submitted.

Sugitani, K., Fukui, Y., Ogawa, H., and Kawabata, K., 1986, *Astrophys. J.*, **303**, 667

Uchida, Y., Kaifu, N., Shibata, K,. Hayashi, S.-S., and Hasegawa, T., 1987a, in *Star Forming Regions*, ed. Peimbert, M. and Jugaku, J. (D. Reidel), p287

Uchida, Y., Kaifu, N., Shibata, K., Hayashi, S.-S., Hasegawa, T., and Hamatake, H., 1987b, *Publ. Astron. Soc. Japan*, **39**, 907

Uchida, Y., Mizuno, A., and Fukui, Y., 1989a, *Publ. Astron. Soc. Japan*, submitted.

Uchida, Y., Pudritz, R., and Shibata, K., 1989b, in preparation

Uchida, Y. and Shibata, K., 1985, *Publ. Astron. Soc. Japan*, **37**, 515

Vrba, F.J., Strom, S.E., and Strom, K.M., 1976, *Astron. J.*, **81**, 958

MONTE CARLO SIMULATION OF H_2 FORMATION BY COSMIC RAYS: EFFECT OF THE GRAIN MODEL

Daniela Averna
Istituto di Astronomia
Università di Catania
95125 Catania
Italy

Valerio Pirronello
Dipartimento di Fisica
Università della Calabria
87036 Rende
Italy

ABSTRACT. It is considered the synthesis of molecular hydrogen induced in the icy mantles of interstellar grains by cosmic rays as due to the recombination of radicals created by the energy loss of the impinging ions. It is in particular investigated, by means of a Monte Carlo simulation, the relevance of the model of grains adopted. Some preliminary results are given.

1 . INTRODUCTION

The energetic balance of an interstellar cloud is strictly related, of course, to the relative importance of processes that, in the unit of time, provide energy with respect to those that subtract it.

In dense clouds the most efficient processes that are able to release energy throughout the cloud are those connected with the energy loss of cosmic rays interacting with the gas inside the cloud itself (e.g. Field, 1974).

Cosmic rays have, by the way, another important and well recognized effect, that of regulating the ionization rate inside interstellar clouds, the parameter that regulates the kinetics of ion-molecule reactions in the gas phase. Such reactions in the gas are probably responsible for the formation of most of the species observed in molecular clouds. Some of these species are relevant for providing a cooling mechanism; they, in fact, once excited collisionally, may emit photons to which the cloud is transparent. Among

G. Belvedere (ed.), Accretion Disks and Magnetic Fields in Astrophysics, 185–190.

these molecules the most important one is H_2 for several reasons and in particular because it is the most abundant by many orders of magnitude.

2. THE FORMATION OF MOLECULAR HYDROGEN

The problem of the formation of molecular hydrogen is not yet completely solved. Several attempts (Gould and Salpeter, 1963; Hollenbach et al., 1971; Barlow and Silk, 1976 and many others) have been made to tackle it because, as it is well known, radiative recombination of hydrogen atoms in the gas phase is not able to produce H_2 at the requested rate. It is believed that molecular hydrogen needs the presence of interstellar dust to form; the generally accepted way uses the grain surface as a catalyst in the sense that on it hydrogen atoms, once adsorbed, can migrate by tunnelling from one site to another until when, encountering each other, can form the molecule, and releasing the formation energy to the grain by "conduction", the just synthesized H_2 molecule can evaporate (e.g. Hollenbach et. al., 1971). Another proposed way consists in the hydrogen abstraction from already formed molecules in the mantle (for instance formaldehyde) by an incoming H atom (Tielens and Hagen, 1982). Both the mentioned mechanisms rely on the high mobility that H adatoms should have even at temperatures of about 10 K, in order to scan the surface of dust grains. The crucial assumption that a sufficiently high mobility takes place is certainly acceptable if the structure of the surface is crystalline: in this case, in fact, the wave packet describing quantum mechanically the hydrogen adatom spreads easily, as is well known (see e.g. Dekker, 1965), in the periodic distribution of potential wells (adsorption sites). This, however, seems not to be the case in the interstellar environment; a comparison between the observed 3.1 μm feature of water ice in astronomical sources with ice samples prepared in the laboratory clearly shows that interstellar ice is almost certainly amorphous (Leger et al., 1979).

Being the structure of the ice mantle on grains noncrystalline, the widely accepted assumption of high mobility does not hold any more and a more proper analysis is needed to treat the transport of adsorbed species on the amorphous surface. This has been done recently with a rigorous quantum mechanical approach by R. Smoluchowski (1983). He used Anderson's theory of localization and obtained that in the amorphous case mobility is reduced by several orders of magnitudes with respect to the crystalline one, because the adsorbed H atoms after few tunnelling remain trapped in the deepest sites existing on the surface. This reduces enormously the recombination rate of adsorbed hydrogen atoms, expecially at temperatures close to 10 K. The same drastic reduction of efficiency should affect the mechanism proposed by Tielens and Hagen (1982).

Other mechanisms are then responsable for the molecular hydrogen synthesis.

In a recent paper Pirronello and Averna (1988) showed by means of a Monte Carlo simulation that the interaction of low energy (E<1GeV) cosmic rays with "dirty ice" mantles on grains, composed by H_2O, CH_4, NH_3 and CO, can be orders of magnitude more efficient than mechanisms requiring the mobility of adsorbed species on amorphous surface. Such numerical simulation used, as an input, empirical values obtained together with W.L. Brown and L.J. Lanzerotti at the Van der Graaf facility of AT&T Bell Laboratories in Murray Hill (New Jersey). Production of molecules in ices and mixtures of them irradiated by energetic ions is now experimentally well established (Brown et al., 1982; Pirronello et al., 1982, Pirronello, 1988; Pirronello et al., 1988) and is possibly due to the recombination of radicals and ions produced by a small fraction of the energy deposited by the impinging ion in the solid. In particular the data used in this simulation are those on the synthesis of H_2 measured as a function of the water ice temperature by Brown and coworkers in 1982. The results of Pirronello and Averna (1988) were obtained under some quite conservative hypotheses, among which the most important one is that the production yield of molecular hydrogen for the mixture of ices is scaled from the measured values for water ice (Brown et al., 1982), while frozen methane seems now to release H_2 at a much higher rate and presumably also frozen ammonia does.

3. EFFECT OF THE GRAIN MODEL: OUR APPROACH

The numerical simulation performed by Pirronello and Averna (1988) has certainly left several unanswered questions. We want to address here one of them: the importance of the type of grains on which the interaction between cosmic rays and dirty ice mantles takes place.

As is well known today, fifty years after the discovery of interstellar dust, we do not have yet a full knowledge of the structure and composition of grains: several models exist. The two of them we chose to use for our comparison are the one introduced by Mathis, Rumpl and Nordisieck (1977) and the other one by J.M. Greenberg and coworkers (see e.g. Hong and Greenberg, 1980; Chlewicki and Greenberg, 1984 and references therein). According to the first model, interstellar grains in dense clouds are made of a refractory core (composed either by silicates or graphite and substantially equal to the grains present in diffuse clouds) that follows a size distribution of the type:

$$a^{-3.5}$$

(where "a" is the radius of the core) on the top of which it is accreted a dirty ice mantle of a thickness, that we assumed roughly equal to the radius of the core in order to agree with observations by Jura (1980). The other model, which is very appealing from the physical point of view and for its evolutionary meaning, describes the grain of a dense cloud as made up by three different layers: the core (made of silicates and having a radius of 500 Å), an organic refractory mantle of thickness between 600 and 2400 Å (which is the residue of an original dirty ice mantle photolized by UV during a stage the grain was in diffuse regions) and, on the top of them, a thin (thickness between 100 and 400 Å) ice mantle accreted inside the dense cloud itself.

In our simulation we have chosen pseudorandomly protons and helium ions, in the relative abundance given by recent evaluations (see e.g. Wefel et al., 1988) with energy between zero and 1000 MeV. The probability of their extraction has been adapted to match the demodulated energy spectra obtained by Morfill et al. (1976). Such ions have been made to impinge on grains of the two types and of selected sizes as if they were at several depths inside an interstellar clouds. The direction of arrival of each ion has been taken also pseudorandomly and the energy loss suffered in the gas before reaching the grain has been taken into account. The model we used for the cloud is one of those developed by Boland and de Jong (1984) that well describes small quiescent spherical clouds with no star formation processes going on, as in the case of L134. The central depth of the cloud, measured in magnitudes, is A_v=8 mag. The silicates used for the core in the two cases have the composition $MgFeSiO_4$. The size of the grains considered in the simulation is between 400 Å and 4000 Å in radius for the grains of Mathis type and between 1200 Å and 3300 Å for those of the Greenberg type. The composition of the ice mantle is in both cases still that one used by Pirronello and Averna (1988), while the refractory mantle of Greenberg grains is tentatively represented by a sort of polymer having an average abundance ratio between "metals" (N+C+O) and H of about 1:3.

4. RESULTS AND CONCLUSIONS

Some of the preliminary results of this investigation,

aiming to try to understand possible effects due to the kind of grains present inside interstellar clouds on the production of molecular hydrogen, are shown in Table 1.

Due to the peculiarities of the two types of structures and size distributions it is not possible to compare in a meaningful way the H_2 production from the two grains size by size in the whole range; such a comparison can be performed meaningfully only for those of the Mathis type having a radius of 800 Å and the Greenberg type having a radius of 3300 Å, because the core and the ice mantle have similar sizes. In the Table it is reported as a function of the depth (measured in the scale of magnitudes) inside the cloud, the ratio "r" between the number of H_2 molecules formed by the investigated processes in Mathis grains and in Greemberg ones. In the third column it is reported the ratio "r_{or}" obtained including the production of molecular hydrogen in the organic refractory mantle in the assumption that it can migrate, escape and be released in the gas phase.

TABLE 1

Depth (A_v)	r	r_{or}
1	.99	.57
2	1.19	.69
4	1.09	.63
6	1.13	.63
8	.96	.58

As it can be seen from the second column of Table 1, the two models seem to give comparable results for the formation of H_2 but, if the migration of molecular hydrogen from the organic refractory mantle is sufficiently high, the Greenberg grains seem to enrich the cloud more efficiently than the Mathis ones. Several other runs have to be performed in order to improve the statistics of the Monte Carlo simulation. However the final answer to the question about what is the effect of the grain model in the production of molecular hydrogen from cosmic rays impinging on interstellar dust, can be obtained only when a reliable size distribution of the two types of grains is taken into account.

REFERENCES

Barlow, M.J., Silk, J.: 1976, Astrophys, J. 207, 131

Boland, W., de Jong, T.: 1984, Astron. Astrophys. 134, 87

Brown, W.L., Augustiniak, W.M., Simmons, E., Marcantonio, K.J., Lanzerotti, L.J., Johnson, R.E., Boring, J.W., Reimann, C.T., Foti, G., Pirronello, V.: 1982, Nucl. Instr. & Methods 198, 1

Chlewicki, G., Greenberg, J.M.: 1984, M.N.R.A.S. 210, 791

Dekker, A.J.: 1965, *Solid State Physics*, Prentice Hall Inc., New York

Field, G.B.: 1974, Astrophys, J. 187, 453

Gould, R.J., Salpeter, E.E.: 1963, Astrophys, J. 138, 393

Hollenbach, D.J., Werner, M.W., Salpeter, E.E.: 1971, Astrophys. J. 163, 165

Hong, S.S., Greemberg, J.M.:1980, Astron. Astrophys. 88, 194

Jura, M.: 1980, Astrophys. J.235, 63

Leger, A., Klein, J., de Cheveinge, S., Guinet, C., Defourneau, D., Belin, M.: 1979, Astron. Astrophys. 79, 256

Mathis, J.S., Rumpl, W., Nordisieck, K.H.: 1977, Astrophys. J. 217, 425

Morfill, G.E., Volk, K.J., Lee, M.A.: 1976 J.G.R. 181, 5841

Pirronello, V., Brown, W.L., Lanzerotti, L.J., Marcantonio, K.J., Simmons, E.: 1982, Astrophys. J. 262, 636

Pirronello, V.: 1988, 'Cosmic Ray Bombarment of Solids in Space' in *Genesis and Propagation of Cosmic Rays*, M.M. Shapiro and J.P. Wefel eds., p. 299, Reidel Publ. Co., Dordrecht

Pirronello, V., Averna, D.: 1988, Astron. Astrophys. 196, 201

Pirronello, V., Brown, W.L., Lanzerotti, L.J., Lanzafame, G., Averna, D.: 1988, 'Formaldehyde Formation in $H_2O/^{13}CD_4$ Frozen Mixtures by Swift Ions', in *Dust in the Universe*, M.E. Bailey and D.A. Williams eds., p.281, Cambridge Univ. Press, Cambridge

Smoluchowski, R.: 1983, J. Phys. Chem. 87, 4229

Tielens, A.G.G.M., Hagen, W.: 1982, Astron. Astrophys. 114, 245

Wefel, J.P.: 1988, 'An Overview of Cosmic Ray Research: Composition, Acceleration and Propagation', in *Genesis and Propagation of Cosmic Rays*, M.M. Shapiro and J.P. Wefel e s., p. 1, Reidel Publ. Co., Dordrecht

MAGNETIC FIELD GENERATION DURING GALAXY FORMATION

Ralph E. Pudritz
Dept. of Physics, McMaster University, Hamilton, ON L8S 4M1

1. Introduction

The detailed study of the interstellar medium in the Milky Way has given rise to a rich history of papers on magnetic field generation in the disks of spiral galaxies (eg Parker 1971). The theoretical models for dynamo action can now be tested because recent advances in radio polarimetry have allowed detailed polarization maps of nearby spiral galaxies to be completed. The observations suggest that galaxies have both symmetric and bisymmetric spiral magnetic field patterns (eg. Sofue et al 1986). Little attention has been given to the nature of magnetic fields in galactic halos however. This arises in part because direct observations of such fields requires that a good subtraction of the background faraday rotation of our own galactic disk be achieved. Recently, Hummel et al (1988) have directly mapped the structure of a halo magnetic field around the edge-on spiral galaxy NGC 4631. This field is as strong as 3.5 μg out to 6 kpc above the plane and is suggestive of a dynamically important halo field. Welter, Kronberg, and Perry (1984) have shown that regions in galaxies as large as 50 kpc harbour magnetic field as strong as 1-10 μG. There is a significant trend of rotation measure with redshift in that the systems at larger redshift have larger rotation measures. This may indicate that young and perhaps protogalactic halos could constitute some of the intervenors.

Recent detections of very broad, damped Ly - α absorption line systems in quasar spectra indicate the presence of possible atomic gas disks of young galaxies (Wolfe 1988). Some of these absorption line systems are found in the quasars used in the Welter et al sample. If the damped Ly systems are young galactic disks, then the data suggest that they are already strongly magnetized by time that they can be detected. The two theoretical possibilities here are that the gas which collapsed out of the halo to form the disk was itself strongly magnetized, or that dynamo action worked very efficiently in young galactic disks.

In this review, I explore the idea that the most important site of magnetic field generation in galaxies may in fact have been in their halos. While the inner portions of galactic halos are now buffetted by local upwellings induced by supernovae in galactic disks or galactic nuclei, they were much more violent places during the epoch of galaxy formation. Current ideas about galaxy formation argue that dissipation may have been a crucial part of galaxy formation (eg review by Efsthathiou and Silk 1983). During that epoch, primordial gas clouds falling into dark matter potential wells (White and Rees 1978) underwent strong shocks during

G. Belvedere (ed.), Accretion Disks and Magnetic Fields in Astrophysics, 191–202.

the course of cloud-cloud collisions. These occurred throughout the volume of the protogalactic halo and magnetic field generation can be very strong, as we (Pudritz and Silk 1988, henceforth PS) have recently shown. Since any theory of galaxy formation that involves dissipation must suppose that shocks in primordial gas play a key role, it behooves us to seriously consider the possibility that magnetic fields were generated there as well.

2. Observations

2.1 FARADAY ROTATION:

Evidence for the existence of strong magnetic fields on very large scales in galaxies comes primarily from the observation of very high Faraday rotation in the direction of some radio loud quasars (Welter, Kronberg and Perry 1984, Kronberg 1988). The technique is to use radio loud quasars as beacons. The propagation of radio waves originating in the quasar through intervening gas serves as a good plasma diagnostic of ionized gas in objects between us and the quasar "lighthouse". In particular, a medium with an electron density n_e and with a component of the magnetic field parallel to the line of sight B_{ll} results in Faraday rotation. The associated rotation measure RM is given by

$$RM = 0.8 \int n_e B_{ll} dl \quad \text{rad} \quad \text{m}^{-2}$$

where the field is measured in microgauss, the density in cm^{-3}, and the path length l is in parsecs. In order to say anything about the magnetic field, the density and size of the intervening gaseous objects must be known. This is accomplished by looking for absorption features in optical spectra taken on these same lines of sight. By choosing only those quasars towards which good optical spectroscopy is avaible, the total column density of the intervening gas can be deduced. By constructing models for the intervenors, one can produce a simple model for the most likely type of object and its associated magnetic field.

This procedure was carried out by Welter et al (1984) and led to some rather startling conclusions. The initial sample consisted of 37 quasars for which both good RM and optical spectroscopy is available (the typical RM being 55 rad m^{-2}). The best fit model involves several intervenors with the properties;

$$l = 50\text{kpc}; \quad n_e = 10^{-3}\text{cm}^{-3}; \quad B_{ll} = 2 \times 10^{-6} \quad \text{gauss}.$$

There is also a weak but clearly discernable trend towards higher RM at larger redshifts.

Interestingly, the intervenors are probably not the usual Ly-α clouds because the magnetic field required from the RM data would have an energy density 24 times larger than the thermal pressure in such clouds. This would threaten to disperse clouds if the field were isotropic. If the Ly-α clouds are in pressure equilibrium with other consituents in IGM, a magnetic energy density so large would be difficult to explain. However magnetic fields can have far greater energy densities than given by thermal pressure if clouds are self-gravitating. An excellent example of this are the magnetic fields in galactic molecular clouds that have measured magnetic energy densities much higher than thermal gas and are comparable to gravity (Myers and Goodman 1988). Therefore, if the Ly-α clouds are not pressure confined but reside in dark matter potential wells, as some cold dark matter scenarios of galaxy formation suggest (eg. Bond, Szalay, and Silk 1988), then the above argument is no longer a constraint.

It turns out that very high magnetic fields are also inferred to exist in cluster gas (eg. Kronberg 1988). Clusters cannot constitute the majority of intervenors however since they have too low a space density.

2.2 LARGE DISKS OR HALOS?

A good candidate for the intervenors is simply galactic halos. If true, this would imply that magnetic fields reside even in the outer reaches of halos and that these fields were generated early in galactic life. However, recent observations of so-called damped Ly-α absorption line systems (Wolfe et al 1986) indicates that very large disks may have formed during the early collapse of galaxies. Wolfe and collaborators conducted a survey designed to try to detect the typical HI disks of galaxies at $z > 2$. For gas at column densities of $N(HI) > 2 \times 10^{20} cm^{-2}$, and with low velocity dispersion $\sigma \simeq 10$ km s^{-1}, the Ly-α lines are broadened by radiation damping. The line width one looks for due to this Voigt damping is several thousands of km per sec wide. The observations indicate a very high detection rate of such objects. Starting with a sample of 68 QSO's with well measured red shifts, 47 of 476 absorption features in the spectra qualify as Ly-α disks. Thus, while 15 truly damped systems are found on these lines of sight, normal disk galaxies would account for about 3 of these systems. In fact, these systems cover more than 20% of the sky at $z \simeq 2 - 3$.

What are they? One possibility is an inflated population of small objects such as dwarf galaxies. It would be difficult to explain the narrowness of the underlying velocity dispersion however since a collection of such objects would have a clump to clump velocity dispersion much broader than 10 km s^{-1}. A more likely possibility is that at the larger red-shifts, *large HI disks* with radii 3-4 times larger than the Holmberg radius are present. If it is true that disks form quickly in protogalaxies, they must also shrink very quickly in size because at redshifts $z \leq 1.5$, one finds a much smaller detection rate of Ly disks.

The damped Ly disks systems are also important from the point of view of magnetic field in protogalaxies because they harbour the strongest magnetic fields, which may be as large as 10 microgauss (Wolfe 1988). Not all the intervenors are Ly disks because there are a few known high RM objects which do not have these absorption features. Thus, there are magnetized intervenors other than protostellar disks, which are very plausibly galactic halos. A halo origin for the magnetic field would be consistent with the data even if all these systems were Ly disks however. The point is that magnetic field formed in the halo would be carried in with gas collapsing to form a disk. When the column density of the collapsed structure becomes high enough, the disk would stand out as a high RM object.

In order to understand why such high field strengths are surprising, consider the strength of seed fields that can be generated in protogalaxies.

3. Seed Fields

I will restrict myself here to a discussion which involves only gas dynamic processes to generate magnetic fields (for exotic possibilities from particle physics see Asseo and Sol 1987).

3.1 COMPTON DRAG

Galaxies aquire their angular momentum by tidal interaction with their neighbours (Peebles 1969, Barnes and Efstathiou 1987). While the density fluctuations that define a protogalaxy at this stage of turn-around from the Hubble flow are irrotational, they possess significant quadrupole moments and therefore exert appreciable torques on neighbouring fluctuations. The specific angular momentum that a protogalaxy aquires at this stage is only an order of magnitude below that observed in the galactic disk today.

Now consider such a rotating protogalaxy embedded in the background radiation field. While the protons do not significantly interact with the radiation field, the electrons undergo significant braking due to Compton drag. At redshifts $z > 1000$, the radiation temperature is high enough to ionize hydrogen and the universe is radiation dominated. Even for redshifts between 1000 and 100, gas if reionized, will remain so and will suffer appreciable drag. Mishustin and Ruzmaikin (1972) were the first to recognize that this was an excellent mechanism for producing magnetic field. The slight charge separation produced by Compton drag on the electrons leads to a small current which in turn generates a small field. One may show that for gas with an angular frequency Ω_{cl}, and for a red-shift of galaxy formation of z_f, the Compton-drag induced magnetic field is

$$B_{compt} = 3.0 \times 10^{-22} \left(\frac{1+z_f}{10}\right)^{5/2} \Omega^{-1/2} \left(\frac{\Omega_{cl}}{10^{-15}\mathrm{s}^{-1}}\right) \quad \text{gauss}$$

where Ω is the cosmological density parameter and for an epoch of galaxy formation of about $1+z_f \simeq 10$. One can do slightly better than this with the Biermann battery mechanism (PS) which we showed, could produce fields of 10^{-19} gauss under these same conditions. The point remains however that incredibly weak seed fields on galactic scales are generated if only the gas dynamics of slowly rotating galaxies is considered. Mishustin and Ruzmaikin identified the basic problem, namely that the existence of strong fields requires some kind of dynamo action to crank up the weak seed field by 13 decades! When one considers what sort of spectacular events could have occurred in the early life of galaxies which could have done this, one's attention soon enough fixes on shocks, for reasons I'll now describe.

4. Shocks

There are two different views of how galaxy formation might have proceeded. If star formation preceeded the formation of the massive dark haloes of galaxies, the resulting collapse to form stellar galaxies would have been dissipationless. The other possibility is that star formation occurs after much of the dark mass of the galaxy has gathered together. The primordial gas clouds collapse into dark matter halos, and undergoe substantial shocking during the course of cloud-cloud collisions. This makes the process of galaxy formation highly dissipative because shocks can radiate away considerable gravitational binding energy of the gas. Therefore, it can efficiently cool and settle into the inner regions of the dark matter potential well (White and Rees 1978).

Typical free-fall speeds in the galactic halo are of order 100 kms^{-1}, which corresponds to a virial temperature of order 10^6 K. Unless the clouds are sufficiently self gravitating, strong shocks will disrupt the primordial clouds rather than creating merged flattened objects. The condition that two infalling clouds not disrupt on colliding is that their internal gravitational binding energy must exceed the ram pressure of the collision, ρv_s^2. This constrains the surface density of the clouds to be

$$\mu_c > (\rho/G)^{1/2} v_s \simeq 8.5 n_3^{1/2} (v_s/100 \quad \mathrm{kms}^{-1}) \quad \mathrm{M_\odot pc^{-2}}.$$

Another condition which must be satisfied by clouds if they are to survive is that they not be tidally disrupted in the background gravitational potential of the dark matter halo. This implies that their volume density should exceed the protogalactic density at the appropriate radius. From the Virial theorem, the total surface density of a galaxy is

$$\mu_{tot} = \frac{\sigma_g^4}{4\pi G^2 M_{tot}} = 1.8 \frac{(\sigma_g/100 \quad \mathrm{kms}^{-1})^4 (\Omega_b/0.1)}{(M_{tot}/10^{12} M_\odot)} \quad \mathrm{M_\odot pc}^{-2}$$

where Ω_b is the baryon mass fraction and σ_g is the velocity dispersion of the gas in the protogalactic potential.

4.1 VORTICITY GENERATION

Generic shocks are oblique which from the point of view of magnetic field generation, is the crucial step. Where there is vorticity, there is the possibility for vigorous dynamo action. Vorticity generation is possible because while the gas velocity tangential to the shock is unchanged, the normal velocity component is diminished. To see this, consider an idealized collision of two self-gravitating isothermal spheres (each with density profiles $\rho \propto r^{-2}$). Unless the collision is exactly head-on, the shock is oblique and the resulting dense sheet of shocked gas has a radial density gradient. For convenience, suppose the shock lies in the $x - y$ plane. The incoming fluid has a non-zero tangential speed which is conserved in the shock, $u'_y = u_y \neq 0$. The incoming fluid is taken to be irrotational, as discussed in the introduction. Therefore, it's vorticity vanishes, ie. $\omega_x \equiv (\partial u_y/\partial z) - (\partial u_z/\partial y) = 0$. In a strong shock, the post shock normal velocity component is reduced to $u'_z = (1/4)u_z$. The density is increased by 4 so that the z-component of the distance between particles is reduced by this same factor (eg Zel'dovich and Novikov 1983). For the above conditions, it is easy to show that the post-shock vorticity is nonzero, ie

$$\omega'_x = (15/4)(\partial u_y/\partial z) \neq 0.$$

This component of the vorticity vector is parallel to the direction of the density gradient in the sheet which means that the fluid has non-zero helicity after being run through the oblique shock.

Numerical simulations of colliding clouds suggest that flattened bar like objects form rather than sheets. A great deal is known about fluid flow and shocks in self gravitating bars (eg Roberts et al 1979). In spite of these differences, a very similar scaling for vorticity generation in bars can be obtained, giving $\omega \simeq 2.5\Omega$ (see PS for details).

4.2 SHOCK COOLING: MOLECULAR HYDROGEN FORMATION

While vorticity generation in shocks is crucial for the generation of magnetic fields, the radiative cooling of these layers is fundamental to the process of star formation in galaxies. Gas of primordial composition radiates very poorly, and in LTE conditions, does not cool much below 10^4 K. The Jeans length for gas this hot implies that only masses of order globular cluster mass could fragment out of the soup by gravitational instability. In shocks however, it is possible that an excellent coolant can be produced in sufficient quantity that the primordial gas can cool to 100 K. The coolant is H_2 and it can be formed in shocks because under non-LTE conditions, a primordial gas has more electrons around than normal (in a gas at 10^4 K, the molecular fraction in equilibrium is merely $n_{H2}/n_H = 10^{-6}$). This allows H_2 formation by the reactions

$$H + e = H^-,$$

$$H^- + H = H_2 + e.$$

These reactions proceed at temperatures down to 100 K. Detailed calculations (see eg. Maclow and Shull 1986, Palla 1988) show that the molecular hydrogen fraction reaches 10^{-3}. An important aspect of this chemistry is that it is very robust. For shock velocities spanning 20-200 km s^{-1}, shock densities in the range $10^{-4} - 10^2$ cm^{-3}, and for background UV ionization fluxes from quasars as large as 3×10^{-21} erg cm^{-2} s^{-1}, the molecular fraction is about the same. The cooling rate that this theory predicts is given by

$$\Lambda = 2.8 \times 10^{-30} T_2^{5/2} \quad \mathrm{erg} \quad \mathrm{cm}^3 \quad s^{-1}$$

per hydrogen atom where T_2 is the temperature in units of 100 K.

The prediction of significant molecule formation in shocks can be tested by looking for emission lines from the vibrational transitions in molecular hydrogen. The observations are difficult and have been somewhat inconclusive so far. One system at $z = 2.81$ in the spectrum of the quasar PKS 0528-250 was detected to have H_2 in the amount predicted (Levshavov and Varshalovich 1985), while in another damped Ly disk, only an upper limit of 10^{-6} could be put on the molecular fraction (Black et al 1987).

4.3 FRAGMENTATION OF SELF-GRAVITATING SHOCK LAYERS

The presence of molecular hydrogen cooling in the shock produced layers stamps a fiducial gas surface density ($\equiv \mu$) on the layers. The balance condition is that radiative cooling balances the heating rate due to the shock. The resulting condition is that

$$\mu > \mu_c \equiv \frac{3kTm_p\sigma}{2\Lambda}. = \mu_o \sigma^{-2} = 12.6 T_2^{-1} \qquad \mathrm{M}_\odot \quad \mathrm{pc}^{-2}.$$

which shows that the cooling pancake are strongly self gravitating objects (see section 4.1).

Thin layers of self gravitating gas are unstable to a variety of instabilities. The fragmentation of an isothermal self-gravitating gas layer is well known to have a dominant mode, which will determine the scale and size of the fragments. Another condition to consider is the stability with respect to the growth of spiral waves in shearing disks (Goldreich and Lynden-Bell 1965, henceforth GLB; Toomre 1981 (stellar disks)).

First I consider the latter stability condition. Jeans' instability occurs for waves perpendicular to the axis of rotation whose wavelength is less than the critical scale $\lambda_c = 4\pi^2 G\sigma_o/\kappa^2$ where κ is the epicyclic frequency. Amplification is possible those modes for which the epicyclic motions in the gas or stars resonate with the wave swinging through the gas. This "lingering" of particles in the wave leads to considerable density perturbations. The subsequent fate of these growing, shearing waves is determined by differential rotation (GLB) which wraps them up sufficiently so that shocks and heating result. The point here is that since differential rotation diminishes the wave-length of the perturbations, Jeans' instability is ultimately discouraged. GLB conjectured that eventually shock waves would form, which would heat the disk up again. They argued that all parts of a disk self regulate to the state that $\pi G\rho/4\kappa^2 \sim 1$ by this process. When pressure effects are included a related instability parameter Q is given by

$$Q = \frac{2^{1/2}\Omega\sigma}{\pi G\mu}.$$

Gas layers are gravitationally unstable if $Q \leq 1$. As already discussed, the growing modes will eventually damp out and reheat the disk to $Q \geq 1$. Detailed numerical simulations of stellar galactic disks allowed to "cool" via star formation (Sellwood and Carlberg 1984) show that Q does indeed self regulate to a value of ~ 1.7 and is nearly independent of both time and radius. Toomre's (1981) showed that the instability shuts off very sharply as Q increases. Henceforth I'll assume that the marginal stability condition $Q \simeq 1$ at all radii in our primordial shocked clouds.

Assuming that the fragments will be disk-like, centrifugal support requires

$$\Omega^2 r^2 = \frac{GM}{r}.$$

The surface density and scale height of self-gravitating isothermal layers are well known, ie.

$$\mu = \frac{\Omega^2 r}{4\pi G},$$

$$H = \frac{\sigma^2}{\pi G\mu}.$$

These results together with $Q \simeq 1$ in an isothermal cloud ($\sigma = const$) therefore prescribe the cloud rotation law. This follows since $Q = (2^{1/2}4\sigma/\Omega r)$ implies that $\Omega r =$ constant. The physical import of this result is that the dissipation time-scale for the waves is tuned to the instability time $\simeq \Omega^{-1}$.

The cloud size is calculated using the standard result for the fastest growing mode in self-graivitating gas layers. This procedure gives

$$r_c = \frac{8\sigma^4}{\pi\mu_o Q^2 G} = 160 T_2^2 \quad \text{pc}$$

The calculations (see PS for details) show that the fiducial cloud mass, scale height, density and rotation rate for the stable fragments which appear in the self-gravitating shock layers are given by

$$M_c = \frac{256\sigma^6}{\pi\mu_o Q^4 G} = 4.0 \times 10^6 T_2^3 \quad \mathrm{M}_\odot,$$

$$H_c = \frac{\sigma^4}{\pi\mu_o G} = 20T_2^2 \quad \text{pc},$$

$$n_c = \frac{\pi\mu_o^2 G\sigma^6}{2m_p} = 2.2\times10^4 T_2^3 \quad \text{cm}^{-3},$$

$$\Omega_c = \frac{\pi\mu_o Q G}{2^{1/2}}\sigma^{-3} = 2.2\times10^{-15} T_2^{-3/2} \quad \text{s}^{-1}.$$

These fragments are the self-gravitating entities in which the first stars must form. Since the fragments are only 100 K, a much more reasonable Jeans' mass for star formation has emerged. The density and mass of the fragments is also similar to the densest most massive star forming clouds in our own galaxy (eg Orion). It appears therefore that primordial star forming clouds are very similar to those in our own interstellar medium. These very fragments turn out to be the objects in the protogalaxy which are resoponsible for vigourous field generation.

5. Dynamos in Protogalactic Halos

On what spatial scales did dynamo action in protogalactic halos occur? The dynamo time scale must be fast enough that a seed field of $\simeq 10^{-19}$ gauss can be amplified up to 10^{-6} gauss in an initial protogalactic free-fall time, which is about 10^9 years. Therefore, approximately 30 e-folding times would have been required to bring the field up to the observed value, and the clumps in which this field amplification occurred must have had internal dynamical (or more precisely, dynamo) time scales of less than 10^8 years. This requirement is indeed satisfied by the self gravitating clouds described above since $\Omega_c^{-1} \simeq 10^8$ yr.

Dynamo action is particularly simple to study at this epoch because the seed magnetic field is orders of magnitude lower in energy density than the thermal, rotational, or gravitational energy densities of the cloud. The dynamo is therefore kinematic, that is, no complicated back reaction of the field upon the fluid motions need be taken into account. Amplification ceases when the magnetic field achieves an energy density comparable to gravity. It is at this point that magnetic braking of the fluid becomes strong. The kinematic dynamo equations for field generation in thin layers are well-known (see, eg. Parker 1971, Moffatt 1978, Pudritz 1981). There are two physical quantities characterizing fluid motions in the stirred cloud which are needed for the dynamo theory. These are the magnitude of the magnetic helicity which is responsible for regenerating the field, α, and the enhanced damping rate of the field due to the "turbulent" diffusivity η_T. The α term must in general be a pseudo-scalar, and one excellent source is a density gradient in a cloud with local vorticity. Under these circumstances, $\alpha \propto (\nabla\rho).\omega$, where ω is the vorticity vector associated with gas motions in the cloud (see, eg. Parker 1971, Moffat 1978). As we saw in section 4.1, cloud-cloud collisions produce flattened bar-like objects (elongated in y direction say) so that $\alpha \propto (\partial\rho/\partial x)'\omega_x'$. The physical picture here is that rising blobs of gas expand along the density gradient. The rising blobs experience a Coriolis force (due to the vorticity) and start to spin. Such gas motions are well known to lead to dynamo action.

The linearized dynamo equations have modes which grow exponentially in time and have a fastest growing mode of the field. The wave-number of this fastest growing mode is calculated in detail in PS. The result is

$$k_m = \left(\frac{\alpha r \Omega'}{32(2\eta_T)^2}\right)^{1/3},$$

this mode growing exponentially as $\exp(t/t_{dyn})$ where the dynamo time is

$$t_{dyn} = \frac{4}{3} \cdot \left(\frac{32\eta_T}{(\alpha r \Omega')^2}\right)^{1/3}.$$

Only modes with sufficiently long wavelengths (small wavenumbers) are amplified since dissipation leads to strong damping on small scales. The critical wavenumber above which only damping occurs is

$$k_{damp} = (16)^{1/3} k_m.$$

Thus, modes with $k > k_{damp}$ are damped, and those with $k < k_{damp}$ are amplified.

To complete this calculation we need to estimate α and η_T in the stirred layers. A reasonable model for these quantities was obtained by Pudritz (1981) assuming that the effective stirring of the gas in a Keplerian disk is due to a superposition of shear waves. In a self gravitating cloud, a superposition of low-amplitude axisymmetric and non-axisymmetric waves in the $Q \simeq 1$ medium has the necessary properties for dynamo action. Thus it is the near onset of Jeans instability which is the wave driver in this problem. Employing this simple picture (see PS), it is easily shown that the length scale which characterises the most rapidly growing field component has a length scale $L_m \equiv (2\pi k_m^{-1}) = L_c(r/r_c)$, where

$$L_c = 10\pi M_T^{2/3} H_c = 0.62 M_T^{2/3} T_2^2 \qquad \text{kpc.}$$

and where the Mach number characterizing the rms wave amplitude is M_T. This can be regarded as the coherence length of the magnetic field associated with disks before any star formation has occurred.

The dynamo time scale characterizing this dominant mode is

$$t_{dyn} = 10.5 M_T^{-2/3} \omega^{-1}.$$

Thus, the dynamo time-scale is directly proportional to the vorticity time scale in the layer. Using the vorticity time calculated for the protogalactic shocks in section 4.1, one finds that the e-folding time for growth of magnetic field is $t_{dyn} = t_{dyn,c}(r/r_c)$, where

$$t_{dyn,c} \equiv 10.5 M_T^{-2/3} \omega_c^{-1} = 6.0 \times 10^7 M_T^{-2/3} T_2^{3/2} \qquad \text{yr.}$$

Therefore, in order to generate a field of order 10^{-5} gauss starting from a seed of about 10^{-19} gauss, the dynamo must have time to undergo 30 e-foldings. The total time required is

$$t_{field} \simeq 1.9 \times 10^9 M_T^{-2/3} T_2^{3/2} \qquad \text{yr}$$

or several free-fall times in the outer halo of the protogalaxy. Thus, the time-scale for field generation is directly proportional to Ω^{-1}, the time scale for gravitational instability. Recall that this was also the time scale characterizing Jeans' instability in the layer. Thus protogalactic field generation draws energy out of marginal gravitational instabilities. These non-axisymmetric disturbances shear and twist magnetic field lines in such a way that dynamo action is stimulated.

When the field begins to have an energy density comparable to the motions which are generating it, then the dynamo must saturate. It is at this point of course that complicated and poorly understood non-linear effects become important in the dynamo equation. Nontheless, the maximum field will certainly be within an order of magnitude of that given by the virial theorem. Given a cloud of surface density μ, the field strength which can support the cloud against collapse is given by the virial theorem (eg Mestel 1965)

$$B_{crit} = (k\pi^2 G)^{1/2}\mu = B_c(r_c/r)$$

where

$$B_c \equiv 3.6 T_2^{-1} \quad \mu\text{G},$$

in agreement with the experimental values for halo fields determined by Kronberg and collaborators (eg. Kronberg 1987).

6. Primordial Star Formation and Galaxy Formation

The processes of primordial molecular cloud and star formation were apparently similar to those occurring in our own interstellar medium. The long cooling time-scale in shocked primordial clouds guarantees inefficient dissipation of gravitational energy, and there is time for dynamo generation of magnetic field. The magnetic field has a crucial role to play in the star formation itself, namely the resolution of the stellar angular momentum problem. Much as in the current interstellar medium, primordial star forming fragments have specific angular momenta 4 decades larger than that of stars. The generation of dynamically significant field in these clouds plays the vital role of resolving this problem by magnetic braking.

6.1 THE FIRST STARS IN GALACTIC HALOS

In order to explain the metallicities of globular clusters, considerable chemical enrichment and thorough mixing must have occurred before the epoch of globular cluster formation. The primordial initial mass function may therefore have been skewed towards higher mass. The evidence for massive star formation in cooling shocks derives mainly from the presence of high star formation rates in merging and strongly interacting galaxies. Larson (1988) has emphasized that high rates of star formation in galaxies requires rapid agglomeration of clouds (as in merging galaxies).

The theory outlined here produces star forming clouds which fragment out of large shock layers in protogalactic halos. Because the densities, masses, and temperatures of the clouds are comparable to clouds in our own interstellar medium which are forming massive star, it is highly likely that O and B star formation will soon enough erupt in these objects. Massive star formation in the disk-like clouds will disrupt them and create large bubbles in the halo gas. This process spreads the magnetic field generated in star forming clouds throughout the halo. The radius of a supershell which is driven by stellar winds and supernovae evolves with time as $t^{3/5}$. The shell continues to expand in the ambient halo gas (taken to have a mean density of 10^{-3} cm^{-3} in the outer halo) until its velocity decreases to the rms ambient speed $v_{rms} \simeq 100$ km s^{-1}. Thus the shell will expand for of order 10^7 yrs reaching a radius of $R_s = 0.52(L_{38}/n_{-3})^{1/2}v_{rms,2}^{-3/2}$ kpc where the rms speed of the medium is in units of 100 km s^{-1} and where the wind and supernovae power L_{38} is in units of 10^{38} erg s^{-1}. The protogalaxy will gradually take on the appearance of a very irregular looking collection of massive HII regions, driven by underlying O-B star formation. Eventually, the halo will be riddled with a collection of supershells turning it into a medium resembling a highly Swiss cheese.

6.2 MAGNETIC FIELDS IN SPIRAL GALAXIES

Gas far out in the halo takes nearly a Hubble time to collapse into the centre of the galaxy, so that this theory predicts the presence of a strong magnetic field in the outer halo. Gas which has not been consumed by star formation in the inner parts of the halo will collapse to form the galactic disk. This gas will have become strongly magnetized so that the disk forms in the presence of a dynamically important, relatively disordered magnetic field whose coherence length is no more than a kpc. The inferred redshift by which these fields could have been produced is (use expression for t_{field}), $z \leq 3$ $(\Omega = 1)$ or $z \leq 7(\Omega = 0.1)$, with $H_o = 75$ km s^{-1} Mpc^{-1}. This is consistent with the expectation that the damped Lyman systems at $z \simeq 2$ seen in absorption against background quasars are protogalactic disks that already contain significant magnetic fields (Wolfe 1988).

Nevertheless a coherence length of several kpc in the halo is too small to support the idea that spiral galaxies originally had strong magnetic fields which were coherent over tens of kpc. Sofue *et al* (1986) have recently reviewed the evidence in favour of possible bisymmetric spiral magnetic fields in galactic disks and suggest that very large-scale primordial fields can explain the formation of such patterns. Recently, Baryshnikova et al (1987) have shown that a non-axisymmetric mode in a thin disk dynamo can be generated that accounts for the existence of bisymmetric field patterns. This suggests that this spiral magnetic field structure does not require a global primordial field ordered on 10 kpc scales.

The focus of this conference has been to develop basic physical paradigms such as dynamo action in accretion disks, which would be useful in understanding a wide variety of astrophysical objects. The theory presented here underscores the importance of that basic philosophy.

It is pleasure to thank Gaetano Belvedere for organizing and orchestrating such a stimulating conference and for providing financial support. I thank Brahm Achterberg, Dick Bond, and Phillip Kronberg, and in particular my collaborator, Joe Silk for many stimulating conversations. This research was supported by grants from McMaster University and NSERC of Canada.

REFERENCES

Asseo, E., and Sol, H. 1987, *Physics Reports*, **148**, 307.

Barnes, J. and Efstathiou, G. 1988, *Ap. J.*, **319**, 575.

Baryshnikova, Y., Ruzmaikin, A., Sokoloff IV, D.D., and Shukurov, A. 1987, *Astron. Ap.*, **177**, 27.

Black, J.H., Chaffee, F.H. Jr., and Foltz, C.B. 1987, **317**, 442.

Bond, J.R., Szalay, A.S., and Silk, J. 1988, *Ap. J.*, **324**, 627.

Efsthatiou, G., and Silk, J. 1983, *Fund. Cosmic Physics*, **9**, 1.

Foltz, C.B., Chaffee, F.H. Jr., and Black, J.H. 1988, *Ap. J.*, **324**, 267.

Goldreich, P., and Lynden-Bell, D. 1965, *M.N.R.A.S.*, **130**, 125.

Hummel,E., Sancisi, R., and Ekers, R.D. 1984, *Astron. Ap.*, **133**, 1.

Kronberg, P.P. 1987, in *Interstellar Magnetic Fields* eds. R. Beck and R. Grave (Springer: Heidelberg), p. 86.

Larson, R.B. 1988, in *Galactic and Extragalactic Star Formation*, R.E. Pudritz and M. Fich eds. (D. Reidel: Dordrecht), p. 459.

Levshakov, S.A., and Varshalovich, D.A. 1985, *M.N.R.A.S.*, **212**, 517.

MacLow, M. M., and Shull, J.M. 1986, *Ap. J.*, **302**, 585.

Mishustin, I.N., and Ruzmaikin, A.A. 1972, *Sov. Phys. JETP*, **34**, 233.

Mestel, L 1965, *Quart J.M.N.R.A.S*, **6**, 161, 265.

Moffatt, H.K. 1978, *Magnetic Field Generation in Electrically Conducting Fluids* (Cambridge University Press, London).

Myers, P.C., and Goodman, A. 1988, *Ap. J. Letters*, **326**, L27.

Palla, F. 1988, in *Galactic and Extragalactic Star Formation*, eds. R.E. Pudritz and M. Fich (D. Reidel: Dordrecht), p. 519.

Parker, E.N. 1971, *Ap. J.*, **163**, 225.

Peebles, P.J.E. 1969, *Ap. J.*, **155**, 393.

Pudritz, R.E. 1981, *M.N.R.A.S.*, **195**, 897.

Pudritz, R.E., and Silk, J. 1989, Ap. J., in press.

Roberts, W.W., Jr., Huntley, J.M., and van Albada, G.D. 1979, *Ap. J.*, **233**, 67.

Sellwood, J.A., and Carlberg, R.G. 1984, *Ap. J.*, **282**, 61.

Sofue, Y., Fujimoto, M., and Wielebinski, R. 1986, *A.R.A.A.*, **24**, 459.

Toomre, A. 1981, in *Structure and Evolution of Normal Galaxies*, S.M. Fall and D. Lynden-Bell eds., (Cambridge University Press:), p. 111.

Welter, G.L., Perry, J.J., and Kronberg, P.P. 1984, *Ap. J.*, **279** 19.

White, S.D.M., and Rees, M.J. 1978, *M.N.R.A.S.*, **183**, 341.

Wolfe, A.M. 1988, in *QSO Absorption Lines: Probing the Universe*, in press.

Wolfe, A.M., Turnshek, D.A., Smith, H.E., and Cohen, R.D. 1986, *Ap. J.*, **61**, 249.

Zeldovich, Ya. B., and Novikov, I.D. 1983, in *Relativistic Astrophysics, Volume 2*, (Chicago: University of Chicago Press), p. 377.

NUMERICAL SIMULATION OF WEAKLY MAGNETIZED PROPAGATING SLAB JETS

Philip E. Hardee
Department of Physics and Astronomy
University of Alabama
Tuscaloosa, AL 35487

Michael L. Norman
National Center for Supercomputing Applications and
Department of Astronomy
University of Illinois
Champaign, IL 61820

ABSTRACT. Simulations of weakly magnetized propagating slab jets have been performed for a range of jet velocities. The magnetic field is introduced in the form of closed field loops and a local emissivity is computed in a cross section through the infinite slab. The use of slab geometry allows study of non-axisymmetric effects. The jet is driven by a periodic sinusoidal oscillation in time of the angle of the jet's velocity relative to the z-direction at the inlet. This driving mechansim excites the sinusoidal modes. These modes are the 2-D analog of helical modes. At early times the jet maintains approximate axisymmetry within a backflowing cocoon. At later times when the jet has penetrated farther into the external medium the symmetry is broken by sideways oscillation which destroys the jet's directionality and the jet "flaps" within a growing lobe.

INTRODUCTION

Stability analysis of thermal pressure confined, supersonic cylindrical jets has shown that these jets are always unstable to the growth of perturbations (Ferrari, Trussoni and Zaninetti 1978). On a cylindrical jet perturbations to the jet consist of harmonic components associated with pinching, helical and fluting modes (Hardee 1979). Each mode contains a fundamental (surface) and reflection (body) modes. On a supersonic jet each fundamental and reflection mode has a resonant frequency and wavelength at which the growth rate is a maximum (Ferrari, Trussoni and Zaninetti 1981; Cohn 1983; Birkinshaw 1984; Payne and Cohn 1985; Hardee 1987).

At the present time, it is not feasible to perform numerical simulations of the nonaxisymmetric 3-dimensional cylindrical jet of an accuracy sufficient to make a comparion with perturbation theory

G. Belvedere (ed.), Accretion Disks and Magnetic Fields in Astrophysics, 203–206.

meaningful. On the other hand, it is quite feasible to conduct numerical simulations of sufficient accuracy which allow for nonaxisymmetric motions of a 2-dimensional slab jet. Hardee and Norman (1988) have performed a spatial stability analysis of the slab jet which shows that the only qualitative difference between slab and cylindrical geometries is that the fluting modes do not have a counterpart on the slab jet. See also the temporal stability results of Ferrari, Massaglia and Trussoni (1982). Thus, the spatial stability of a supersonic slab jet provides a reasonable analog for the spatial stability of a supersonic cylindrical jet.

Results from numerical simulations of a supersonic slab jet initially in static pressure balance with an external medium have been compared with the linearized spatial stability analysis (Norman and Hardee 1988). In general, results from the numerical simulations compare favorably with predictions of the linearized analysis. In all simulations the slab jet is disrupted by sinusoidal oscillation at or near the resonant wavelength of the fundamental sinusoidal mode. The jet then propagates behind a jet front and forms a lobe and the jet flaps within the growing lobe. The simulations also show weak oblique shock waves in the jet which can be identified with the first and second sinusoidal reflection modes predicted by the linearized analysis.

It has been assumed that some of the observed behavior of jets and the morphology of extragalactic radio sources produced by the jets can be understood as resulting from instabilities associated with supersonic flow. We now report on results from simulations designed to more closely emulate the conditions in extragalactic radio sources in which a jet propagates through an initially quiescent medium.

THE NUMERICAL TECHNIQUE

We have used the 2-D MHD code ZEUS discussed by Norman, Clarke and Burns (1987). The code solves the equations of motion explicitly in time on an Eulerian grid. The upwinded, second order accurate, monotonic van Leer interpolation algorithm is used for the advective terms in the equations of motion except for the vector potential. A third order accurate scheme was used to advect the vector potential. Shock fronts are handled by the usual Von Newmann-Richtmyer artificial viscosity, which will typically smear a shock over several grid zones. We used 600 uniform grid zones in the z direction. The slab jet had half thickness R with 40 uniform grid zones across the jet in the r direction. We employed 140 grid zones in the external medium on either side of the jet in the r direction. The grid outside the jet was stretched as distance increased away from the jet in the r direction with ratio $\Delta r_{j+1}/\Delta r_j = 1.05$. The slab jet was driven by a periodic sinusoidal oscillation in time of the angle of the jet's velocity relative to the z direction at the inlet on the left edge of the grid. The maximum amplitude was $u_r/u = 0.01$. Reflection boundary conditions were used on the jet inflow side of the grid and outflow boundary conditions were used elsewhere.

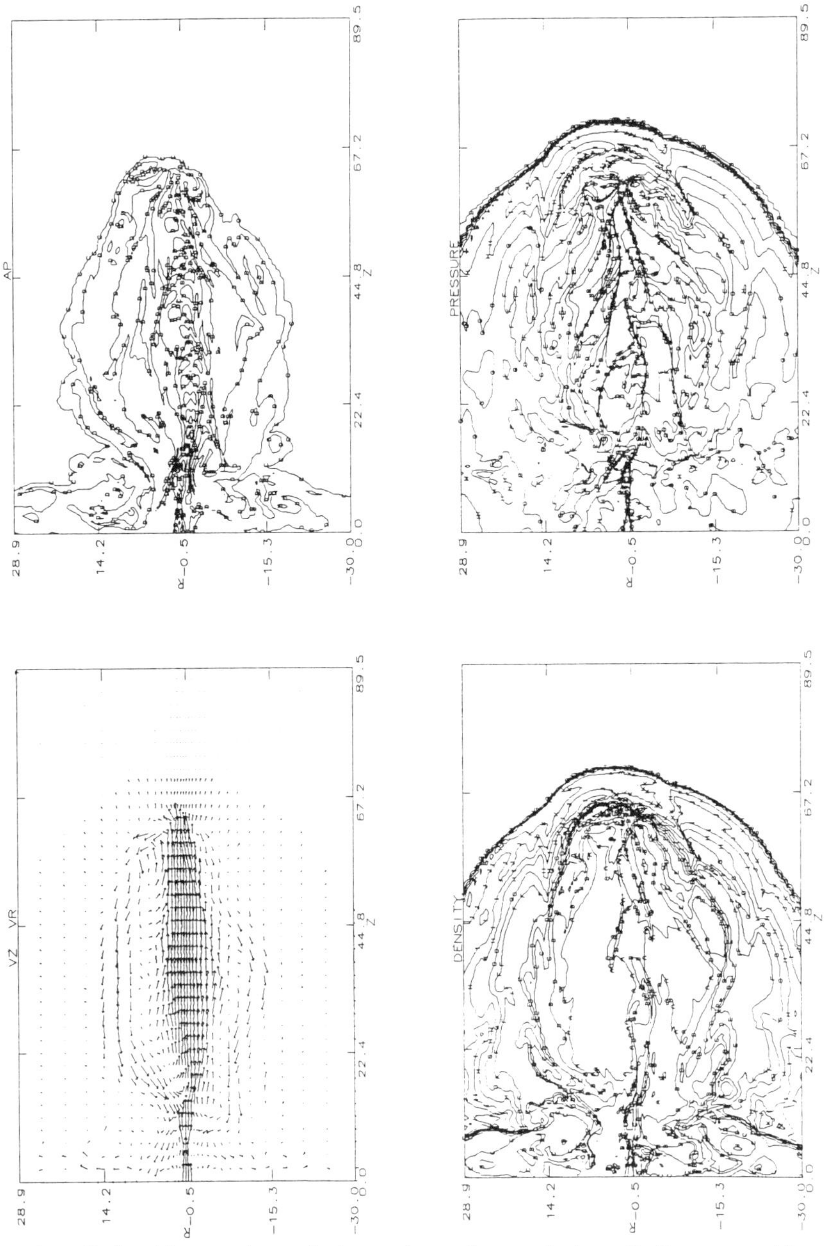

Figure 1. Velocity vector plot, and contour plots of the magnetic vector potential, density and pressure at time $\tau \equiv tR/c = 8$ ($2R$ = slab thickness and c = sound speed in the undisturbed external medium). The jet has an internal Mach number of 8.2 at the inlet, emerges over-pressured relative to the undisturbed external medium and has an external Mach number $\equiv u/c = 45$.

RESULTS

Density of the jet in all simulations was one tenth the density of the undisturbed medium and the flow velocity was three to fifteen times the sound speed in the external medium. In one simulation the jet was subsonic behind the bow shock and the jet broke up into eddies within a growing lobe. Three other simulations remained supersonic behind the bow shock. Figure 1 shows velocity vectors and contours of the density, pressure and vector potential (AP) about three quarters of the way through a simulation in which the jet remained supersonic. At early times the jet maintained approximate axisymmetry within a backflowing cocoon. At later times the symmetry was broken by sideways oscillation which destroyed the jet's directionality and the jet "flapped" within a growing lobe. The oscillation wavelength and disruption length scaled approximately with the Mach number and the jet's radius as predicted.

The simulations showed evidence for "hot spot" like features after the symmetry was broken. These features included shocks and jet bending inside the lobe as well as protrusions at the leading edge of the lobe. These features occured further into the simulation and at larger distances from the inlet as the Mach number increased. In all simulations the magnetic field developed filamentary structure in the growing lobe. The slab jet was strongly influenced by vortical motions associated with backflow in the lobe. We would expect a cylindrical jet to be less strongly influenced by vorticies in the lobe and more sensitive to jet stability properties than the slab jet.

ACKNOWLEDGEMENTS

The work was supported by the National Science Foundation through grant AST85-16921 and allotment of 100 hours of time on the Cray at The National Center for Supercomputing Applications, University of Illinois and EPSCoR grant RII8610669 to The University of Alabama.

REFERENCES

Birkinshaw, M. 1984, *M.N.R.A.S.*, 208, 887.
Cohn, H. 1983, *Ap.J.*, 269, 500.
Ferrari, A., Trussoni, E., and Zaninetti, L. 1978, *Astr.Ap.*, 64, 43.
Ferrari, A., Trussoni, E., and Zaninetti, L. 1981, *M.N.R.A.S.*, 196, 1051.
Ferrari, A., Massaglia, S., and Trussoni, E. 1982, *M.N.R.A.S.*, 198, 1065.
Hardee, P.E. 1979, *Ap.J.*, 234, 47.
Hardee, P.E. 1987, *Ap.J.*, 313, 607.
Hardee, P.E. and Norman, M.L. 1988, *Ap.J.*, 334, in press.
Norman, M.L. and Hardee, P.E. 1988, *Ap.J.*, 334, in press.
Norman, M.L., Clarke, D.A. and Burns, J.O. 1987, in the Fifth Cargese Symposium on *Magnetic Fields and Extragalactic Objects*, eds. E. Asseo and D. Gresillon, p.297.
Payne, D.G. and Cohn, H. 1985, *Ap.J.*, 291, 655.

NUMERICAL SIMULATION OF MASS OUTFLOWS FROM STAR FORMING REGIONS

James M. Stone, Michael L. Norman, and Dimitri Mihalas
Dept. of Astronomy, University of Illinois, Urbana, IL
and National Center for Supercomputing Applications

ABSTRACT

Observational studies of star forming regions in the Galaxy have demonstrated the ubiquity of mass outflows from objects in the late stages of star formation. The most dramatic examples of these flows are molecular bipolar outflows and optical jets. We summarize the results from observational studies of these two classes of objects, and outline the major theoretical problems in understanding how such flows are generated. We then describe a project currently underway to address these problems in detail. This project centers on increasingly sophisticated numerical simulations of mass outflows from star forming regions using a multiphysics hydrocode being developed at the NCSA.

1. INTRODUCTION

Observational evidence has long been accumulating that objects in the late stages of star formation (protostars) can be the source of vigorous mass outflows. Such evidence includes the broad, double-peaked, optical emission line profiles of T Tauri stars (which suggest that strong, steady mass loss is occurring from these stars), and the large radial velocities and proper motions of both Herbig-Haro (HH) objects and H_2O masers associated with star forming regions (indicating these objects are participating in a high velocity flow). However, the magnitude and significance of these outflows has become apparent only with the discovery of two remarkable classes of objects. The first is energetic bipolar molecular outflows, whose enormous kinetic energies ($\sim 10^{43} - 10^{47}$ $ergs$) pose interesting problems not only in understanding their internal dynamics, but also for understanding the dynamics and evolution of the parent molecular cloud as a whole. The second, more recently discovered, class is optical protostellar jets. How these highly collimated, high velocity flows are generated, and how they may be related to the large scale molecular outflows to which they are often associated, are further interesting puzzles. It is now clear, though, that a period of vigorous, and often highly collimated, mass loss is an important phase in the formation of most, if not all stars. Furthermore, the fundamental questions regarding this phase of star formation can now be clearly defined and formulated.

G. Belvedere (ed.), Accretion Disks and Magnetic Fields in Astrophysics, 207–221.

This paper describes a theoretical study underway to address some of the fundamental issues of mass outflow from star forming regions in the Galaxy. We are motivated by observations of sites of recent star formation in nearby molecular clouds; therefore strictly speaking this study will be applicable to these regions alone. We will not be able to address the interesting problems encountered in star formation in other environments or epochs.

We begin in section 2 by summarizing some of the latest observational data on energetic molecular outflows and protostellar jets, and the constraints that these observations place on theoretical models. We continue in section 3 by outlining a theoretical project to study the most important, unresolved problems in the generation of these flows. Due to the complex physical nature of these regions, detailed answers can be arrived at only with numerical simulations, and we have therefore begun the development of a multiphysics hydrocode for this problem. This development is described in section 3, along with the progression of numerical simulations which will be performed with the code. Finally, in section 4, we present our conclusions, and describe the significance of this work and its applicability to future studies.

2. SUMMARY OF THE OBSERVATIONS

2.1 Energetic Molecular Outflows (EMOs)

When molecular emission line profiles are observed in molecular clouds, it is generally found that the line widths at half intensity are typically between $1-3km/s$. However, in localized areas around active star forming regions (identified by the prescence of deeply embedded protostars and young stellar objects), the line widths are often much larger, typically in the range $6-8km/s$. Initially, the increase in line width was interpreted as infall of molecular gas into the star forming regions. However, over the last decade, sensitive radio observations of star forming regions has demonstrated that the line emission can extend over an enormous range of velocities often giving line wings extending over $150km/s$ or more (full width at zero intensity). Since the typical sound speed in the gas in these regions is a few tenths of a kilometer per second, these velocities represent highly supersonic bulk motions of the gas superimposed on the already supersonic, random motions in the ambient medium. The large velocities implied by these observations indicate the gas is not gravitationally bound (unless the mass exceeds $10^4 M_\odot$), and therefore is actually participating in an *outflow*. This supposition is supported by observations of the distribution of shock excited H_2 emission, and the proper motion of both HH objects and H_2O masers associated with the flow (see Lada (1985) or Welch *et al.* (1985) for recent reviews of molecular outflows).

We can therefore define high velocity outflow sources as regions in which the observed molecular line width in the wings greatly exceeds that of the ambient medium (typically $\Delta v \geq 10km/s$ is used as a selection criterion). With this definition, more than 60 sources have been identified, all of which are associated with deeply embedded protostars with radiative luminosities $L \sim 1-100L_\odot$. The large range in luminosities of the central source indicates that high velocity outflows can occur from both high and low mass protostars. Moreover, the frequency of occurrence of molecular outflow sources is comparable to the formation rate of 1 $M_\odot$ stars, suggesting that most, if not all, stars undergo a period of high velocity outflow during their formation.

With the advent of high resolution millimeter interferometers, imaging of the outflow regions has become possible. The remarkable result of such studies is that the flows are often collimated into two, oppositely directed (bipolar) streams. The degree of collimation varies widely from source to source, to such an extent that the variation must be intrinsic to the sources themselves, and not merely a projection effect.

High velocity molecular outflows are thus a clearly defined and commonly observed phenomena. From catalogues of currently known sources, we can compile the following list of parameters as characterizing these flows (Lada, 1985):

- temperature: $T_{EMO} \sim 10 - 90K$
- velocity: $v_{EMO} \sim 5 - 20km/s$
- number density: $n_{EMO} \sim 300 - 3000cm^{-3}$
- mass: $M_{EMO} \sim 0.1 - 100M_{\odot}$
- morphology: usually bipolar, with size $R \sim 0.1 - 4pc$
  poorly collimated, with $\overline{R_{coll}} \equiv \frac{length}{width} = 2.4$

From these observed values, we can estimate some of the important dynamical properties of the flow. In particular, from the observed masses and velocities we find that the kinetic energy in the flow is enormous ($\sim 10^{43} - 10^{47} ergs$), thus the label energetic molecular outflows (EMOs). Although the energy contained in the flow is small compared to that in, say, a supernova explosion, it is rather impressive when one considers it has often emanated from a low luminosity (and therefore low mass) protostar. Furthermore, such large energies put severe demands on the driving mechanism for the flow. We defer discussion of models of the dynamics of EMOs until section 2.3.

2.2 Protostellar Jets

Deep CCD images of star forming regions taken in low excitation optical emission lines ($H\alpha$, SII, NII, OIII) have revealed the prescence of elongated, optical nebulae. Spectroscopic and proper motion studies have clearly demonstrated that these nebulae are produced in high velocity mass outflows, and thus they have been termed protostellar jets (Mundt and Fried, 1983). To date, 20 such objects have been discovered, and of these nearly half are associated with known EMOs (Mundt *et al.*, 1987). Morphologically, these jets are bipolar and highly collimated ($R_{coll} \approx 10$, but it may be higher since the jets are often not resolved radially). Often they show knots and clumps of HH objects along their length, especially at their ends. Only a few examples of jets are perfectly straight, most show wiggles or kinks along their axis.

Mundt *et al.* (1987) has concluded that the morphology and characteristics of optical jets and their associated HH objects can be completely understood if the nebulosity is generated by a high velocity, highly collimated, ionized outflow from the embedded protostar. In this context, the HH objects and knots represent either the working surface of the jet, or radiative oblique internal shocks in the flow. Spectroscopic measurements of the flow velocity in HH objects located at the end of the jets (the working surface) have indeed measured a substantial decrease in velocity, as expected. Internal shocks are expected to form when the jet is pressure confined, thus regions where the jet seems to disappear may be areas in which the flow is freely expanding. One can even attribute the wiggles and kinks observed in some jets as due to fluid instabilities in the flow.

Based on the 20 known examples, typical parameters of protostellar jets have been estimated as (Mundt, 1985):

- size: $R_j \sim 0.01 - 1pc$
- velocity: $v_j \sim 200 - 400km/s$
- velocity dispersion: $\delta v_j \sim 50 - 100km/s$
- Mach number: $M_j \sim 10 - 40$
- opening angle: $\theta_j \sim 2^\circ - 10^\circ$
- number density: $n_j \sim 15 - 150cm^{-3}$
- mass flux: $\dot{M}_j \sim 10^{-9} - 10^{-8} M_\odot/yr$
- kinetic luminosity: $L_j \sim 0.01 - 0.2L_\odot$
- dynamical age: $\tau_j \sim 200 - 3000yrs$

All observed jets are associated with moderate luminosity, highly obscured protostars ($L \sim 1 - 10L_\odot$), which all show evidence for activity and stellar winds (at least one line exhibits a P Cygni profile). From the above jet parameters, and from those typical of an EMO, it is clear that optical protostellar jets differ as a class from EMOs in their physical size, flow velocities, densities, and compositions. How these objects are related dynamically is unknown. Furthermore, how these jets are collimated and driven is a mystery.

2.3 Observational Constraints on EMO and Jet Dynamics

Understanding how star forming regions are able to produce molecular bipolar flows and ionized jets is an important problem. Detailed observations of these objects have been accumulating in the last decade, and have provided estimates and constraints on important dynamical variables. Unfortunately, however, the theoretical understanding of these objects is still in a rudimentary stage. Even the resolution of the most fundamental aspects of these flows remains uncertain, with many competing ideas. Until these issues are addressed, the detailed modelling of specific features common to the flows (e.g. the kinks, knots and wiggles of optical jets) is premature.

The two most fundamental issues concerning mass outflows from star forming regions are regarded as:

1. How are these flows collimated?

2. How are these flows driven?

In principle, collimation can be achieved either by an initially anisotropic flow (e.g. if the flow is driven from an initially anisotropic mass distribution such as a disk), or by external collimation of an isotropic flow by the ambient medium. In support of the first idea, direct observations of protostellar disks have been claimed for some objects (Rodriguez, 1986). Indirect evidence for such disks is also provided by the high degree of polarization of the embedded protostars. However, the size of these disks is generally much greater than the scale over which collimation occurs, especially in the case of protostellar jets. Collimation of an isotropic flow by the external medium could occur by external density gradients, or by large scale, ordered magnetic fields. Königl (1982) has considered the collimation of molecular flows by density gradients. However, as stated above, density gradients on a scale comparable to the collimation scale of the flow are not supported by current observations. To date, the collimation of an isotropic wind by a magnetic medium has not been considered in the literature.

The question of how mass outflows from star forming regions are driven is even more uncertain. Observations, however, can help to constrain the possibilities. For instance, it seems certain the radiation field from the central protostar cannot drive the flow. This is demonstrated by the fact that the momentum observed in EMOs is 2 to 3 orders of magnitude larger than the momentum contained in the radiation field, thus the flow cannot be momentum-driven by the radiation field. Furthermore, although the radiation can provide enough energy to drive the flow (i.e. the radiant luminosity is larger than the kinetic luminosity), the kinetic energy is sufficiently high as to require an unusually strong coupling between the radiation field and the outflowing material, so that a significant fraction of the radiant energy is deposited. This fact argues that it is also unlikely that the flow is energy-driven by the radiation field.

With the elimination of the radiation field as the driving force for mass outflows, it seems the next most likely candidate is the (observed) ionized wind from the central protostars. However, observations of these winds have demonstrated that they contain neither enough momentum nor kinetic luminosity to drive the molecular flows. It would seem, therefore, that in order for protostellar winds to drive the flows, there must be a considerable (unobserved) neutral component. How such a neutral wind could be generated by the protostar is an enigma. Nonetheless, detailed spectroscopic and kinematic studies of a few objects have found that the properties of EMOs are entirely consistent with the molecular gas being swept up into a shell located around a cavity evacuated by a wind from the protostar. Currently, this is the most widely accepted scenario (Lada, 1985).

Models using magnetic fields to drive the flows have also been proposed. Pudritz and Norman (1983,1986) have suggested EMOs might result from a hydromagnetic wind driven off a protostellar disk. The rotational energy of the disk provides the energy source for the flow. Uchida and Shibata (1985) have simulated a similar scenario, however the disk is initially allowed to collapse (it is not in radial equilibrium). Although mass outflow is generated in their simulation, it does not reproduce the observed parameters of EMOs. Such models are intriguing, however, as they provide a plausible mechanism by which an EMO and an optical jet can be produced simultaneously. Thus, the molecular flow is driven off the (molecular) protostellar disk, whereas the ionized jet can arise in the inner regions of the disk near the accretion shock. More recently, Shu *et al.* (1988) have proposed a model in which the outflow is driven magnetocentrifugally from the equatorial regions of a protostar rotating near break-up. Detailed modelling of these scenarios, involving the complex interplay of a rotating molecular disk threaded by a magnetic field, and an inner disk heated by radiation from the protostar would be challenging, though not impossible. In particular, the self consistent treatment of the interaction of the radiation field from the protostar with the disk is essential.

Finally, we conclude that the questions of how outflows from star forming regions are collimated and driven still remain largely unanswered. The collimation of an isotropic wind by conditions in the ambient medium typical to star forming regions has not been fully investigated. Furthermore, complex models in which an intrinsically collimated wind is driven from a protostellar disk requires a more in depth treatment to ascertain if internally consistent models can reproduce the observed parameters of mass outflows.

3. NUMERICAL SIMULATIONS

It is clear that mass outflows from star forming regions is a topical and important area for theoretical modelling. Much progress needs to be made in understanding the fundamental physical processes responsible for these flows. In this section we describe a project to investigate some of these issues.

Due to the complex physical nature of mass outflows from star forming regions, we can model them in detail only with self consistent numerical simulations. Based on the observations, we can identify the major characteristics that a computer code used to model these flows must have:

1. Since these flows are far from spherical symmetry, the code must be at least two-dimensional.
2. Magnetic fields may play an important role in the dynamics, therefore the code must incorporate magnetic fields self consistently.
3. In some models, the radiation field from the protostar plays an important role energetically, thus a 2-D radiation transfer module is highly desirable.

We have already begun the development of a new code which incorporates these capabilities. Below we describe this development work , and how the new code will be used to model mass outflows with a progression of simulations.

3.1 The ZEUS Code

This project represents an integral part of the development of a multiphysics code being headed by one of us (MN) at the NCSA. This development is based on ZEUS, a working two dimensional hydrocode which in turn is based on the code described by Norman and Winkler (1986). The new code which will be used for this project will be derived largely from ZEUS.

Currently, ZEUS solves the usual equations of inviscid fluid dynamics including pressure, self gravity and magnetic fields (assuming flux freezing):

$$\frac{D\rho}{Dt} + \rho \nabla \cdot \mathbf{v} = 0 \tag{1}$$

$$\rho \frac{D\mathbf{v}}{Dt} = - \nabla p - \rho \nabla \phi + \frac{1}{4\pi} (\nabla \times \mathbf{B}) \times \mathbf{B} \tag{2}$$

$$\rho \frac{D}{Dt} \left(\frac{e}{\rho} \right) = -p \nabla \cdot \mathbf{v} \tag{3}$$

$$\frac{\partial \mathbf{B}}{\partial t} = \nabla \times (\mathbf{v} \times \mathbf{B}) \tag{4}$$

Here, the $\frac{D}{Dt}$ denotes the Lagrangean or comoving derivative

$$\frac{D}{Dt} \equiv \frac{\partial}{\partial t} + \mathbf{v} \cdot \nabla \tag{5}$$

The fluid equations are closed with Poisson's equation which determines the gravitational potential

$$\nabla^2 \phi = 4\pi G \rho \tag{6}$$

and an equation of state $p = p(\rho, e)$. Thus, the dependent variables evolved in ZEUS are the fluid density ρ, the three components of both the velocity $\mathbf{v}$ and the magnetic field $\mathbf{B}$, and the internal energy density e. These are computed as functions of the three independent variables, namely the time t and two spatial coordinates. Note that all three components of vector field variables are evolved as functions of two spatial coordinates. This approximation is often referred to as 2-$\frac{1}{2}$ MHD, and is necessary for the proper treatment of all MHD wave modes.

The numerical scheme used in ZEUS is an explicit, multistep solution procedure based on a staggered grid. Shocks are mediated with a von Neumann-Richtmyer artificial viscosity, which smears the shock over several grid zones. Numerical tracking of fluid interfaces, using the method of LeBlanc, is possible. Grid zoning in both dimensions is arbitrary and independent, and the grid can be allowed to move in an arbitrary manner to follow global features of the flow (such as collapse or expansion) by rezoning after each time step. The grid must remain orthogonal at all times, though. When interpolation of a field variable between zones is required, an upwinded monotonic van Leer algorithm is used (van Leer, 1977), except in the case of the magnetic field. In this case, in order to guarantee that $\nabla \cdot \mathbf{B} = 0$ is satisfied to within round off error during the simulation, it is necessary to evolve the magnetic vector potential $\mathbf{A}$, rather than the magnetic field itself, where the two are related via $\mathbf{B} = \nabla \times \mathbf{A}$. Thus, in order to maintain first order accuracy in the Lorentz force term in the momentum equation, it is necessary to use a third order accurate advection scheme for the vector potential (since the Lorentz force involves second derivatives of the vector potential). Therefore, the third order accurate piecewise parabolic method (PPM) of Collela and Woodward (1984) is used.

To date, ZEUS has been subjected to a wide battery of test problems, including reproducing the 1-D MHD shock jump conditions, advection of a wide variety of waveforms, Bennet pinch equilibrium and 2-D spherical wave propagation. Furthermore, ZEUS has been, and continues to be, used for a wide range of physical problems and in all cases it has proven to be a reliable and robust code. It therefore provides an excellent and well tested foundation on which to develop the hydrocode used in this project.

3.2 Extensions to ZEUS

Most of the work involved in this project will be associated with the development a new hydrocode, based on ZEUS, which will be particularly suited to simulating mass outflows from star forming regions. Much of this development work has already begun; it involves making both modifications and additions to ZEUS. To date, work that has been completed includes rewriting the code in a completely coordinate independent fashion, using metric scale factors. This allows the new code to compute in Cartesian, cylindrical, spherical, or in principle any orthogonal coordinate system without changes. The single coordinate dependent step is the computation of the scale factors in the desired geometry, which can be performed, once and for all, at the beginning of the simulation. We have also updated the numerical solution package to compute the gravitational potential from Poisson's equation, incorporating fast iterative matrix solvers.

Future substantial improvements that will be incorporated in the new code include the constrained transport scheme of Evans and Hawley (1988) for the magnetic field. This scheme incorporates the constraint $\nabla \cdot \mathbf{B} = 0$ directly into the magnetic evolution equation, removing the necessity of computing with the magnetic vector potential. Thus, if a third order accurate advection scheme is used for the magnetic field, second order accurate Lorentz force values will be achieved in the momentum equation. This improvement stimulates the conversion of the interpolation algorithm for *all* field variables to PPM, ultimately giving the new code an entirely third order accurate advection step. Other technical changes include ensuring the new code conforms to the Fortran 77 standard, allowing it to be developed, and even tested, entirely on a local workstation under UNIX^{TM}. The transportability, modularity and multitasking capability of the code will therefore necessarily be improved as well.

Perhaps the most ambitious additional capability that will be incorporated in the new code as part of this project will be the addition of a 2-D radiation transport algorithm. This will allow the self consistent calculation of the evolution of an embedded radiation field concurrent with the hydrodynamics, effectively making the new code a 2-D radiation magnetohydrodynamics code. The interaction of radiation with matter is incorporated by adding additional terms to the momentum and energy equations (Mihalas and Mihalas (1984), chapter 7), so that

$$\rho \frac{D\mathbf{v}}{Dt} = -\nabla p - \rho \nabla \phi + \frac{1}{4\pi}(\nabla \times \mathbf{B}) \times \mathbf{B} + \frac{1}{c}\int_0^\infty \chi(\nu_\circ)\mathbf{F}_\circ(\nu_\circ)d\nu_\circ \tag{7}$$

$$\rho \frac{D}{Dt}\left(\frac{e}{\rho}\right) = -p\nabla \cdot \mathbf{v} - \int_0^\infty [4\pi\eta_\circ(\nu_\circ) - c\chi_\circ(\nu_\circ)E_\circ(\nu_\circ)]d\nu_\circ \tag{8}$$

where $\chi_\circ(\nu_\circ)$ and $\eta_\circ(\nu_\circ)$ are the opacity and emissivity of the material, $E_\circ(\nu_\circ)$ the radiation internal energy density (zeroth angular moment of the radiation field), and $\mathbf{F}_\circ(\nu_\circ)$ the radiation flux vector (first angular moment of the radiation field). The subscript " o " on all radiation quantities denotes they are measured in the comoving frame of the fluid, the importance of which is described below. Equations (7) and (8) can be easily cast into a form involving only the total (frequency integrated) variables $E_\circ$ and $\mathbf{F}_\circ$. For instance, if we assume strict LTE, so that the source function $S_\circ \equiv \eta_\circ/\chi_\circ$ is identical to the Planck function $B_\circ$, then we find

$$\rho\frac{D\mathbf{v}}{Dt} = -\nabla p - \rho\nabla\phi + \frac{1}{4\pi}(\nabla\times\mathbf{B})\times\mathbf{B} + \frac{1}{c}\chi_F\mathbf{F}_\circ \tag{9}$$

$$\rho\frac{D}{Dt}\left(\frac{e}{\rho}\right) = -p\nabla\cdot\mathbf{v} - 4\pi\kappa_P B_\circ + c\kappa_E E_\circ \tag{10}$$

where χ_F , κ_P, and κ_E are the flux mean, Planck mean, and energy mean opacities respectively

$$\chi_F \equiv \frac{1}{\mathbf{F}_\circ}\int_0^\infty \chi_\circ(\nu_\circ)\mathbf{F}_\circ(\nu_\circ)d\nu_\circ \tag{11}$$

$$\kappa_P \equiv \frac{1}{B_\circ}\int_0^\infty \chi_\circ(\nu_\circ)B_\circ(\nu_\circ)d\nu_\circ \tag{12}$$

$$\kappa_E \equiv \frac{1}{E_\circ}\int_0^\infty \chi_\circ(\nu_\circ)E_\circ(\nu_\circ)d\nu_\circ \tag{13}$$

To calculate $E_\circ$ and $\mathbf{F}_\circ$ self consistently requires solution of the dynamical equations of the radiation field, namely the frequency integrated zeroth and first moments of the transfer equation (the total radiation energy and momentum equations respectively). In the comoving frame, and to order unity in (v/c) , these equations are

$$\rho\frac{D}{Dt}\left(\frac{E_\circ}{\rho}\right) + \nabla\cdot\mathbf{F}_\circ + \nabla\mathbf{v} : \mathsf{P}_\circ = 4\pi\kappa_P B_\circ - c\kappa_E E_\circ \tag{14}$$

$$\frac{\rho}{c^2}\frac{D}{Dt}\left(\frac{\mathbf{F}_\circ}{\rho}\right) + \nabla\cdot\mathsf{P}_\circ = -\frac{1}{c}\chi_F\mathbf{F}_\circ \tag{15}$$

Dropping terms of $O(v/c)$ and higher in equations (14) and (15) is justified by the fact that the magnitude of the flow velocities encountered in mass outflows is small compared to c. Moreover, this approximation considerably simplifies the computations. An important feature of this work, however, has been to ensure that only terms which are $O(v/c)$ in *all* regimes of interest (including the optically thin, and the static and dynamic diffusion regimes) are dropped. This step is essential to ensure that the calculations remain self consistent over the entire computational mesh whenever the physical conditions in different parts of the flow become widely disparate. In addition, in writing equations (14) and (15), we have exploited an important advantage of the comoving frame, namely that the material properties are isotropic in this frame. Thus, the angular moments involving these quantities are particularly simple. Furthermore, using the comoving frame for the radiation variables conforms to the scheme used for the hydrodynamics, where all material variables are really Lagrangean variables measured in an Eulerian coordinate system.

Equations (14) and (15) are closed with the variable Eddington tensor which relates the total radiation stress tensor $\mathsf{P}_\circ$ (second angular moment of the radiation field) to the total radiation internal energy density

$$\mathsf{f} = \frac{\mathsf{P}_\circ}{E_\circ} \tag{16}$$

Generally, the Eddington tensor can be found only from a formal solution of the full angle and frequency dependent transfer equation to compute the radiation field. Appropriate angle and frequency quadratures then yield the Eddington tensor from (16), in addition to the mean opacities (11) - (13), thus closing the problem.

Implementing the radiation hydrodynamics problem in the form described above in a multi-dimensional numerical algorithm is computationally expensive. Since solving the transfer equation represents the most expensive step in this procedure, most previous work on 2-D radiation hydrodynamics has generally made use of the non-equilibrium diffusion approximation, in which the Eddington approximation ($\mathsf{f} = \frac{1}{3}$) is used. This circumvents the need for a formal solution of the transfer equation. However, although this method results in considerable reduction of the computational demands of the problem, it is applicable only to optically thick flows, and breaks down if the radiation becomes streaming *anywhere* on the grid (and there is no *a priori* way to know whether a particular flow will become optically thin somewhere on the grid). Applying the diffusion approximation to calculations of optically thin flows leads to the classic problem of flux limiting. Numerical fix-ups of this problem, such as flux limiters, are of unknown generality and reliability, and therefore are probably unsatisfactory. Only a full transport solution can treat correctly many situations we expect to occur in mass outflows, such as direct irradiation of a local (optically thin) region (the envelope of a protostellar disk) by a remote source of intense radiation (the protostar itself). For this reason, we are implementing a full transport solution in the new code being developed for this project. We outline below the major steps in the solution procedure we are developing.

The first step is to solve the dynamical equations of the radiation field coupled to the fluid equations. We proceed by adding the fluid and radiation internal energy equations (10) and (14) to derive the first law of thermodynamics of the radiating fluid

$$\rho \frac{D}{Dt}\left(\frac{e + E_\circ}{\rho}\right) + \nabla \mathbf{v} : \mathsf{P}_\circ = -p \nabla \cdot \mathbf{v} - \nabla \cdot \mathbf{F}_\circ \tag{17}$$

Notice that the material-radiation interaction terms have cancelled on the RHS of (17), this ensures our numerical scheme will be accurate when the fluid is near radiative equilibrium. The complete set of moment equations to be solved are now (14), (15), and (17). We continue to integrate the fluid equations explicitly in time, however, on a fluid flow timestep Δt, the solution of the moment equations must be implicit for stability. In 2-D, we must therefore solve four coupled, implicit equations for the four variables e, $E_\circ$, and the two non-zero components of $\mathbf{F}_\circ$. We can, however, reduce this number (and thereby greatly simplify the problem) by using the "automatic flux limiting" method of Mihalas and Weaver (1982). Here, the radiation momentum equation is integrated analytically, albeit in an approximate way. Expanding the Lagrangean derivative, equation (15) can be rewritten as

$$\frac{\partial \mathbf{F}_\circ}{\partial t} + \nabla \cdot (\mathbf{F}_\circ \mathbf{v}) = -c\chi_F \mathbf{F}_\circ - c^2 \nabla \cdot \mathsf{P}_\circ \tag{18}$$

The second term on the RHS is the advection term. With the explicit multistep numerical scheme implemented in ZEUS, this advection term is accounted for in a logically independent step (the transport step) from the rest of the equation. Thus, for the discussion below, we drop the advection term with the understanding that it is computed in the transport step.

Thus, we can rewrite (18) as

$$\frac{\partial \mathbf{F}_\circ}{\partial t} + c\chi_F \mathbf{F}_\circ = -c^2 \nabla \cdot (\mathsf{f} E_\circ) \tag{19}$$

The integrating factor for this equation is $e^{c\chi_F t}$, thus integrating over a fluid flow timestep gives

$$\mathbf{F}_\circ(t+\Delta t)e^{c\chi_F(t+\Delta t)} - \mathbf{F}_\circ(t)e^{c\chi_F t} = -c^2 \int_t^{t+\Delta t} \nabla \cdot (\mathsf{f} E_\circ) e^{c\chi_F t} dt \tag{20}$$

If we now choose some approximate, "average" value over the timestep Δt for χ_F and $\nabla \cdot (\mathsf{f} E_\circ)$ on the RHS, then (20) becomes

$$\mathbf{F}_\circ(t+\Delta t) = e^{-x}\mathbf{F}_\circ(t) - (1 - e^{-x})\frac{c}{\overline{\chi_F}}\overline{[\nabla \cdot (\mathsf{f} E_\circ)]} \tag{21}$$

where $x \equiv c\overline{\chi_F} \triangle t$. Equation (21) represents an exponentially attenuated, stable prescription for evaluating the flux; it can handle the numerical stiffness that arises when $x = \left(\frac{photon\ free\ flight\ distance\ in\ \Delta t}{photon\ mean\ free\ path}\right) \gg 1$. The accuracy of the method will depend on how well the time averaged values $\overline{\chi_F}$ and $\overline{\nabla \cdot (\mathsf{f} E_\circ)}$ can be approximated. Equation (21) can now be inserted algebraically into (14) and (17), yielding only two coupled, implicit equations for the two variables e and $E_\circ$. Note, however, that these equations will be highly nonlinear due to the temperature dependence of the opacities and the Planck function. Nonetheless, numerical methods for solving such equations in 2-D are well known and readily available.

Of course, solution of this coupled set still requires knowledge of the variable Eddington tensor, which can be found only from a formal solution of the full angle-frequency dependent transfer equation for the intensity. We can greatly simplify this step if we assume grey opacities (independent of frequency). Then, the frequency dependence of the radiation variables is removed, and we need solve only the total (frequency integrated) transfer equation. In 2-D, formal solutions for the total transfer equation for a *static* medium using the method of direct integration along short characteristics (Kunasz and Auer, 1988) are well developed. Of course, in hydrodynamics the medium is not static. However, to a high degree of approximation, static solutions will be accurate as long as the flow is not evolving on a timescale which is short compared to the radiation flow timescale (*i.e.* so long as $\lambda/c \ll l/v$, where λ is a photon mean free path, l is a characteristic gradient length, v is the fluid velocity, and c the speed of light). Except in very unusual circumstances, this condition is always satisfied, certainly in the protostellar mass outflows considered here. Our working philosophy is therefore to take a "snapshot" of the quasistatic radiation field at each time step (or perhaps less frequently) via a formal solution of the static transfer equation, and use this solution to compute the variable Eddington tensor.

In Figure 1 we summarize the whole solution procedure for the radiation module being implemented in the new code. The development of this module is currently well underway.

3.3 A Hierarchy of Simulations

The theoretical considerations involved in understanding protostellar mass outflows are complex, and will challenge the capabilities of the multiphysics code being developed for this problem. Our current understanding of the importance and relationship of the processes involved in these objects is limited. Detailed modelling must therefore begin with very idealized, simple scenarios. Most ideas which involve complex physical processes to explain the flow (*e.g.* magnetic reconnection, dynamos, effects of departures from LTE and non-grey (monochromatic) radiation transport) will therefore remain only ideas, unaccessible to detailed modelling at least in the near future.

We therefore propose a hierarchy of numerical simulations, outlined below, which will address key questions regarding the problems of mass outflows for simple, idealized situations. Such a method will help us isolate the effects of the addition of increasingly more complex physics to the problem. Our philosophy is not to be an advocate of a particular model or process, but merely an investigator, sorting through the competing ideas and identifying the most promising.

With this in mind, we shall begin by addressing the question of collimation of an isotropic mass outflow from a protostar. Simulations in three different regimes are envisioned.

1. Collimation occurs by magnetic forces in a uniform ambient medium. Realistic choices for the parameters of the medium include $n \sim 10^7 cm^{-3}$, $T \sim 10K$, and $B \sim 10^{-3}G$.

2. Collimation occurs by pressure gradients in a protostellar disk in equilibrium. Dimensions for the disk would be $R \sim 10^{16} - 10^{17} cm$, $a/b \sim 10$, $M \sim 10 - 100 M_{\odot}$, and $T \sim 10K$.

3. Collimation occurs by a magnetic protostellar disk in equilibrium, combining the physics of the first two simulations.

In all cases, we will choose the wind to be steady and isotropic, with sufficient momentum to drive a typical EMO. Realistic dimensions for the protostar are $M \sim 0.5 M_{\odot}$, and $R \sim 4R_{\odot}$ (Shu *et al.* , 1987). More advanced models might include the effects of radiative heating of the region near the protostar. These simulations may provide important clues as to the whether conditions in the ambient gas around a protostar (as best as can be determined from current observations) are sufficient to collimate an isotropic wind.

With regard to understanding what drives the outflows, the simulations become more difficult. If the flows are initiated by a massive neutral wind from the protostar (as will be assumed above), then the question becomes what drives the neutral component of the wind, a formidable problem indeed. Current ideas are not accessible to detailed modelling. If a hydromagnetic wind is driven from a protostellar disk, then the problem is more tractable. A detailed, though complex simulation along the lines of those of Uchida and Shibata (1985) is possible. The scenario is a protostar surrounded by a rotating disk which is threaded by a magnetic field. Initially, the disk is not in radial equilibrium, and it contracts, with some material being accreted directly onto the protostar. Realistic choices for the dimensions of the problem are a protostar of $M \sim 0.5 M_{\odot}$, $R \sim 4R_{\odot}$, and luminosity determined from by the accretion rate, a disk with $M \sim 10 - 100 M_{\odot}$, $R \sim 10^{16} - 10^{17} cm$, $a/b \sim 10$, and $T \sim 10K$, and a magnetic field strength of $B \sim 10^{-3}G$. Eventually, magnetic forces in the disk will drive a hydromagnetic wind, removing angular momentum and driving further contraction. Whether a fully self consistent model which accounts for the observed parameters of mass outflows from star forming regions can be constructed in such a scenario is an intriguing question.

Results obtained during the course of these simulations may pose new questions and generate new ideas. Regardless of the outcome of individual simulations, deep insight into the dynamics of these outflow is expected by the conclusion of this study.

4. CONCLUSIONS

Understanding mass outflows from star forming regions is an important problem in modern star formation theory. Detailed models are needed not only to understand the observed manifestations of outflows (EMOs and protostellar jets), but also to examine the effects that mass loss may have on the dynamics of the parent molecular clouds, and ultimately on the star formation process itself. Furthermore, the knowledge gained by modelling protostellar jets, guided by the wealth of observational data available for these objects due to their proximity, may prove an enormous aid in understanding all types of jets in space.

We have therefore begun a theoretical study of mass outflows from star forming regions. This study requires the development of a multiphysics hydrocode capable of simulating the complex physical environment of mass outflows. Development of this code is already well underway. A hierarchy of simulations will be performed with the new code, starting with simple models and adding progressively more complex physics. In this manner, the relationship and importance of each effect on the dynamics of the outflows will be isolated. It is hoped that this process will allow us to identify the most promising physical scenario operating in mass outflow regions, and the areas in which further work is needed.

Of course, we cannot expect the hydrocode developed in this project to be able to describe mass outflows from star forming regions completely. There are many areas in which future development could prove important. For instance, it might be important to incorporate a more advanced radiation transfer module, allowing more realistic opacities and equations of state to be used, and perhaps even frequency dependent transport. The interface tracking capabilities of ZEUS would prove indispensable in this case, as it would allow the numerical treatment of sharp variations in material properties (such as opacity, temperature, *etc*). Or perhaps we will find the assumption of flux freezing (perfect coupling between the fluid and magnetic field) is not justified, and therefore future work might concentrate on allowing the magnetic field to diffuse through the fluid (ambipolar diffusion).

As stated previously, this project represents an integral part of the entire multiphysics hydrocode development project underway at NCSA. The techniques and numerical methods developed in this project will therefore see wide ranging applications in many problems. Hopefully, they will ultimately prove useful as a foundation for the development of increasingly more sophisticated hydrocodes in the future.

REFERENCES

Colella, P. and Woodward, P.R. 1984. *J. Comp. Phys.*, 54:174
Evans, C.R. and Hawley, J.F. 1988. Preprint.

Königl, A. 1982. *Ap. J.*, 261:115

Kunasz, P. and Auer, L.H. 1988. *JQSRT*, 39:67

Lada, C.J. 1985. *Ann. Rev. Astron. Astrophys.*, 23:267

Mihalas, D. and Mihalas, B.W. 1984. *Foundations of Radiation Hydrodynamics*, Oxford University Press.

Mihalas, D. and Weaver, R. 1982. *JQSRT*, 28:213

Mundt, R. 1985. In "Jets from Young Stars and Galaxies", *Canadian J. Phys.*, 64:407

Mundt, R. and Fried, J.W. 1983. *Ap. J.*, 274:L83

Mundt, R., Brugel, E.W. and Bührke, T. 1987. *Ap. J.*, 319:275

Norman, M.L. and Winkler, K.-H. 1986. In *Astrophysical Radiation Hydrodynamics*, edited by K.-H. Winkler and M.L. Norman, D. Reidel

Pudritz, R.E. and Norman, C.A. 1983. *Ap. J.*, 274:677

Pudritz, R.E. and Norman, C.A. 1986. *Ap. J.*, 301:571

Rodriguez, L. 1986. In *Masers, Molecules and Mass Outflows in Star Forming Regions*, edited by A.D. Haschick

Shu, F.H., Adams, F.C. and Lizano, S. 1987. *Ann. Rev. Astron. Astrophys.*, 25:23

Shu, F.H., Lizano, S., Ruden, S.P., and Najita, J. 1988. *Ap. J.*, 328:L19

Uchida, Y. and Shibata, K. 1985. *Publ. Astron. Soc. Japan*, 37 :515

van Leer, B. 1977. *J. Comp. Phys.*, 23:276

Welch, W.J., Vogel, S.N., Plambeck, R.L., Wright, M.C.H. and Beiging, J.H. 1985. *Science*, 228:1389

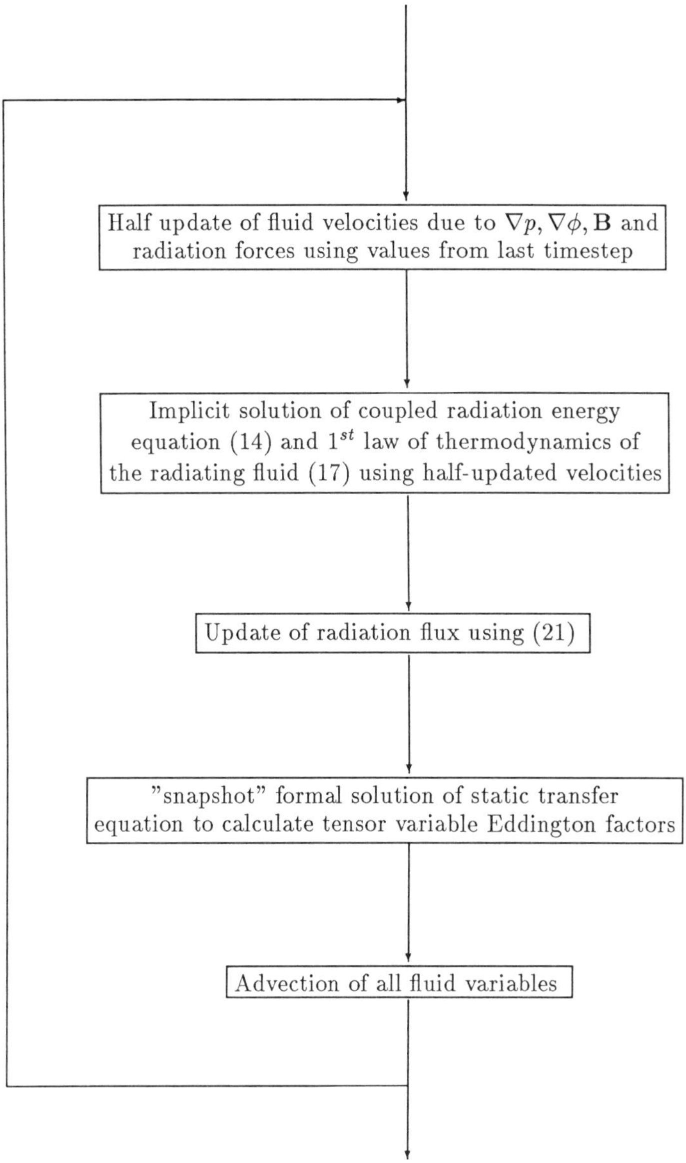

Fig. 1. Schematic flow chart for the numerical solution of the 2-D radiation dynamical equations coupled to an explicit, multistep hydrocode as described in section 3.2.

OUTFLOWS FROM ACTIVE GALACTIC NUCLEI

Attilio Ferrari

Osservatorio Astronomico di Torino, and
Istituto di Fisica Matematica, Università di Torino, Italy

Abstract. In this review I discuss some aspects of the physics of collimated outflows emerging from accretion disks in connection with the modelling of Active Galactic Nuclei (AGN). Analytical and numerical studies of outflow propagation and instabilities are used to interpret structures detected at radio (non-thermal continuum radiation) and optical (continuum and emission line radiation) frequencies. The present situation is that good qualitative knowledge of the several dynamical effects which may shape AGN's structures has already been reached; however the quantitative estimate of their relative importance and conditions of occurrence still awaits a better study of the nonlinear processes in astrophysical conditions and more observations to determine the values of the physical parameters.

1. Introduction.

Galaxies are generally interpreted as self-gravitating collections of stars and gas. However in 1954 radioastronomy provided the first clues that galaxies are more than this, as their nuclear regions are the seats of powerful dynamical and radiative processes which have been later discovered to be peculiar to all compact, dynamically active astrophysical objects. For instance the discovery of pulsars in 1967 and X-ray binaries in 1970 has extended this scenario to the case of collapsed stars; more recently it has been realized that stellar formation regions and luminous blue variables also show similar processes.

In all these years a rather compelling evidence has grown that very-high-energy sources (on both stellar and galactic scale) are powered by accretion flows towards collapsed objects. The gravitational energy liberated in these inflows is large enough (and can be concentrated to high enough particle/photon energy density) that the energy budget can be easily accounted for. In addition the trend of observations suggests that supersonic plasma outflows are basic physical elements by which AGNs build up the various morphologies displayed on several length scales and in different regions of the radiation spectrum.

However how this energy is liberated and channeled into the various forms of activity (radiative and dynamical) still remains to be understood, especially because the information on physical parameters, time and length scales, etc. are still rather uncertain. Future space missions will clarify these questions, but we have also to prepare the theoretical background for interpeting such exotic physical conditions.

G. Belvedere (ed.), Accretion Disks and Magnetic Fields in Astrophysics, 223–232.

In this review I discuss some of the physical processes devised in the study of astrophysical outflows, especially adopting methods and techniques already tested in better studied phenomena. In particular I address the subject of how outflows can be originated from hot accretion disks using the standard wind theory proposed by Parker in 1958. Moreover I also suggest effects that may account for the formation of morphologies observed in the various frequency bands.

The basic observational scenario for the model discussed is summarized in Sec. 2; then, physical interpretations of the results of numerical simulations of jet propagation presented by Mike Norman in his review in this Conference are suggested (Sec. 3). The two following Secs. 4 and 5 attempt a morphological discussion of the phenomena displayed in simulations and their application to astrophysical objects.

2. Observations of Extragalactic Outflows.

The distinctive feature of AGNs is their bright, "star-like" optical core, indipendently of their specific morphological and spectral morphologies.

Radio, optical, infrared and X-ray data combine in indicating an enormous energy budget, up to 10^{63} ergs over their lifetime; in addition extended radio structures (collimated jets, lobes, hot-spots) and emission line profiles (in Broad-Line Regions, Narrow-Line Regions, Extended Emission-Line Regions) suggest a strong, persistent dynamical activity characterized by supersonic outflows (Fosbury 1987). The detection, by VLBI technique, of "superluminal" motions in the radio cores of several AGNs is generally interpreted as an indication that outflows can reach relativistic velocities.

Often AGNs are classified in two large groups: radio-quiet AGNs and radio-loud AGNs. These two classes are found both in nearby galactic objects (Active Galaxies and Radiogalaxies) and in objects at cosmological distances (radio-quiet QSOs and radio-loud QSOs). Extended structures are only seen through nonthermal radio emission; this fact has led some authors (Orr and Browne 1982) to propose that in general nonthermal radio emission arises only when plasma jets can break out of the deep AGN's cores. In a scenario in which outflows from the deep cores are present in all AGNs, the difference between radio-loud and radio-quiet AGNs may be routed to the physical conditions leading these outflows to emerge from the cores under the form of more or less collimated radio jets, or to be stopped by some process which entails dissipation of the kinetic energy into plasma heating around the cores.

Radio data display a large variety of morphologies in extended, well-developed outflows: collimation of outflows with different regimes of confinement (by thermal and/or magnetic pressure), knots of enhanced brightness, wiggles and bendings in the jet propagation direction, interaction with the external medium yielding diffuse radio lobes and hot spots.

Supersonic outflows are observed in galactic objects also: stellar winds are one well-known example, but more recently collimated bipolar outflows have been observed in star formation regions and in active compact objects like SS 433 and Cyg X-3.

We may conclude that outflows and in particular bipolar structures must be ubiquitous in all astrophysical environments; thus we should learn a lot from comparing morphologies and physical processes in different situations (see for instance Ferrari 1987).

3. Outflow Topologies from the Wind Theory.

Leaving aside the process of outflow acceleration from AGN cores (Begelman *et al.* 1984, Rees 1984, Begelman 1985), I address here the results of simulations of jet propagation, namely their interaction with gravitational fields, surrounding plasma, external pressure and velocity fields.

In the theoretical study of jets a fundamental role is beeing played by numerical simulations, which must be regarded as experiments in hydro- and magneto-hydro- dynamics. They should be used as such, and their results should always be cautiously adopted taking into considerations the many assumptions about the physical parameters. In fact most of these parameters are very poorly known in astrophysical conditions and therefore, although simulations may beautifully compare with observations, this is not necessarily a test of their applicability.

Numerical codes have been developed by Norman, Smarr and Winkler (1982), Falle *et al.* (1985), Arnold and Arnett (1986), Woodward and Colella (1984), Uchida and Shibata (1986) and other authors. Most codes use 2-D hydrodynamics for axially symmetric jets, but some also include 3-D and MHD effects. A large variety of features are displayed in the results of simulations that can usually be associated with observational morphologies. However it should be pointed out that most of them do in fact arise from the assumed boundary and initial conditions and from the specific integration methods. In particular "artificial" viscosity is used to follow the formation of shocks; although mathematical, this viscosity correspond to a very high nonlinear interactions of the plasma particles, which is a completely arbitrary condition on physical grounds.

Similarly, the development of structures is followed for limited integration time, as numerical calculations are already very lengthy. It has been proven in other types of simulations that in fact many of these structures are transient and disappear in long range, stationary configurations.

Actually we must realize that, before accepting the simulation results, it is essential to understand the basic physics of the nonlinear processes they involve; in this sense they are a real "experiment" in hydrodynamics, because they give evidence of physical effects that we must analytically investigate in order to evaluate their dependence from the adopted physical parameters.

A simple, powerful tool for this analysis of physical effects has been recently presented by Ferrari and Tsinganos (1986); the method is based on the use of the wind theory proposed by Parker since 1958, after proper modelling to the problem of collimated outflows from AGNs. I stress that adopting the wind theory does not at all imply that outflows are necessarily accelerated by thermal pressure as in the case of the solar wind. Any acceleration process can be accepted, and any collimation: we are here investigating the hydrodynamics of the jet propagation only.

The simplest representation of flow dynamics can be derived following Liepman and Roshko (1957). It leads to a set of equations in one space variable, r, along the propagation direction, and time, t; stresses transverse to r are included by integrating across streamlines and they appear in the definition of the flow-channel cross-sectional area. In this sense the method is technically called "one-and-a-half" dimensional.

The system consists of three equations, the equations of continuity, of motion and of state. After algebric manipulations they are:

$$\frac{\partial}{\partial t}[\rho A(r,t)] + \frac{\partial}{\partial r}[\rho A(r,t)v] = 0 \tag{1a}$$

$$\frac{\partial}{\partial t}[\rho A(r,t)v] + \frac{\partial}{\partial r}[\rho A(r,t)v^2] =$$

$$= -A(r,t)\frac{\partial P}{\partial r} - \rho A(r,t)\frac{Gm(r)}{r^2} + \rho A(r,t)D(r,t) \ , \tag{1b}$$

$$P = K\rho^{\Gamma} \tag{1c}$$

where v is the flow velocity along r, $A(r,t)$ the flow-channel cross-sectional area, ρ the density, P the pressure, Γ the plasma polytropic index, m the gravitational mass distribution of the galaxy around the AGN and $D(r,t)$ the generic nonthermal momentum deposition rate. Following Kopp and Holzer (1976), one can also write $A(r,t) = r^2 f(r,t)$, where the function $f(r,t)$ introduces the spatial and temporal profile of the non-spherically symmetric geometry.

3.1 Stationary topologies, multiple solutions, shocks.

Stationary solutions consistent with the above system of equations provide necessary conditions for the existence of steady flows, and allow the analysis of the effects which may be produced by various physical "ingredients". Our discussion will exclude relativistic velocities, although this aspect has been extensively treated in other papers (Ferrari *et al.* 1985).

The stationary topologies can be directly found from equations (1) by setting $\partial/\partial t \equiv 0$. Thus, if we use the flow Mach number $M = v/v_s$ (v_s is the local sound speed), we obtain (Kopp and Holzer 1976):

$$\frac{M^2-1}{M^2}\frac{\mathrm{d}M^2}{\mathrm{d}r} = \left[1 + \left(\frac{\Gamma-1}{2}\right)M^2\right] \times$$

$$\times \left[\frac{2}{r} - \frac{\Gamma+1}{2}\frac{Hr_0 m(r)/m(r_0)}{r^2\left[1+(\Gamma-1)/(2M^2)\right]} + \frac{1}{f(r)}\frac{\mathrm{d}f(r)}{\mathrm{d}r} + \frac{D(r)}{v_{s0}^2}\right] \tag{2}$$

where r_0 is the coordinate of the base of the flow, to which all quantities with index "0" are referred; H is a numerical factor depending on the AGN mass and temperature (Ferrari *et al.* 1985, 1986).

The wind topologies are those connecting $r \approx 0, M \approx 0$ with $r \to \infty, M \to \infty$. In the standard Parker (spherically symmetric, isothermal) wind, only one such topology exists and it possesses an X-type critical point where pressure forces and gravitational pull come into balance (nozzle) and from then on the outflow becomes supersonic.

The system (2) allows to test how the mass distribution, confinement (and/or boundary instability) and various nonthermal acceleration forces affect the jet propagation through the action of the 2nd, 3rd and 4th terms of the right-hand side of Eq. (2) respectively.

For instance, when nonthermal momentum is added upstream the Parker critical point ($D > 0$), the wind topology is initially distorted, but when the momentum addition is strong enough a pair of new critical point appear, one X-type and one O-type (so-called because topologies do not cross it, but encircle it). In this last case three wind solutions are possible in principle: one continuous solution through the inner X point, and two discontinuous solutions that, after going supersonic at the inner X point, become subsonic where shock jumps connect them to another topology going again supersonic through the outer X point. When the added momentum is very large, only the inner X point survives and the discontinuous solutions are not allowed. These various cases are discussed in detail in Ferrari *et al.* (1985, 1986).

In general discontinuous solutions are very important from the point of view of the flow hydrodynamics as they correspond to steady shocks which modulate the flow. These shocks are not due to boundary perturbations (or initial conditions), as displayed in numerical simulation experiments, but are related with real physical agents: actually momentum addition can represent both nonthermal pressure added to the flow or a sudden expansion of the flow channel cross-section (3rd term of Eq.(2)).

Another type of shock has interesting implications for our astrophysical problem, the so-called "terminal shock". The wind solutions we have just discussed imply that for $r \to \infty$ the plasma pressure vanishes. In AGNs we must expect instead the presence of a non-zero pressure confining the region; therefore the standard wind solution is not applicable in the regions in which the internal flow pressure becomes comparable with the external pressure. This problem was already discussed by Parker himself showing the possibility for the flow to join with a different asymptotic subsonic branch through a shock; after this shock the flow velocity slows down toward $M \to 0$.

It is possible to calculate the position of the terminal shock as a function of the external pressure (see Holzer and Axford 1970). Whether this position is stationary or it moves up- or downstream, it depends on the physical conditions of the plasma surrounding an AGN core. In any case the shock tends to be relatively strong and corresponds to regions of enhanced thermal energy dissipation.

3.2 Flow instabilities, multiple shocks.

Collimated flows are unstable to Kelvin-Helmholtz modes at the surface discontinuity with the confining medium or magnetic field. This instability can be very important for supersonic flows also, as shown in the context of astrophysical jets (see for instance Ferrari 1985). In particular body modes can arise which affect the whole jet structure (Bodo *et al.* 1989). Nonlinear effects are likely to saturate the growth of the instability and avoid the disruption of the outflow, but unstable modes do certainly modulate the flow pattern and provide substructures responsible for the formation of observable features. For instance wiggles in the jet direction of propagation may be amenable to the development of helical modes, and enhanced brightness spots to pinching modes (Ferrari 1985).

The wind model predicts that transverse shocks can form when the flow surface becomes unstable to shears, as, for instance, in the case of Kelvin-Helmholtz or resistive instabilities. It has been calculated by Ferrari *et al.* (1986) that a sinusoidal perturbation of the surface of a cylindrical jet allows multiple discontinuous solutions associated with shocks. In a sense these solutions represent the nonlinear evolution of linear instability perturbations.

These fluid instabilities can be easily connected with the morphologies produced in numerical simulations. However other, more subtle types of instabilities are likely to develop along jets. For instance, resistive instabilities (Ferrari 1985) and thermal synchrotron instabilities (Bodo *et al.* 1988) can also develop over time scales shorter than convective time scales. In terms of the linear analysis they both lead, for a large range of physical conditions, to the formation of filamentary structures aligned with the magnetic field component helically wrapped around the jet. These modes appear to be related to the morphologies observed in the jet associated with M87 and possibly in other still not well resolved jets. Filaments are therefore a direct consequence of the presence of magnetic field in jets, whose presence on the other hand is confirmed by the nonthermal synchrotron emission at radio frequencies.

The simulation of resistive and thermal instabilities is at present far from reach because of the limitations of our computational capabilities.

3.3 Time-dependent solutions, transient shocks.

When a perturbation is created along the propagation, either in the confinement conditions or in the interaction with external medium, multiple solutions will appear, some of them with shocks. Actually it is important to show which are the parameters defining how and when each solution is adopted by the flow, and whether such a solution is in fact accessible and stable. These questions require a time-dependent analysis of the problem by solving Eqs. (1).

A study of time-dependent wind solutions has been performed recently by Trussoni *et al.* (1988). In particular it has been shown that solutions with a shock downstream of the O-type critical point are not accessible (unstable); the adoption of one or the other of the remaining solution depends on the way momentum is added to the flow. Fast momentum addition always leads to continuous solutions without shocks, slow and weak addition allows the formation of the upstream stationary shock.

As a matter of fact, for a while after momentum is added and before the relaxation to one or the other of the accessible solutions, non-stationary shocks are formed along the flow which are convected (about sonically) up- or downstream. These transient shocks are not simple numerical accidents: they actually represent transient perturbations in the flow and, due to their slow damping, give rise to variabilities in the flow physical conditions (possibly connected with variabilities of the radiation characteristics).

3.4 Asymptotic flow velocities.

An important jet parameter is the propagation velocity; in fact this determines the energy flux required to support the associated extended radiosources. In general this velocity cannot be directly measured from observations, because the radio spectrum does not display lines. The only direct evidence of an outward motion (in addition to the above mentioned energetic arguments) are the so-called "superluminal motions", corresponding to proper motions of bright knots in the VLBI cores of several radio-loud AGNs, with estimated apparent velocities larger than c.

The common interpretation of this phenomenon is that such bright knots move at relativistic velocity almost along the observer's line of sight, so that relativistic corrections increase the apparent velocity to $\approx \gamma c$. The predicted real velocities are larger than $0.5c$.

In wind models velocities larger than $\approx 0.3c$ are not allowed because of the radiation drag inside AGN cores where the photon density is necessarily very high. On the other hand the observed proper motions may in fact represent the propagation of disturbances on reflecting screens inside the AGN cores; the observed velocities would be phase velocities. More data are required to solve this problem.

4. Correlation of Theoretical Results and Observations.

The study of topologies of stationary and non-stationary outflows has shown, beyond specific details, that any slight modification of the classical Parker problem produces a complicated pattern of critical points in the solution topologies and yields wind solutions with shocks.

These shocks certainly represent a wider class than those found in numerical simulations which, as discussed by Mike Norman in this Conference, have been restricted to the propagation of collimated jets in pressure equilibrium with the surrounding medium. In that limit the morphologies are due to the nonlinear development of surface instabilities at the contact discontinuity between the jet and the external plasma (or confining magnetic field). Instead, in addition to surface instabilities, other physical effects also lead to shocks.

Therefore the morphologies so beautifully displayed in simulations can in fact be produced by a wider choice of effects than those tested numerically. All these effects are potentially responsible for the structures observed in AGNs: for instance, they all produce shocks eventually that heat thermal plasma and accelerate suprathermal particles and are thence possibly associated with knots, wiggles and hot spots in extended jets, with moving clouds of ionized plasma (BLR, NLR) or with extended regions of diffuse ionized plasma (EELR).

Hiwever different effects yield different combination of morphologies for different physical paremeters. Thus, the existence of structures in jets is not, unfortunately, a unique way to derive these basic physical parameters.

It may be interesting to propose a cross-correlation list of physical processes and observational characteristics which can be used to propose tests for future numerical experiments and to make predictions for observations of AGNs.

(a) Mass distribution in AGNs.

The gravitational field profile has to satisfy precise conditions to allow the breaking out of extended jets; in particular for $M(r) \approx r^{-\alpha}$ with $\alpha \geq \alpha(\Gamma), (\alpha(1) = 1)$, jets are trapped. This means that, although all AGNs can accelerate outflows, these do not develop on large scales in all cases. When outflows are trapped, their kinetic energy is released as thermal heating, perhaps in the moving plasma of the BLR.

Such interpretation predicts an anticorrelation between the presence of jets and BLR, which in fact has been suggested (Norman and Miley 1984), but still needs confirmation.

Trapping can develop on larger scales also, and this may yield small jets associated with extended ionized regions (NLR, EELR).

(b) Fluid instabilities in collimated outflows.

This problem has been extensively analysed in the attempt to understand jet morphologies. In particular Kelvin-Helmholtz instabilities allow to work out a scenario in which the excitation of fast growth rate modes, instead of disrupting the collimated outflow, can be saturated by nonthermal particle acceleration which in turn dissipate energy via synchrotron radiation (e.g. Ferrari 1985). In addition acceleration (and radiation) is enhanced at the position of shocks as nonlinear development of instabilities. This suggestions are supported by the results of numerical simulations.

Instabilties may also be linked to the formation of knots and wiggles along jets, and of transients structures responsible for variabilities (Ferrari *et al.* 1986, Trussoni *et al.* 1988).

(c) Transverse shocks.

Geometries, confinement, non-thermal momentum deposition processes all produce stationary and/or non-stationary transverse shocks. Their morphological effects are equivalent to those discussed for the instability case; however they mark now specific situations during the outward jet propagation.

Transverse shocks are obviously strong ionizing agents: in addition to modulate the large scale nonthermal radiation from jets and outflows, they are strong ionizing local sources for BLR, NLR, EELR.

Actually stationary shocks are relatively exceptional, depending on well-defined boundary conditions; as shown by numerical calculations, in most cases shocks move along the flows as transient structures. It will be important to follow the evolution of observed knots to detect these proper motions.

(d) Terminal shocks.

The asymptotic behaviour of any outflow has to match to an external pressure, which although negligible is typically nonzero. Correspondingly the outflow reaches equilibrium against the external medium through a shock, whose position and strength is fixed by the overall set of boundary conditions. Some AGNs may have this terminal transition close to the core: then extended radio jets cannot break out and the shock is associated with BLRs. In other AGNs the terminal shock is responsible for the hot spots at the working surface of radio jets in the external medium; however the part of the outflow which is not collimated may feed an EELR.

(e) Thermal instabilities.

Thermal instabilties can arise in outflows when radiative losses favor the collapse of dense, cold clouds or filaments aligned to the magnetic field lines. This process is possibly associated with the filamentation and disruption of the homogeneous flow. A model of this filamentation has been proposed for extragalactic jets and radio lobes (Bodo *et al.* 1988).

6. Conclusions.

This review was intended to show that, although the study of outflows from AGNs is still in its infancy, the dynamical effects associated with the rich variety of observed morphologies have probably already been basically understood. However which processes operate in each specific sources is still undefined.

Further analytic studies of the fundamental physics must be combined with numerical experiments to test the models proposed. At the same time observations must proceed in such a way to organize a statistically complete picture which takes into account all the possible morphologies.

Acknowledgements. Support by the Italian Ministero della Pubblica Istruzione and Consiglio Nazionale delle Ricerche is gratefully acknowledged.

REFERENCES.

Arnold C.N., and Arnett W.D., 1986, *Astrophys. J. Lett.*, **305**, L57.

Begelman M.C., 1985, *Ann. N.Y. Acad. Sci.*, **470**, 51.

Begelman M.C., Blandford R., and Rees, M.J., 1984, *Rev. Mod. Phys.*, **56**, 255.

Bodo G., Ferrari A., Rosner R., and Knobloch E., 1989, *Astrophys. J.*, in press.

Bodo G., Ferrari A., Massaglia S., and Trussoni E., 1988, in *Plasma Astrophysics*, Proc. of Varenna Intl. School (ESA Scient. Publ.).

Falle S.A.E.G., and Wilson M.J., 1985, *Mon. Not. R. Astron. Soc.*, **216**, 971.

Ferrari A., 1985, in *Unstable Current Systems and Plasma Instabilities in Astrophysics* (Reidel: Dordrecht).

Ferrari A., 1987, in *Mass Outflows from Stars and Galactic Nuclei* (Reidel: Dordrecht).

Ferrari A., Trussoni E., Rosner R., and Tsinganos T., 1985, *Astrophys. J.*, **294**, 397.

Ferrari A., Trussoni E., Rosner R., and Tsinganos T., 1986, *Astrophys. J.*, **300**, 577.

Ferrari A., and Tsinganos K., 1986, *Can. J. Phys.*, **64**, 456.

Fosbury R.A.E., 1987, *Proc. 10th I.A.U. European Regional Meeting*, vol. IV, 351.
Holzer T.E., and Axford W.I., 1970, *Ann. Rev. Astron. Astrophys.*, **8**, 31.
Kopp R.A., and Holzer T.E., 1976, *Solar Phys.*, **49**, 43.
Liepman H.W., and Roshko R., 1957, *Elements of Gasdynamics* (New York: John Wiley).
Norman C., and Miley G.K., 1984, *Astron. Astrophys.*, **141**, 85.
Norman M.L., Smarr L., and Winkler K.H., 1982, *Astron. Astrophys.*, **113**, 285.
Orr M.J.L., and Browne I.W.A., 1982, *Mon. Not. R. Astron. Soc.*, **200**, 1067.
Parker E.N., 1958, *Astrophys. J.*, **128**, 664.
Rees M.J., 1984, *Ann. Rev. Astron. Astrophys.*, **22**, 471.
Trussoni E., Ferrari A., Rosner R., and Tsinganos K., 1988, *Astrophys. J.*, **325**, 417.
Uchida Y., and Shibata K., 1986, *Can. J. Phys.*, **64**, 507.
Woodward P.R. and Colella P., 1984, *J. Comp. Phys.*, **54**, 115.

FORMATION OF RADIO JETS AND LOBES, AND ENHANCED ENERGY LIBERATION AT AGN'S

Yutaka UCHIDA, and Hisashi HAMATAKE
Department of Astronomy, and Institute of Astronomy, University of Tokyo

ABSTRACT. A magnetodynamic model for the formation of jets and lobes of radio galaxies and quasars is presented with a tentative picture for the formation of galaxies from a magnetized medium. Interaction of the large scale magnetic field with the gravitationally contracting disk, which is rotating at the throat of an hourglass-shaped magnetic field, twists up the magnetic field, and the $j \times B$ force in the relaxing twist (non-linear torsional Alfven wave) drives the gas in the disk atmosphere into helical flows in the bipolar directions along the large scale magnetic field. The twist pinches the poloidal field along the axis on one hand, and, on the other hand, tuckes up the magnetic field into the shape of an umbrella with increasing pinch strength as the low Alfven velocity region near the edge of the "cavity" is approached. These explain the observed jets, lobes, hot spots, etc., in a natural way. The back-reaction of these on the disk means an efficient extraction of the angular momentum from the disk, allowing a high rate of accretion, and may explain the enhanced energy liberation at AGN's.

1. Introduction

Certain radio galaxies and quasars are known to have double radio lobes associated with them at a distance of a few hundred kpc on both sides of the central compact source (eg., Miley 1980). It is known that there exist stem-like parts (jets) connecting the lobes to the central object. High resolution radio observations by interferometers revealed recently that there are extremely thin jets connecting the lobes to the central source (Dreher *et al.* 1987) even in the cases in which the jets had not been observed. The Faraday rotation-corrected direction of the magnetic field is parallel to the jets in the high energy sources (Fanaroff-Riley type II), while it is parallel to the jets near the central source but tilted from a certain distance on in the low energy ones (Fanaroff-Riley type I) (Bridle and Perley 1984). The magnetic field in the lobes is complicate, but it is aligned along the periphery of them (Dreher *et al.* 1987), suggesting that the magnetic field is confining the lobes in some way or other.

In the earlier days the magnetic models were the most natural ones to conceive for the explanation of these radio features because the radio emission from these objects was known to have spectra of synchrotron radiation, and a variety of magnetic models, such as the expanding magnetic loop model, the magnetic cloud sling-shot model, etc, were proposed (Sturrock 1969, Piddington 1970, Ozernoy and Somov 1971). The paradigmatic model in the recent years, however, may be said to have been the one in which the initially *isotropic* powerful wind from AGN's is collimated by the polar funnels opened up along the axis of the accretion disk (Blandford and Rees 1974, Konigl 1982, see Ferrari and Pacholczyk 1983, Begelman *et al.* 1984). In this model, the wind is assumed to be due to some unknown activity at the central AGN, and the flow is collimated and accelerated through the de Laval nozzle which is thought to be formed in the funnel due to the cross-sectional as well as gravitational effects. There are a lot of works done along this line, some including relativistic effects (eg., Fukue 1982). Behavior of the thus-accelerated non-magnetic flows in the intergalactic medium, namely, the

G. Belvedere (ed.), Accretion Disks and Magnetic Fields in Astrophysics, 233–239.

confinement of them in a medium with much lower density, or, the stability of the flow against the Kelvin-Helmholz instability, has been discussed both analytically (eg., Ferrari *et al.* 1984, see Ferrari and Pacholczyk 1983, Begelman *et al.* 1984), and by numerical simulations (Norman *et al.* 1981).

A revival of the magnetic models was given by Blandford and Payne (1982) by making use of the centrifugal effect to accelerate the flow as in the magnetic sling-shot model. They tried to explain the fast outflows into the bipolar directions by first centrifugally accelerate the flows which is originated as a wind from the disk to higher velocities along the rigidly rotating flux tube whose direction is finally bent to the polar direction of the disk by the magnetic effect. Uchida and Hamatake (1988), based on the sweeping-magnetic-twist model of the star-formation bipolar flow proposed by Uchida and Shibata (1985) combined with the similarity consideration (Uchida and Shibata 1986), proposed a magnetic model which explains the formation of the jets and lobes of radio galaxies. In the following, we concentrate ourselves on this model and its extension.

2. Proposed Magnetodynamic Model for the Formation of Jets and Lobes of Radio Galaxies

Our standpoint is to extend our magnetodynamic model proposed for the problem of the bipolar flows from the forming stars (Uchida and Shibata 1985, 1986, Shibata and Uchida 1986) to the problem of the jets and lobes from the galaxies in the formation by using the following similarity rule in the MHD equation system. Namely, in the non-dimensionalized scale-free equation system which we solve, there appear three non-dimensional coefficients, ratios of the acoustic, Alfven, and rotation velocities to the Kepler velocity, all taken at a typical point in the coordinates scaled by the size of the system. Similar solution will be rendered in two different situations *if* the initial normalized distributions of physical quantities and force fields in the scaled coordinates are the same, *and if* these three non-dimensional coefficients appearing in the non-dimensionalized system of equations are in common, respectively (Uchida and Shibata 1986). No exact similarity is expected in the stellar and galaxy cases, of course, but a very rough "similarity" in the following sense may hold: namely, the initial normalized distribution of physical quantities and force fields are more or less like each other in the scaled coordinates in the sense that both may have an hourglass shaped large scale magnetic field squeezed by the contracting disk at its throat, rotating with the axis parallel to the field. The ratios (v_s^2/v_K^2 , v_A^2/v_K^2 , v_r^2/v_K^2) naturally do not have exactly the same values in these two cases, but the values of them may fall in the same domain fulfilling

$$v_s^2/v_K^2 \ , \ \ v_A^2/v_K^2 \ll 1 \ , \ \ v_r^2/v_K^2 \approx 1 \tag{1}$$

in the disk and its atmosphere, and

$$v_A^2/v_K^2 \gg 1 \ , \ \ v_s^2/v_A^2 \ll 1 \tag{2}$$

at certain scaled distance along the axis outside the disk.

The physical meaning of these are that the gravitational force is roughly balanced by the centrifugal effect of rotation, and these dominate the pressure and magnetic effects in the disk, while the magnetic effect dominates that of gravity and other effects from certain distance on outside the disk. In order to get the initial state, we tentatively assume a primordial state of $n_0 \sim 10^{-4}$ cm^{-3} , $T \sim 10^6$ K, and homogeneous poloidal magnetic field of $B_{op} \sim 10^{-9}$ G as considered by some researchers (Thomson and Nelson 1982). The contraction of the gas in a sphere with a radius ~ 100 kpc can provide the mass of the galaxy ~ 10^{11} M_o . Then, the density and the intensity of the magnetic field collected in a body of galactic scale will be $n_G \sim 10^{0-1}$ cm^{-3} and $B_{Gp} \sim 10^{-7}$ G, respectively, where the subscript p denotes poloidal. This value B_{Gp} may be consistent with the toroidal field of $B_{Gt} \sim 10^{-6}$ G to be produced and amplified by the galactic rotation. It should be noted that a "cavity" with a lowered density is left behind in the process of condensation of the central body. Next, we assume that the large part of the gas in this protogalaxy is freed from the original poloidal magnetic field by being turned into the stars (plus the part of the gas involved in the life cycle of the stars afterwards), and a few percent of it may drift down to the central part with the trapped poloidal magnetic field due to the magnetic braking which we discuss in more detail later. If 10^9 M_o gas contracts to a disk with a scale of, say, 100 pc, the density will become $n_C \sim 10^4$ cm^{-3} and the poloidal field becomes $B_{Cp} \sim 10^{-3}$ G . In this stage, we may very roughly estimate velocities (all in cm/s) as $v_A \sim 10^{6.5}$, $v_s \sim 10^6$, $v_r \sim v_K \sim 10^{7.5}$ in the disk, and $v_A \sim 10^{8.5-9}$, $v_s \sim 10^7$, and $v_K \sim 10^6$ at a distance outside the disk along the pole. Conditions (1) and (2) are fulfilled with these values, and flows like that in the star-forming bipolar flows may be expected in the present case (a "soft jet" according to Uchida and Shibata 1986), though the similarity is very crude.

Figure 1 shows a tentative example of the result of 2.5D MHD simulations by Uchida and Hamatake(1989).* It is seen from Figure 1a that the poloidal magnetic field near the axis is strongly bunched as the relaxing B_ϕ-packet propagates up (called by Uchida and Shibata as the sweeping pinch effect). This effect becomes strongest as the packet of B_ϕ approaches the edge of the "cavity" due to the accumulation of B_ϕ by the decrease in v_A there (Figure 1b). It is also seen that the weak field lines are tucked up in that phase into the shape of an umbrella. This is due to the enhancement of v_A due to the pinch effect near the axis even in the outer part of the "cavity" where v_A decreases to lower values otherwise. We identify the bunched axial field with the *jet*, the strongest pinch at the hub of the "umbrella" with the *hot spot*, and the part of the field tucked up in the form of an "umbrella" with the *lobe* of the radio galaxies, when they are lit up by synchrotron radiation by the injection of high energy electrons. The re-acceleration of the injected high energy electrons may be executed in our present picture by Fermi-I mechanism between the consecutive B_ϕ-packets which may be caused by the non-steady accretion of the inhomogeneous disk, because the late-coming packets have higher velocities than the ones approaching the edge of the cavity.

If we proceed further and assume that a blackhole is formed at the center since the magnetic braking effect continues to feed the mass to the central body so that the central body grows to a blackhole of 10^9 M_o surrounded by a compact disk of ~ 1 pc scale. If we retain 10% of the mass in the disk, then, $n_B \sim 10^{12}$ cm^{-3} and $B_{Bp} \sim 10^1$ G . In this stage, $v_A \sim 10^{7.5}$, $v_s \sim 10^{6-7}$, $v_r \sim v_K \sim 10^{8.2}$ in the disk, and $v_A \sim 10^{10}$, $v_s \sim 10^7$, and $v_K \sim 10^6$ outside the disk. These again fulfill conditions (1) and (2), and a similar process will take place, though can be pretty different in the appearance. The jets are expected to be very thin and energetic (a "hard jet" according to Uchida and Shibata 1986).

It should be remarked that the formation of the jets and lobes is not a mere ad-hoc phenomenon but quite intrinsic in the formation of the galaxies or blackholes in our present picture. More careful discussion of the formation of radio jets and lobes is given in Uchida *et al.*(1989), and the effect of the extraction of angular momentum to the enhancement of energy liberation of the central object is discussed in Uchida and Norman (1989).

3. Explanation of the Enhanced Energy Liberation at AGN's

In the previous models with the polar funnels of the disk collimating the otherwise isotropic wind from the central blackhole, the collimated flow cannot have a large angular momentum due to the point-like geometry of the source of the wind. The mechanism driving the energetic wind in these models was *unknown*, and the disk in these models did not play very positive roles but providing a rigid pipe (with

* It should be noted that a simulation with realistic scale ratios of ~ $10^{-3\sim-5}$, of the central body to the jets and lobes from AGN's is not possible by the present-day computers. Also the term ratios (v_A^2/v_K^2, etc) cannot be made extreme in order to avoid the dominance of numerical errors of a term over other terms. Thus, we should concentrate our attention only on characteristic behaviors in a qualitative way with appropriate scale ratios and term ratios in equations.

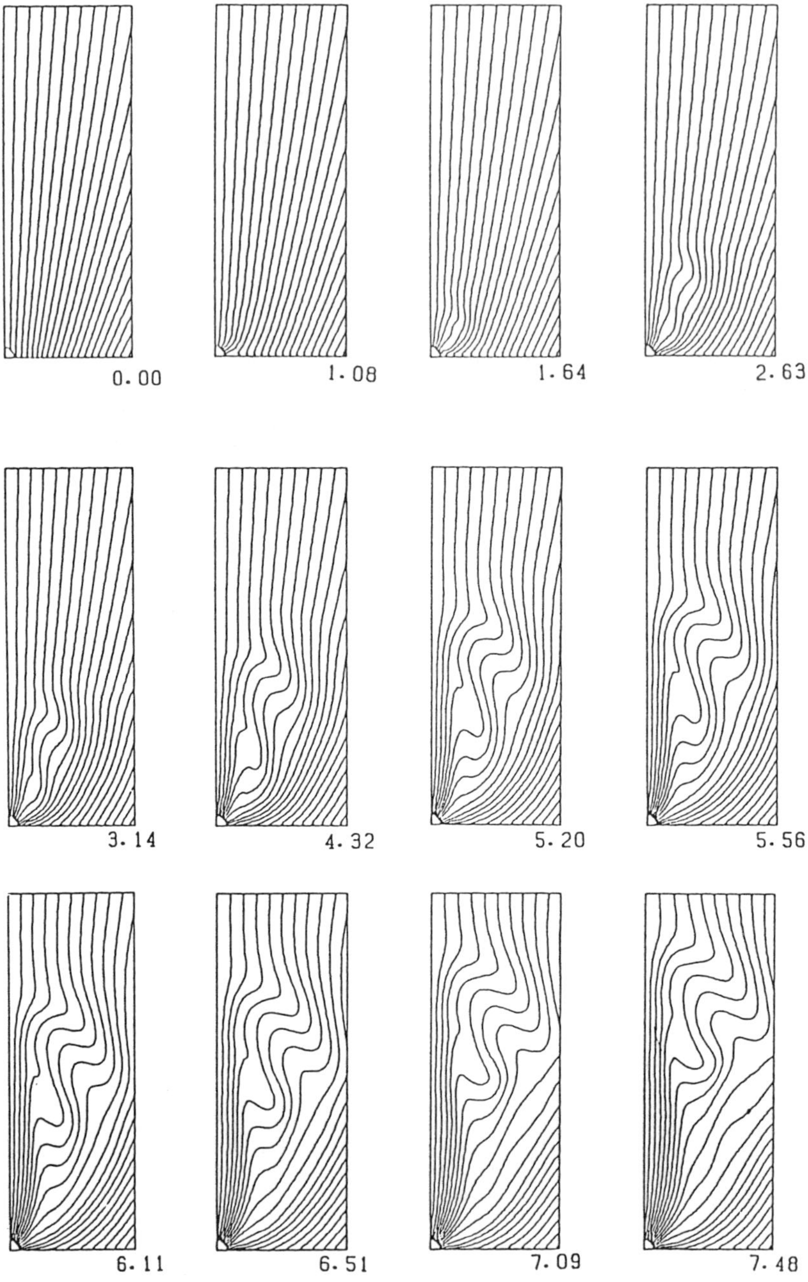

Fig.1(a) An example of the results of the 2.5D MHD simulations using appropriate parameters (Uchida and Hamatake 1989). Evolution in time of the poloidal magnetic field lines.

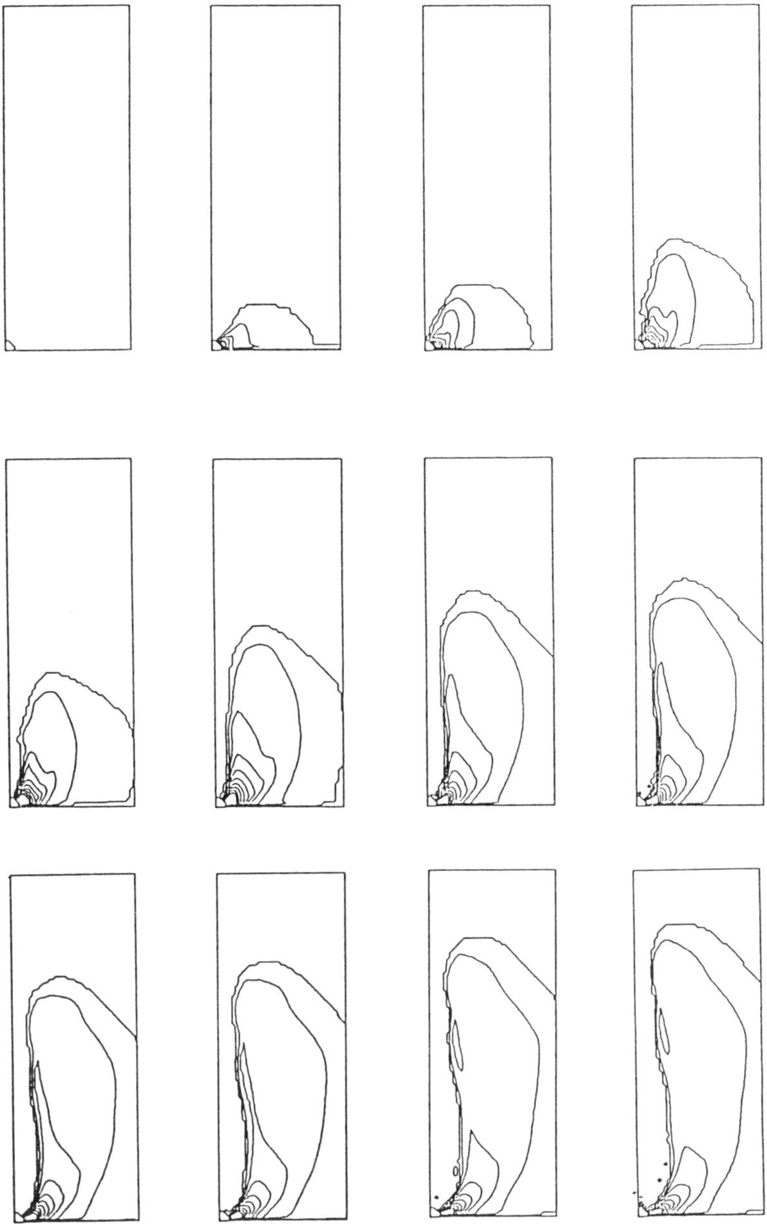

Fig.1(b) Same for the toroidal magnetic field corresponding to the magnetic twist. It is seen that the poloidal field along the axis is bunched as the packet of the toroidal field is produced and relaxes along the poloidal field, and the surrounding weak field lines are tucked up as the sweeping-pinch progresses and the low-V_A region at the edge of the "cavity" is approached.

a de Laval point) for the collimation of the flow.

In contrast, the flow does not come from the central source in our present model. It comes from the *disk* with the $j \times B$ force as the driving mechanism. This is caused by the interaction of the large scale poloidal field which we newly introduced in our proposed model with the (differentially) rotating disk. The magnetic field is passively carried around in the disk by the rotation of the dense part of the disk near the equatorial plane, and is twisted up. When the accumulated twist is relaxed along the poloidal field into the bipolar directions, it drives the material of the atmosphere of the disk into helical bipolar flows spinning in the same direction as that of the rotation of the disk. The angular momentum is transferred to the shells of much larger radius than that of the flow-driving part of the disk through the hourglass-shaped large scale magnetic field by the magnetic stresses. This point merited the magnetic models* when the indication of the spinning motion was found in the star-formation bipolar flow case (Uchida *et al.* 1987a, b). In the present case of galactic situation, it is seen from the result that the action of the foregoing twist wavepacket, rather than the mass flow, does play an important role.

If there is a transfer of angular momentum by magnetic stress, there must be a magnetic braking effect back-reacting on the disk material and this is estimated to be rather effective (Uchida and Norman 1989). It may largely exceed the α-viscosity of the so-called standard disk model (Shakura and Sunyaev 1973). If this is the case, it enhances the accretion rate and contributes in providing a large rate of conversion of gravitational energy into kinetic energy, and eventually into thermal energy near the central object. There are various aspects of the detailed way of energy conversion, for example, into radiation and further into pair creation of electrons, and so on, but it is of a prime importance to identify the main energy supply. In our picture, it may be the release of the gravitational energy due to enhanced angular momentum loss which back-reacts on the disk in the form of magnetic braking. We do not have to recur to unknown mechanism of energy release in the central object, but rather, we may be able to explain it by the present picture. Thus, the production of the jets and lobes is not an ad-hoc event with the AGN's, but it may be closely related to the essential problem of the energy release at AGN's.

4. Discussion

In the above, we have discussed the applicability of the magnetodynamical picture which was first applied for the bipolar jets around the forming stars (Uchida and Shibata 1985) and extended to the jets and lobes of AGN's. The analogy of the phenomena suggests that AGN's may have, at least, a major subclass consisted of the newly formed nuclei of galaxies. This will be consistent with the supposition that quasars are newly formed galaxies in the past. Our model takes the role of the interaction of the primordial large scale magnetic field with the disk into account to explain the enhanced energy liberation of the AGN's through the enhancement of accretion by magnetic braking.

We regard the presence of the wiggles in the jet part of radio galaxies like 3C449 (Perley *et al.* 1979) as evidence supporting the magnetic model (Hamatake *et al.* 1989, Tohdoh *et al.* 1989) rather than their precessional origin. The argument is that the helix should lie on the surface of a cone with a constant apex angle in a precessional model, but this is apparently not the case, eg., in 3C449. The jet part in 3C449 is straight up to certain point and then begins to wiggle, and the shapes of the jets and lobes do neither lie on the surface of a cone, nor are point symmetric as should be expected in the precession model. They are mirror symmetric! In our magnetic model, these are naturally explained as

* Of the magnetic models proposed thus far, the centrifugal wind model (Blandford and Payne 1982, Pudritz and Norman 1983) assumes a part of the flux tube rigidly rotating. The gas slides outwards along the tubes of force fixed in the frame rotating with the angular velocity of the footpoint. In contrast, in our model (Uchida and Shibata 1985), the deformation of the field line (magnetic twist) produced in the differentially rotating disk propagates dynamically as a non-linear torsional wavepacket driving material into a nonsteady flow. The velocity field in both models differs from each other characteristically, and will be discriminated by the future observations.

due to helical instability of a flux tube which may develop when the magnetic twist accumulates in a bounded region. In our case, the effect may be expected when the Alfven velocity decreases with distance, eg., on approaching the boundary of the "cavity" from which the mass has been evacuated to the central body.

Lastly, it may be worth pointing out that the mechanism we described in relation to the generation of the magnetic twist may be considered to be a variation of a non-steady astrophysical dynamo but now having open boundaries in our case through which a pair of *coaxial* current systems is supplied continuously into the bipolar directions. The non-force-free field characteristics are important in our dynamical model (cf. Lovelace *et al.* 1987). These coaxial current systems (packet of B_ϕ field) exert pinching effect which sweeps up a long distance until the region of low Alfven velocity (weak field and/or high density) is encountered. This is the case when the packet of B_ϕ encounters the edge of the "cavity" left after the galaxy formation in the radio galaxy case, and the wiggling of the jet or the tucking up of the field lines may result by the accumulation of the twist in our sweeping-magnetic-twist model when a criterion corresponding to the Kraskal-Shafranov criterion for the helical instability is violated.

References

Begelman, M.C., Blandford, R.D., and Rees, M.J., 1984, *Rev. Mod. Phys.*, **56**, 255.

Blandford, R.D. and Payne, D.G., 1982, *Mon. Not. Roy. Astron. Soc.*, **199**, 883.

Blandford, R.D. and Rees, M.J., 1974, *Mon. Not. Roy. Astron. Soc.*, **169**, 395.

Brindle, A.H. and Perley, R.A., 1984, *Ann. Rev. Astron. Astrophys.*, **22**, 319.

Dreher, J.W., Carilli, C.L., and Perley, R.A., 1987, *Astrophys. J.*, **316**, 611.

Ferrari, A., Habbal, S.R., Rosner, R., and Tsinganos, K., 1984 *Astrophys. J.*, **227**, L35.

Ferrari, A. and Pacholczyk, A.G. (eds.), 1983, *Astrophysical Jet* (D. Reidel).

Fukue, J., 1982, *Publ. Astron. Soc. Japan*, **34**, 163.

Hamatake, H., Uchida, Y., Ferrari, A., and Sato, T., 1989, in preparation.

Konigl, A., 1982, *Astrophys. J.*, **261**, 115.

Lovelace, R.V.E., Wang, J.C.L., and Sulkanen, M.E. 1987, *Astrophys. J.*, **315**, 504.

Miley, G., 1980, *Ann. Rev. Astron. Astrophys.*, **18**, 165.

Norman, M., Smarr, L., Wilson, J.R., and Smith, M.D., 1981, *Astrophys. J.*, **247**, 52.

Ozernoy, L.M. and Somov, B.V., 1971, *Astrophys. Sp. Sci.*, **11**, 264.

Perley, R.A., Willis, A.G., and Scott, J.S., 1979, *Nature*, **281**, 437.

Pudritz, R. and Norman, C.A., 1983, *Astrophys. J.*, **274**, 677.

Piddington, J.H., 1970, *Mon. Not. Roy. Astron. Soc.*, **148**, 131.

Shakura, N.I. and Sunyaev, R.A., 1973, *Astron. Astrophys.*, **24**, 337.

Shibata, K. and Uchida, Y., 1986, *Publ. Astron. Soc. Japan*, **38**, 631.

Sturrock, P.A., 1969, in *Quasars and High Energy Astronomy*, eds. K.N. Douglas et al. (Gordon and Breach), p147.

Thomson, R.C. and Nelson, A.H., 1982, *Mon. Not. Roy. Astron. Soc.*, **201**, 365.

Tohdoh, Y., Uchida, Y., Rosner, R., and Sato, T., 1989, in preparation.

Uchida, Y. and Hamatake, H., 1989, in *Big Bang, Active Galactic Nuclei and Supernovae*, eds. Hayakawa, S. and Sato, K. (Universal Academy Press), in press.

Uchida, Y., Hamatake,H., Norman,C., and Shibata,K., 1989, in preparation.

Uchida, Y., Kaifu, N., Shibata, K,. Hayashi, S.-S., and Hasegawa, T., 1987a, in *Star Forming Regions*, ed. Peimbert, M. and Jugaku, J. (D. Reidel), p287.

Uchida, Y., Kaifu, N., Shibata, K., Hayashi, S.-S., Hasegawa, T., and Hamatake, H., 1987b, *Publ. Astron. Soc. Japan*, **39**, 907.

Uchida, Y., and Norman, C., 1989, in preparation.

Uchida, Y. and Shibata, K., 1985, *Publ. Astron. Soc. Japan*, **37**, 515.

Uchida, Y. and Shibata, K., 1986, *Can. J. Phys.*, **64**, 507.

RELATIVISTIC RADIATIVE TRANSFER USING MOMENT FORMALISM

ROBERTO TUROLLA and LUCIANO NOBILI
Department of Physics, University of Padova
Via Marzolo 8, 35131 Padova, Italy

ABSTRACT. An analysis of Thorne's (1981) structure equations governing radiation field moments is presented in the case of stationary, spherically symmetric flows. It is shown that the truncated system of moment equations must be solved subject to boundary conditions and a number of critical points equal to the truncation order l_{max} arises. The moment formalism can be made tractable once a suitable closure assumption for moments of order higher than l_{max} is introduced. A sample numerical integration of the moment equations is performed for a spherical, accreting flow with an assigned velocity and temperature profile.

1. INTRODUCTION

The increasing observational evidence for high–speed flows in astrophysical objects (jets, accretion processes both in galactic and extragalactic sources) has prompted in the past few years new interest in relativistic radiation hydrodynamics. Any investigation in this field suffered, however, the lack of an efficient technique for handling properly the transition region where the photons mean–free path becomes comparable to relevant macroscopic lengthscales. All the major theoretical efforts spent to obtain a fully relativistic description of radiative transfer in differentially moving media, valid also outside the diffusion limit (e.g. Anderson & Spiegel 1972; Schimd–Burgk 1978; Thorne 1981; Udey & Israel 1982; see also Schweizer 1982, 1984a,b for a discussion of the diffusion regime), led in fact to very complex formalisms, so complex that no one of them seemed to be of any practical use in astrophysical applications. Among them, Thorne's PSTF moment formalism (Thorne 1981) seems nevertheless worthy of further investigations for a number of reasons. Its complexity is in fact reduced by far for stationary flows with spatial symmetries and using the frequency integrated version. In this case the radiation moments are just scalars and, for planar and spherical flows (Thorne, Flammang & Żytkov 1981 hereafter TFZ), the evolution equations are linear ordinary differential equations. Moreover, source moments were already computed by Thorne for a variety of radiative processes of astrophysical relevance, so any interesting input physics will automatically fit into the formalism.

Here we show that PSTF moment formalism is indeed tractable and works correctly at all optical depths once a convenient closure approximation is used. An

G. Belvedere (ed.), Accretion Disks and Magnetic Fields in Astrophysics, 241–246.

analysis of the moment equation is carried out and shows that l_{max} critical points are present in the system of differential equations, l_{max} being the truncation order. Singular points are encountered when the flow velocity equals the zeros of Legendre polynomial $P_{l_{max}+1}(v/c)$. The boundary conditions and the choice of the closure approximation are then discussed and the numerical integration of moments equations is presented in the specific case of a spherical infall on to Schwarzschild black hole with assigned velocity and temperature profiles; source terms include electron scattering and bremsstrahlung. Results indicate that $6 \div 8$ moments are enough to give a satisfactory description of the radiation field.

2. MOMENT EQUATIONS

PSTF moment expansion of the radiation field can be regarded, in many respects, as a modern, highly sophisticated version of old moment formalism (Chandrasekhar 1960 and references therein), introduced more than forty years ago for studying spherically symmetric, static, scattering atmospheres. In treating relativistic radiative transfer in a differentially moving medium, complexities are, of course, much greater, expecially if a completely general formalism, which could be applied to situations without particular symmetries, is developed. Thorne succeed in building such a formalism by using projected, symmetric, trace–free (PSTF) tensors to describe radiation moments. The evolution equations are then derived by evaluating the moments of the relativistic transfer equation. In the general case the l–th moment is a PSTF tensor of rank l: $\mathcal{M}_{\nu}^{\alpha_1 \dots \alpha_l}(x^{\beta})$, $\beta = 0, 1, 2, 3$; moments of order 0, 1, 2, have a precise physical meaning and represent the radiation energy density, energy flux and shear stress in the local rest frame of the fluid.

We shall deal in the following with stationary, spherical flows in a Schwarzschild spacetime, treating only the frequency independent problem. For this particular case all moments are scalars and depend only on the radial coordinate. The corresponding equations were already contained in Thorne's paper and are discussed in more details in TFZ. The l–th moment of the transfer equation reads

$$\begin{aligned}
&w'_{l+1} - vw'_l + a_l w'_{l-1} + v(l-1)\left(\frac{v'}{v} + \frac{y'}{y} - \frac{1}{r}\right) w_{l+2} + \\
&\left[\frac{l+2}{r} - (l-2)\frac{y'}{y}\right] w_{l+1} - v\left[\frac{4}{3}\left(\frac{v'}{v} + \frac{y'}{y} + \frac{2}{r}\right) + \right. \\
&\left.\frac{5l(l+1)}{3(2l-1)(2l+3)}\left(\frac{v'}{v} + \frac{y'}{y} - \frac{1}{r}\right)\right] w_l - a_l\left[\frac{l-1}{r} - (l+3)\frac{y'}{y}\right] w_{l-1} - \\
&v a_l a_{l-1}(l+2)\left(\frac{v'}{v} + \frac{y'}{y} - \frac{1}{r}\right) w_{l-2} + \frac{\varrho_0}{y}(\kappa_l w_l - \epsilon_l) = 0,
\end{aligned} \tag{1}$$

where the prime denotes derivation with respect to r and $a_l = l^2/(4l^2 - 1)$, $y = [(1 - 2M/r)/(1 - v^2)]^{1/2}$. The quantity $-\varrho_0(\kappa_l w_l - \epsilon_l)$ is the l–th moment of source function and describes all matter–radiation interactions; κ_l and ϵ_l are the opacity and the emissivity for the l–th moment of radiation, respectively. Here, and in the following, units in which $c = G = 1$ are used, M denotes the mass of the central source and v is the flow speed as measured by a stationary observer; v is positive for inward motions. Of course the moment expansion must be truncated at some maximum order l_{max}, introducing at the same time a closure hypothesis, which involves both $w_{l_{max}+1}$ and $w_{l_{max}+2}$. The infinite hierarchy of equations (1) reduces consequently to a finite system of $l_{max} + 1$ linear, ordinary differential equations, where $w_{l_{max}+1}$ and $w_{l_{max}+2}$ are to be regarded as known functions of r.

It can be proved (see Turolla and Nobili 1988 for details) that the determinant of coefficients is proportional to the Legendre polynomial of order $l_{max}+1$ in v and since $P_{l_{max}+1}(v)$ vanishes $l_{max} + 1$ times in the interval $-1 \leq v \leq 1$, the system (1) contains $l_{max} + 1$ singularities in correspondence to the points where the flow speed equals the zeros of $P_{l_{max}+1}(v)$. If the actual flow velocity spans a range which contains all the zeros of $P_{l_{max}+1}$, the radiation field is uniquely determined by the $l_{max}+1$ regularity conditions that must be imposed at the critical points. However, when not all the critical points are encountered, some of, or even all, the $l_{max} + 1$ conditions must be specified elsewhere in the flow and their form will depend on the particular problem under examination.

3. BOUNDARY CONDITIONS AND CLOSURE

For the sake of simplicity we shall deal in the following only with isotropic emission processes, so that $\epsilon_l = 0$ for $l \geq 1$, and assume that κ_l and ϵ_0 are regular functions of r which reduce to pure scattering at radial infinity. We shall refer moreover, to a flow which becomes optically thick with respect to true emission–absorption at small radii, that is to say, we imagine a situation in which LTE always holds near the central source.

We start the discussion about boundary conditions noticing that in any case some of the free conditions should be given at radial infinity where optical depth τ vanishes , since we know that the physical solution of the transfer equation is characterized by the vanishing of all moments, together with their derivatives, there. Moreover, if LTE holds when $r \to 0$, moments must behave in such a way that $w_0 > w_1 > w_2 \ldots > w_{l_{max}}$ there. By assuming, for definiteness, that l_{max} is even it is possible to prove that the only solution of moment equations which satisfies the above mentioned boundary conditions must obey $(l_{max}+1)/2$ constraints when $r \to \infty$ and $(l_{max} + 1)/2$ ones when $r \to 0$, if no critical points are present. The former must be imposed to ensure that even moments go smoothly to zero at radial infinity, while the latter are required to prevent odd moments to grow unbounded when $\tau \gg 1$. In general, therefore, the solution of moment equations is a genuine boundary value problem in which, if a velocity field is present, some of the conditions can be transferred from the endpoints to critical points.

As we already mentioned, moment equations can be solved only once a *ad hoc* assumption for $w_{l_{max}+1}$ and $w_{l_{max}+2}$ is prescribed. Clearly the simplest closure consists in asking that

$$w_{l_{max}+1} = w_{l_{max}+2} = 0. \tag{2}$$

Equation (2) represents, moreover, the correct closure in photons diffusion regime but becomes utterly inadequate when the flow becomes optically thin or even marginally thick. At radial infinity, where $\tau \to 0$, photons stream radially and all moments must be proportional to the energy flux $w_1 \sim r^{-2}$:

$$w_{l\infty} = \frac{l!(2l+1)}{(2l+1)!!} w_1 \tag{3}$$

(TFZ), while, by using (2), even moments do not follow the inverse square law. The correct asymptotic behaviour can be recovered just asking that both $w_{l_{max}+1}$ and $w_{l_{max}+2}$ are given by (3), but such a closure would work however only at large radii, where optical depth is much smaller than unity, since in the diffusion limit it would produce $w_{l_{max}+1} > w_1$. A reasonable compromise can be therefore to assume that (3) holds for $w_{l_{max}+1}$ and $w_{l_{max}+2}$ when $\tau \lesssim 1$ and then both moments drop rapidly to zero for increasing optical depth, so that the correct asymptotic behaviour of the solutions is preserved in both the limits $\tau \gg 1$ and $\tau \ll 1$.

4. A NUMERICAL EXAMPLE

In order to test the applicability of moment formalism to real astrophysical problems we have solved numerically moment equations in the particular case of spherical accretion onto a Schwarzschild hole with assigned density and temperature profiles :

$$\varrho_0 \propto r^{-n}, \qquad T \propto \varrho_0^m. \tag{4}$$

All the other quantities characterizing both the hydrodynamics of the flow and emission–absorption processes are then easily derived, if only bremsstrahlung is taken into account in addition to electron scattering. The integration was carried out in the range $\tau_\infty \le \tau \le \tau_h$, $\tau_\infty \ll 1$, using electron scattering optical depth as the independent variable. As it was discussed, half of the boundary conditions need to be specified at $\tau = \tau_\infty$ and can be cast in the form

$$\frac{d\ln w_0}{d\ln\tau} = \frac{2}{n-1}, \qquad \frac{d\ln w_{l+1}}{d\ln\tau} = \frac{d\ln w_{l+1}}{d\ln\tau} \tag{5}$$

(l odd), while the remaining $(l_{max}+1)/2$ constraints must be necessarily imposed at the critical points, since the same number of zeros of $D_{l_{max}}$ is crossed for $0 \le v \le 1$. Since all singular points are met in the inner, optically thick region critical point conditions have the same form of boundary conditions at $\tau \gg 1$:

$$\frac{w_l'}{w_l} = \frac{w_{l-1}'}{w_{l-1}} - \frac{y}{\tau}\frac{\kappa_{es}}{\kappa_l}\left(\frac{\tau}{y}\frac{\kappa_l}{\kappa_{es}}\right)', \qquad \frac{w_{l+1}'}{w_{l+1}} = \frac{w_l'}{w_l} - \frac{v'}{v} \tag{6}$$

(1 odd, $' \equiv d/d\ln\tau$). The result is shown in figure 1, which illustrates the run of radiation moments vs. optical depth for $l_{max} = 5$. The present case refers to $n = 1.5$, $m = 0.2$ and $T = 10^7$ °K, $\varrho_0 = 5$ gcm^{-3} at the horizon. An estimate of the errors, introduced both by the truncation and by our particular choice of the closure, can be obtained comparing results of integrations with different values of l_{max}. In going from $l_{max} = 1$ to $l_{max} = 5$ the maximal relative change in w_0 and w_1 reduces form $\sim 11\%$ to $\sim 1\%$. This result indicates that convergence is quickly reached and eight moments are enough to describe correctly the radiation field in all regimes, with an accuracy $\sim 1\%$.

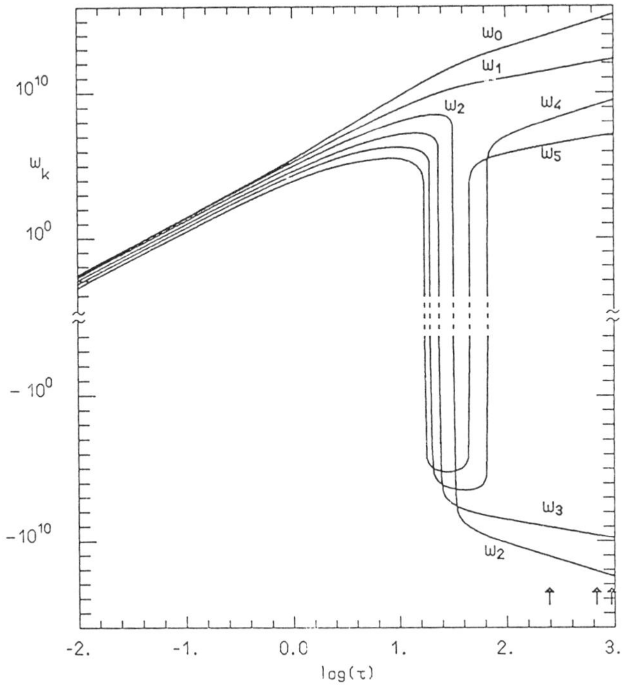

Figure 1. The run of radiation moments vs. optical depth for $l_{max} = 5$. Arrows mark the location of critical points. Moments are in units of $erg\,cm^{-3}$.

5. REFERENCES

Anderson, J.L. & Spiegel, E.A., 1972. *Astrophys. J.*, **171**, 127

Chandrasekhar, S., 1960. *Radiative transfer* (New York: Dover)

Schmid-Burgk, J., 1978. *Astrophys. Space Sci.*, **32**, 403

Schweizer, M.A., 1982. *Astrophys. J.*, **258**, 798

Schweizer, M.A., 1984a. *Astrophys. J.*, **280**, 809

Schweizer, M.A., 1984b. *Mont. Not. R. astr. Soc.*, **210**, 303

Thorne, K.S., 1981. *Mont. Not. R. astr. Soc.*, **194**, 439

Thorne, K.S., Flammang, R.A. & Żytkov, A., 1981. *Mont. Not. R. astr. Soc.*, **194**, 475
Turolla, R., Nobili, L., 1988. *Mont. Not. R. astr. Soc.*, in the press
Udey, N. & Israel, W., 1982. *Mont. Not. R. astr. Soc.*, **199**, 1137

FLUID MODELS FOR RELATIVISTIC WARM PLASMAS

A.M.Anile
Dipartimento di Matematica
Universita' di Catania
(Italy)

Relativistic warm plasmas are of great interest in several areas of astrophysics (e.g.stellar winds[1], pulsar magnetospheres[2], collisionless accretion onto neutron stars and black holes [3]) as well as in laboratory plasma physics[4]. The models utilized for describing warm plasmas are based on a kinetic approach starting from the relativistic Vlasov - Maxwell system[4]. Whereas for linear problems a kinetic approach is analytically tractable, in the non linear case one has to resort to numerical simulation. However for many problems a reduced description in terms of fluid quantities could be adequate, leading to problems which are analytically tractable. For this reason several fluid models have been introduced starting from the simplest cold fluid one (zero temperature)[4]. Warm plasmas consisting of finite temperature electron beams have been modeled among others, by Toepfer [5] assuming that the electrons have a thermal distribution. More adequate models have been obtained by Siambis[6] and by Newcomb[7] by taking the moments of the relativistic Vlasov equation and truncating in an appropriate way .

The beam is then modeled as a relativistic fluid with a so-called generalized state equation . The models of Siambis and of Newcomb, however, are not manifestly covariant, and in a relativistic theory this is unsatisfactory. A relativistic covariant model along these lines has been proposed by and investigated by Amendt and Weitzner[8,9]. Amendt[10] has also made a detailed study of the linear wave modes.

The Amendt and Weitzner model however has the drawback that the constraints on the moments arising from the requirement that the moments derive from a distribution function are satisfied only approximately.

In this paper we introduce a fluid model which solves the above drawback of the Amendt and Weitzner one in that the constraint equations are *satisfied exactly* . As a bonus of our method we find that our model leads to a *symmetric*

G. Belvedere (ed.), Accretion Disks and Magnetic Fields in Astrophysics, 247–254.

hyperbolic system (for symmetric hyperbolic systems one has a much more satisfactory mathematical theory [11]).

Our approach is based on the entropy principle method which was originally expressed and utilized in the theory of relativistic extended thermodynamics by Liu, Müller and Ruggeri [12]. The entropy principle method consists in assuming that a supplementary conservation law exists for the truncated system of moment equation and that such a law is a mathematical consequence of the moment equations. This requirement puts severe restrictions on the possible forms of closure for the truncated moment equations (and in fact, jointly with other reasonable assumptions, almost determines a unique closure).

Let u^{μ} denote the four velocity of a particle, which, in inertial coordinates (x^{μ}), writes

$$u^{\mu}=\Gamma(v^{i},1)$$

where $\Gamma = (1-v^{i}v_{i})^{-1/2}$ is the Lorentz factor and v^{i} the three velocity. Let $u^{i}=\Gamma v^{i}$ and introduce the one particle invariant distribution function $f(x^{\mu}, u^{i})$. Then the Lorentz covariant Vlasov equation writes [8]

$$u^{\mu}\frac{\partial f}{\partial x^{\mu}} + \frac{e}{m}F^{i\mu}u_{\mu}\frac{\partial f}{\partial u^{i}} = 0 \qquad (1)$$

with e the electric charge, m the particle mass . The following quantities related to the moments of the distribution function f can be defined

$$h=\int f\omega \qquad (2)$$

$$hw^{\mu}=\int fu^{\mu}\omega \qquad (3)$$

$$\Theta^{\mu\nu}=\int f(u^{\mu}-w^{\mu})(u^{\nu}-w^{\nu})\omega \qquad (4)$$

$$S^{\mu\nu\lambda}=\int f(u^{\mu}-w^{\mu})(u^{\nu}-w^{\nu})(u^{\lambda}-w^{\lambda})\omega \qquad (5)$$

where $\omega=\dfrac{du^{1}du^{2}du^{3}}{u^{0}}$ is the invariant measure on the velocity space.

The quantity h can be interpreted as an invariant particle density, w^{μ} as an average four-velocity.

From the definitions (4) and (5) the following relationships are obtained

$$\Theta^{\mu}_{\ \mu} = - h\ (1 + w^{\mu} w_{\mu}) \tag{6}$$

$$S^{\mu\nu}_{\ \ \nu} = - 2\ \Theta^{\mu\nu} w_{\nu} \tag{7}$$

By taking the moments of the Vlasov equation (1), assuming that f vanishes sufficiently fast at infinity in the velocity space, we obtain

$$\frac{\partial h w^{\mu}}{\partial x^{\mu}} = 0 \tag{8}$$

$$\frac{\partial}{\partial x^{\mu}} (h\ w^{\mu}\ w^{\nu} + \Theta^{\mu\nu}) = \frac{e}{m}\ h\ F^{\mu\nu}\ w_{\mu} \tag{9}$$

$$\frac{\partial}{\partial x^{\mu}} (h\ w^{\mu} w^{\nu} w^{\alpha} + w^{\mu}\Theta^{\nu\alpha} + w^{\nu}\Theta^{\mu\alpha} + w^{\alpha}\Theta^{\mu\nu} + S^{\mu\nu\alpha}) =$$

$$= \frac{e}{m} (\ F^{\nu\mu}\Theta^{\alpha}_{\ \mu} + F^{\alpha\mu}\Theta^{\nu}_{\ \mu} + F^{\nu\mu} w_{\mu} h w^{\alpha} + F^{\alpha\mu} w_{\mu} h w^{\nu}) \tag{10}$$

The set of equations (8),(9),(10) obviously is not a closed system. In order to close it one must introduce constitutive equations for $S^{\mu\nu\lambda}$ on the basis of some approximation scheme and determine their range of validity.

The basic physical approximation will be that the spread in momentum space is very small, expressing the concept of warm plasma. More precisely we introduce the following quantity

$$\varepsilon^2 = -(1 + w^{\mu} w_{\mu}) \tag{11}$$

and define a warm plasma by the requirements : i) $\varepsilon \ll 1$ ii) in the local rest frame of the plasma, in which

$$w^{\mu} = (0,0,0\overline{w}^4) \quad \text{with} \quad \overline{w}^4 = \sqrt{1+\varepsilon^2},$$

the following representation holds

$$\frac{\overline{\Theta}^{ij}}{h} = [\overline{K}^{ij} + \frac{1}{3}\ (1 - \overline{K}^{l}_{\ l})\delta^{ij}]\varepsilon^2 + \frac{1}{3}\ \overline{K}^{44}\ \delta^{ij}\varepsilon^4$$

$$\frac{\overline{\Theta}^{i4}}{h} = \overline{K}^{i4}\varepsilon^3 \quad , \quad \frac{\overline{\Theta}^{44}}{h} = \overline{K}^{44}\varepsilon^4$$

$$\frac{\bar{S}^{ijk}}{h} = [\bar{K}^{ijk} - \frac{3}{5}\,\delta^{(ij}\,\bar{K}^{k)l}{}_{l}]\varepsilon^{3} + \frac{6}{5}\delta^{(ij}\bar{K}^{k)4}\varepsilon^{3}\sqrt{1+\varepsilon^{2}} + $$

$$+\frac{3}{5}\,\delta^{(ij}\bar{K}^{k)44}\varepsilon^{5} \tag{12}$$

$$\frac{\bar{S}^{ij4}}{h} = [\bar{K}^{ij4} - \frac{1}{3}\,\bar{K}^{4l}{}_{l}\delta^{ij}]\varepsilon^{4} + \frac{2}{3}\,\delta^{ij}\bar{K}^{44}\varepsilon^{4}\sqrt{1+\varepsilon^{2}} + $$

$$+ \frac{1}{3}\,\delta^{ij}\bar{K}^{444}\varepsilon^{6}$$

$$\frac{\bar{S}^{i44}}{h} = \bar{K}^{i44}\varepsilon^{5}, \qquad \frac{\bar{S}^{444}}{h} = \bar{K}^{444}\varepsilon^{6}$$

where overbars denote components in the plasma local rest frame and $K^{\mu\nu}$, $K^{\mu\nu\lambda}$ are dimensionless tensors at most of order zero in ε, such that $K^{\mu\nu} = K^{(\mu\nu)}$, $K^{\mu\nu\lambda} = K^{(\mu\nu\lambda)}$. It is easy to check that the above ordering satisfies the constraints (6) and (7) exactly.

It is obvious that the ordering (12) is based upon the physically intuitive idea that, for a sufficiently collimated beam, the higher order off-center moments are of decreasing order of magnitude with respect to the *spread parameter* ε. It is convenient to introduce also another set of variables.

Let $p^{\mu}=mu^{\mu}$ be the four momentum of a particle. Then, we define

$$V^{\mu} = \int fp^{\mu}\,\pi \tag{13}$$

$$T^{\mu\nu} = \int fp^{\mu}p^{\nu}\pi \tag{14}$$

$$A^{\mu\nu\alpha} = \int fp^{\mu}p^{\nu}p^{\alpha}\pi \tag{15}$$

where $\pi = \dfrac{dp^{1}dp^{2}dp^{3}}{p^{4}}$ and f is the distribution function.

The relationship with the former variables is

$$V^{\mu} = m^{3} h \; w^{\mu} \tag{16}$$

$$T^{\mu\nu} = m^{4} \left(\Theta^{\mu\nu} + h \; w^{\mu} w^{\nu} \right) \tag{17}$$

$$A^{\mu\nu\alpha} = m^{5} \left(S^{\mu\nu\alpha} + 3 \; w^{(\nu} \Theta^{\mu\alpha)} + h w^{\mu} w^{\nu} w^{\alpha} \right) \tag{18}$$

and the constraints (6), (7) read

$$T^{\mu}{}_{\mu} = -hm^{4}, A^{\mu\nu}{}_{\nu} = -m^{2} V^{\mu}.$$

The energy-momentum tensor $T^{\mu\nu}$ can be decomposed in the usual way

$$T^{\mu\nu} = E \; U^{\mu} U^{\nu} + p h^{\mu\nu} + 2 U^{(\mu} q^{\nu)} + t^{\langle\mu\nu\rangle} \tag{19}$$

where

$$U^{\alpha} = V^{\alpha} \left(-V^{\mu} V_{\mu} \right)^{-1/2}$$

$$n = \sqrt{-V^{\mu} V_{\mu}}$$

is the rest frame particle number density,

$$h^{\mu\nu} = g^{\mu\nu} + U^{\mu} U^{\nu}$$

is the projection tensor onto the 3-space orthogonal to U^{α},

$$p = \frac{1}{3} h_{\mu\nu} T^{\mu\nu}$$

is the isotropic pressure,

$$E = T^{\mu\nu} U_{\mu} U_{\nu}$$

is the total rest frame energy density,

$$q^{\alpha} = -h^{\alpha}_{\mu} U_{\nu} T^{\mu\nu}$$

is the heat flux vector,

$$t^{\langle\alpha\beta\rangle} = \left[h^{\alpha}_{\mu} h^{\beta}_{\nu} - \frac{1}{3} h^{\alpha\beta} h_{\mu\nu} \right] T^{\mu\nu}$$

is the stress tensor.

In terms of the former variables we have

$$p=\frac{1}{3}\, m^4 h_{\mu\nu}\theta^{\mu\nu}$$

$$E=hm^4+m^4 h_{\mu\nu}\theta^{\mu\nu}$$

$$q^{\alpha}=-\frac{1}{\sqrt{-w^{\mu}w_{\mu}}}\, m^4 h_{\mu}w_{\nu}\theta^{\mu\nu}$$

$$t^{\langle\alpha\beta\rangle}= m^4\left[h^{\alpha}_{\mu}h^{\beta}_{\nu}-\frac{1}{3}\, h^{\alpha\beta}h_{\mu\nu}\right]\theta^{\mu\nu}$$

$$n=m^3 h\sqrt{-w^{\mu}w_{\mu}}$$

$$U^{\alpha}=\frac{1}{\sqrt{-w^{\mu}w_{\mu}}}\, w^{\alpha}$$

The closure method of Amendt and Weitzner[8] consists in neglecting $S^{\mu\nu\alpha}$ altogether and this forces the constraint (7) not to be satisfied exactly (with their closure the truncated moment equations constitute, with reasonable restrictions on the range of variables, a hyperbolic system). Instead we look for $S^{\mu\nu\alpha}$ as function of the previous moments h, w^{α}, $\theta^{\mu\nu}$ such that it *satisfies the constraint* (7) exactly and *preserves the physical ordering* (*12*). Therefore,we shall look for a "constitutive function"

$$S^{\mu\nu\alpha}=\hat{S}^{\mu\nu\alpha}(h,\ w^{\beta},\ \theta^{\sigma\beta}) \tag{20}$$

such that:

i) the constraint (7) is satisfied exactly,

ii) the ordering (12) is preserved.

It is more convenient to use the variables V^{μ}, $T^{\mu\nu}$, $A^{\mu\nu\alpha}$ and seek "constitutive functions"

$$A^{\mu\nu\alpha}=\hat{A}^{\mu\nu\alpha}(V^{\lambda},\ T^{\lambda\rho}) \tag{21}$$

such that the constraint: i) $A^{\mu\nu}{}_{\nu}=-m^2V^{\mu}$ is verified and, ii) the corresponding ordering (12) for $S^{\mu\nu\alpha}$ is preserved.

We shall consider the case when the plasma is near thermal equilibrium . Then if we take into account only condition i) the problem is exactly the same as in the case of relativistic extended thermodynamics (apart from a minor difference arising from the coupling to Maxwell's equations) and we can apply the constitutive functions obtained by Liu, Müller and Ruggeri[12], which yields

$$A^{\mu\nu\alpha}=(C_1'^{0}+C_1^{\pi}p)\left[U^{\alpha}U^{\mu}U^{\nu}+\frac{1}{2}g^{(\alpha\mu}U^{\nu)}\right]-$$

$$-\frac{nm^2}{2}g^{(\alpha\mu}U^{\nu)}-3C_3[g^{(\alpha\mu}+6U^{(\alpha}U^{\mu}]q^{\nu)}+3C_5t^{\langle\langle\alpha\mu\rangle}U^{\nu)} \quad (22)$$

where the quantities $C_1'^{0}$, C_1^{π}, C_3, C_5 depend only on the thermal equilibrium variables n, T.

The coefficients in eq. (22) are given in the article by Liu, Müller & Ruggeri[12].

From the corresponding expression for $A^{\mu\nu\alpha}$ we can now obtain $S^{\mu\nu\alpha}$ and finally we can check condition ii), i.e. that the ordering (12) holds.

From the definition of total energy density $E=hm^4+m^4h_{\mu\nu}\theta^{\mu\nu}$ we obtain

$$G-\frac{1}{\gamma}=\frac{h^{-1}h_{\mu\nu}\theta^{\mu\nu}+1}{\sqrt{-w^{\mu}w_{\mu}}}\,. \quad (23)$$

Now, assuming the ordering of eq. (12) *only for* $\theta^{\mu\nu}$ we get

$$G-\frac{1}{\gamma}=\left[1+\varepsilon^2\right]^{-1/2}(1+\varepsilon^2+\varepsilon^4\bar{K}^{44}) \quad (24)$$

whence we see that the limit $\varepsilon\longrightarrow 0$ corresponds to $\gamma\longrightarrow\infty$.

A long series of calculations[13] shows that by assuming the ordering (12) only for $\theta^{\mu\nu}$. then the ordering for $S^{\mu\nu\alpha}$ follows. Therefore expression (22) for $A^{\alpha\beta\gamma}$ provides the sought for closure for the system of moment equations (8),(9),(10).

With the closure given by eq. (22), one also obtains that, near thermal equilibrium, the resulting system is equivalent to a symmetric hyperbolic one[12] which is, from the mathematical view point, a much stronger result than mere hyperbolicity.

An explicit expression for the leading term in ε of $S^{\alpha\beta\gamma}$ in terms of the variables h, w^{μ}, $\theta^{\mu\nu}$ can also be obtained and reads

$$
\begin{aligned}
S^{\alpha\beta\gamma} = & -2(1+w^{\tau}w_{\tau})(-w^{\delta}w_{\delta})^{-1/2}\theta^{\mu\nu}\left[h^{\langle\alpha}{}_{\mu}h^{\beta}{}_{\nu} - \frac{1}{3}h^{\langle\alpha\beta}h_{\mu\nu}\right]w^{\gamma\rangle} + \\
& +4(1+w^{\tau}w_{\tau})(-w^{\delta}w_{\delta})^{-3/2}\left[\frac{1}{5}h^{\langle\alpha\beta}(-w^{\lambda}w_{\lambda}) + w^{\langle\alpha}w^{\beta}\right]h^{\gamma\rangle}{}_{\mu}\theta^{\mu\nu}w_{\nu} - \\
& -(1+w^{\tau}w_{\tau})(-w^{\delta}w_{\delta})^{-3/2}\left\{\frac{h}{9}(1+w^{\lambda}w_{\lambda})^{2} + \left[\theta^{\mu\nu}w_{\mu}w_{\nu}(-w^{\lambda}w_{\lambda})^{-1} - \right.\right. \\
& \left.\left. - \frac{h}{6}(1+w^{\lambda}w_{\lambda})^{2}\right]\left(\frac{2}{15}\alpha_{1}+2\right)\right\}(w^{\langle\alpha}w^{\beta} - w^{\sigma}w_{\sigma}h^{\langle\alpha\beta})w^{\gamma\rangle} - \\
& -2h^{\langle\alpha\beta}\left(\frac{w^{\gamma\rangle}w_{\delta}}{w^{\lambda}w_{\lambda}} - \frac{3}{5}h^{\gamma\rangle\delta}\right]\theta_{\delta\mu}\quad w^{\mu} \qquad (25)
\end{aligned}
$$

where α_1 is an arbitrary function of h. Details of the calculations will be published elsewhere.

REFERENCES

(1) S.Cuperman, I.Weiss & M.Dryer, Ap. J. 239, 345 (1980).
(2) D.B.Melrose, *Plasma Astrophysics, vol. I-II*, Gordon and Breach Science Publishers, New York, (1980).
(3) Shapiro, L.S. & Teukolsky, S.A., *Black Holes, White Dwarfsand Neutron Stars*, Johm Wiley & Sons, New York (1983).
(4) R.B.Miller. *Intense Charged Particle Beams*, Plenum Press, N.Y. (1985).
(5) A.J.Toepfer, Phys.Rev.A3, 1444 , (1971).
(6) J.G. Siambis, Phys.Fluids, 22, 1372, (1979).
(7) W.A Newcomb , Phys. Fluids, 25, 846 , (1982).
(8) P.Amendt and H.Weitzner, Phys.Fluids, 28, 949, (1985).
(9) P.Amendt and H.Weitzner Phys.Fluids 30, 1814 (1987)
(10) P.Amemdt, Phys.Fluids, 29, 1458 (1986)
(11) A.Majda, *Compressible Fluid Flow and Systems of Conservation Laws in Several Space Variables*, Springer, Berlin (1984).
(12) I-Shih Liu, I.Müllerand T.Ruggeri Annals of Physics, 159, 191 , (1986).
(13) Anile, A.M. and Pennisi, S. (1988). An improved relativistic warm plasma model, to be published in A.M.Anile and Y.Choquet-Bruhat (Eds.), Relativistic Fluid Dynamics, Proceedings of the C.I.M.E Symposium on Relativistic Fluid Dynamics, (Noto, 1987) .

MAGNETIC FIELDS IN THE CENTRAL ENGINES OF ACTIVE GALACTIC NUCLEI

MITCHELL C. BEGELMAN
Joint Institute for Laboratory Astrophysics
University of Colorado and National Institute of Standards and Technology
Boulder, Colorado 80309-0440 USA

ABSTRACT. Magnetic fields play a crucial role in a variety of physical processes thought to occur in the central engines of active galactic nuclei. I will examine three classes of phenomena in which magnetic fields are probably important: the plasma physics of hot tenuous accretion flows; the production of nonthermal continuum radiation; and the radiative manifestations of hydromagnetic jet production.

1. INTRODUCTION

It has generally been assumed that magnetic fields play an important role in the central engines of active galactic nuclei (AGNs). At the very least, a small magnetic field is necessary if the accreting gas is to behave like a fluid (Blandford 1983). More substantial fields, of order 10^3 Gauss, are often invoked in order to explain part of the observed continuum radiation as synchrotron (or cyclotron) radiation, or as a means of accelerating outflows which eventually become the observed jets. Magnetic fields of this order of magnitude (which is $\sim 10^{11}$ times larger than the field needed merely to "fluidize" the plasma) correspond to equipartition in gas accreting onto a supermassive black hole at the rate required to explain the AGN luminosity. Although it is not yet possible to deduce all of the observable properties of the central engine from the black hole accretion paradigm, the associated theory does at least give us a plausible idea of what physical conditions are like in the central engine. In this paper I will discuss several important physical processes which may occur in this environment, all of which rely on the presence of a strong magnetic field. First, it is worth reviewing the main arguments which support the hypothesis that supermassive black holes are the "prime movers" in the central engines of AGNs, as well as some of the major deductions regarding the physical state of the accreting gas.

G. Belvedere (ed.), Accretion Disks and Magnetic Fields in Astrophysics, 255–265.

2. THE BLACK HOLE ACCRETION PARADIGM

Although a number of arguments support the hypothesis that supermassive black holes are the "prime movers" in AGNs, none of these arguments by itself is conclusive. The most direct arguments couple upper limits on the size of the central engine, obtained from variability, with lower limits on its mass. Using variability to estimate size is valid provided that relativistic beaming is not important; thus, the method is questionable for blazars. There are at least three methods of estimating the minimum mass of the central engine: 1) multiply the observed luminosity by a source lifetime, estimated from statistical arguments; 2) use the minimum energy content of relativistic electrons and magnetic fields in radio lobes to estimate the cumulative energy output of the nucleus; 3) compare the observed luminosity to the Eddington limit. The first two methods require one to assume a maximum mass-to-energy conversion efficiency (typically taken to be ~10%), while the first and third methods are subject to uncertainties about the role of beaming. Values of M_{min}/R_{max} typically $\gtrsim 10^{-3}$ c^2/G make it plausible but not essential that a black hole be present. More circumstantial arguments include detections of "dark" masses in the centers of several galaxies (e.g., Kormendy 1988, Dressler and Richstone 1988); evidence of relativistic motion in radio jets; evidence of stable jet axes over time scales 10^{-3} years, indicating a good "gyroscope" (a spinning black hole?) at the center; and the "inevitability" of black hole formation through evolutionary processes in the galactic nucleus (Rees 1984). All of these arguments are discussed in greater detail in reviews by Begelman, Blandford and Rees (1984) and by Rees (1984).

We now know that energy can be released both by tapping the gravitational potential energy of material falling into the black hole (accretion) and by tapping the rotational kinetic energy of the black hole itself (the flywheel mechanism: Phinney 1981). In the most plausible realization of the flywheel mechanism, the region surrounding the black hole must be threaded by magnetic fields generated by currents in the surrounding plasma (Blandford and Znajek 1977, MacDonald and Thorne 1982). Close to the black hole, accreting gas will probably possess too much angular momentum to fall into the hole directly: some form of angular momentum transport is necessary. In standard accretion disk theory (Pringle 1981), this transport is assumed to occur as a result of local viscous processes, which generate heat and ultimately (if the density is high enough) radiation. However, not all gravitational binding energy need be released in this way. Blandford and Payne (1982) demonstrated that organized magnetic fields are efficient at removing the angular momentum of an accretion disk. The excess angular momentum and energy of the accreting gas are carried away in the kinetic energy and Poynting flux of a hydromagnetic wind. Energy released by the flywheel mechanism is likely to emerge in similar forms (Phinney 1983). This is good news: hydromagnetic winds are a promising mechanism for jet production, because they can readily attain relativistic speeds and because they tend to be self-collimating (see articles by Lovelace et al. and by Camenzind, in this volume).

In the standard accretion theory it is usually assumed that the gravitational binding energy released by inflow is transferred to internal energy through local dissipative processes. The structure of the accretion flow depends on whether the internal energy is converted efficiently into radiation, and whether the radiation escapes or not. There are three principal "modes" of accretion, which are outlined below and described in greater detail in the review by Begelman (1985). If the accreting gas radiates efficiently and most of this radiation escapes, then the flow will be a thin accretion disk. When the luminosity approaches the Eddington limit, radiation pressure builds up inside the flow, inflating it into a "radiation torus". In this case radiation is produced efficiently but it is prevented from escaping by the large optical depth. An "ion torus" develops when the density is so low that the hot gas is unable to radiate efficiently, resulting in the buildup of thermal ion pressure (Rees et al. 1982). Thin disks and radiation tori are expected to be luminous thermal sources of UV and soft X-ray radiation, and are natural candidates for producing the "ultraviolet bump" which is observed in the spectra of many AGNs (Malkan 1983, Malkan and Sargent 1982). It should be noted, however, that any optically thick gas which locally intercepts and radiates away a significant fraction of the energy output of the central engine should produce thermal radiation with a temperature of order 10^5 K (Guilbert and Rees 1988, Ferland and Rees 1988). Therefore, the existence of a UV bump should not be interpreted as evidence of an accretion disk or torus per se.

We are not yet able to make detailed predictions about the radiative properties of optically thin, hot ion tori, but the properties of an ion torus which we can predict make it a particularly intriguing type of flow. Gas in an ion torus tends to heat up to the local virial temperature, which exceeds the rest mass energy of an electron at $R < 2000\ R_g$, where $R_g \equiv GM/c^2$ is the gravitational radius of the black hole. Since relativistic electrons cool very efficiently, ion tori can exist only if the ions can be maintained at a temperature much higher than that of the electrons. I will discuss whether or not this is possible in §3 below. By its nature, an ion torus need radiate only a small fraction of the gravitational energy released through accretion. For this reason it was proposed as the dominant type of flow in powerful radio galaxies, which seem to have kinetic energy outputs far greater than their luminosities (Rees et al. 1982).

The three modes of accretion can be classified according to the dimensionless ratio of the accretion rate to the "Eddington accretion rate," $\dot{m} \equiv \dot{M}/\dot{M}_E = \dot{M}c^2/L_E$. If we parametrize the black hole mass in units of $10^8\ M_8$ solar masses, then $L_E = 1.3 \times 10^{46}\ M_8$ erg s^{-1}. A thin disk and an ion torus are both possible at low accretion rates, the thin disk at $\dot{m} \lesssim 1$ and the ion torus at $\dot{m} \lesssim 50\ (v_{inflow}/v_{freefall})^2$ (for Coulomb coupling between ions and electrons: Rees et al. 1982). A radiation torus occurs at $\dot{m} \gtrsim$ a few.

To establish the physical parameters of the environment we will be discussing below, it is worth listing some characteristic scales associated with the accretion theory:

luminosity $\sim 10^{45}\ M_8\dot{m}$ erg s^{-1},
radius $\sim 3 \times 10^{14}\ M_8$ cm,
density $\sim 10^9\dot{m}/M_8$ cm^{-3},
radiation energy density $\sim 3 \times 10^5\ \dot{m}/M_8$ erg cm^{-3},
equipartition magnetic field $\sim 3 \times 10^3(\dot{m}/M_8)^{1/2}$ Gauss.

The actual flow is likely to be much more complicated than the simple models described above. Elements of all three modes may coexist, e.g., an ion pressure-supported corona may overlie an optically thick disk or optically thick clouds may be embedded in a hot intercloud medium. The role of magnetic fields in transporting energy from one region of the flow to another is unknown; the flow may resemble an extreme version of the solar atmosphere with magnetic flares transferring a large fraction of the available energy into low-density regions. The apparently nonthermal nature of much of the continuum luminosity may be telling us that something like this is occurring.

3. THE PLASMA PHYSICS OF ION TORI

The idea of a "two-temperature" state close to a black hole, with $T_i \gg T_e$, was first discussed by Shapiro, Lightman and Eardley (1976) in their model for the bistable behavior of Cygnus X-1. Can such a state really exist in an accreting plasma? The minimal thermal coupling between ions and electrons is through binary Coulomb collisions; consideration of this coupling alone leads to the constraint on $\dot{m}$ given in §2 above and in Rees et al. (1982). Obviously it is important to consider whether collective effects might couple the species more rapidly. Chiueh and I (1988) investigated some of the plasma physical properties of two-temperature gas in an ion torus. The system is characterized by the following unusual ordering of plasma length scales:

$$\lambda_{ge} \ll \lambda_{Di} < \lambda_{gi}, \tag{1}$$

where λ_{ge} and λ_{gi} are the gyroradii of electrons and ions, respectively, while λ_{Di} is the ion Debye length. Because the electron gyroradius is so small, the electrons can respond to a fluctuating electric field with wavelength $\sim \lambda_{Di}$ (supported by ion oscillations) only by sliding along the magnetic field. When the wave vector $\vec{k}$ is nearly (but not exactly) perpendicular to $\vec{B}$, the electron and ion oscillations couple as illustrated in Fig. 1. In this limit, we found three types of waves in which the restoring force comes from ion pressure, while the inertia is supplied by the electrons. In other words, the electrons behave like the more massive particles! The three modes are: 1) an "electron-acoustic" wave, in which the phase speed parallel to $\vec{B}$ is $\sim (kT_i/m_e)^{1/2}$; 2) an electromagnetic wave, in which the

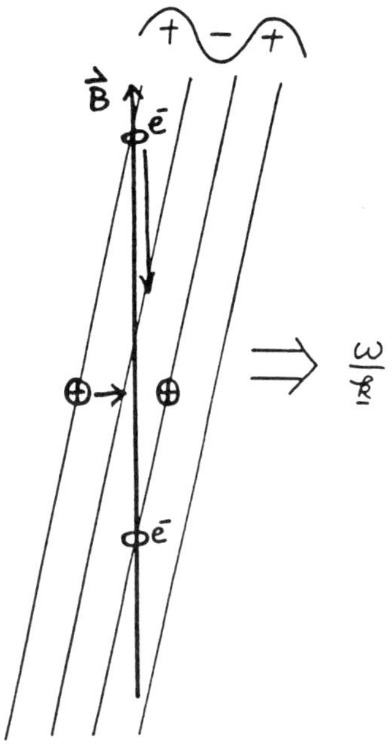

Figure. 1. Coupling of electrons to nearly transverse ion oscillations in an ion torus with $T_i \gg T_e$. Because of their small gyroradii, electrons are constrained to slide along magnetic field lines.

parallel phase speed ~ c; and 3) a drift wave, in which the phase speed perpendicular to $\vec{B}$ is equal and opposite to the diamagnetic ion drift speed in the background plasma.

Provided that $T_i \gg T_e$, the parallel phase speed exceeds the electron thermal speed in all three cases, hence electron Landau damping is unimportant. These waves can be driven unstable by ion drift in the background plasma, caused by density gradients or magnetic inhomogeneities. We were able to show that the wave amplitudes most probably saturate by coupling to the electrons through stochastic $\vec{E} \times \vec{B}$ drifts. The electron heating rate in a saturated wave background is sufficient to destroy the two-temperature state by draining heat away from the ions, under a wide range of conditions. The important question is, under what conditions do the waves grow to saturation?

It turns out that there are some fairly stringent requirements for wave growth. Since the waves are driven unstable by ion drift, the growth rate depends on the size and scale of the inhomogeneities which cause the drift. In order for the instability to have time to grow, the background plasma must be highly turbulent on small scales, with large-amplitude fluctuations on scales as small as a few hundred ion gyroradii. We do not know whether such intense turbulence is likely to occur over large portions of the flow. Even a small region of turbulence, e.g., downstream from a shock or close to a magnetic reconnection site, would be sufficient to couple the ions and electrons locally. We must regard the existence of two-temperature plasma in ion tori as still open to question.

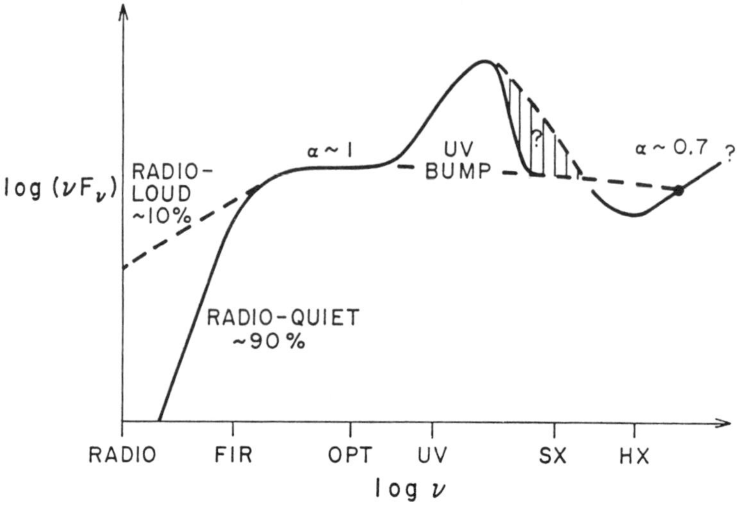

Figure. 2. Generic AGN continuum spectrum. High frequency behavior of UV bump is uncertain because of absorption by intervening material.

4. CLUES FROM THE NONTHERMAL CONTINUUM

It is likely that at least part of the continuum spectrum from an AGN is produced by nonthermal synchrotron radiation and/or inverse Compton scattering. Figure 2 shows a cartoon of a typical AGN continuum spectrum. The vertical scale νF_ν is proportional to the luminosity per decade of frequency. In addition to the UV bump discussed earlier, note the relatively flat spectrum between far-infrared and optical frequencies, the possible correlation of this flat "power-law" with the X-ray flux, and the evidence for a possible upturn in νF_ν towards the hard X-rays. Unfortunately we do not know how much of this radiation comes to us from the vicinity of the black hole. There is some evidence that the variability time scale decreases with increasing frequency, suggesting that the higher frequencies are produced closer to the black hole, whereas the infrared flux might be thermal emission from dust.

If one assumes that the bulk of the continuum emission is nonthermal, however, it becomes possible to place some interesting constraints on the physics of the central engine. A turnover in the far-IR, typically at frequencies $\nu_t \sim 5 \times 10^{12}$ Hz (Fig. 2), is predicted from synchrotron self-absorption. Turning the calculation around, the radius of the "far-IR photosphere" is

$$R_t \sim 3 \times 10^{15}\, B_3^{1/4}\, L_{45}^{1/2} \left(\frac{\nu_t}{5 \times 10^{12}\ \mathrm{Hz}}\right)^{-7/4} \mathrm{cm} \quad , \tag{2}$$

where $B_3 \equiv B/10^3$ G and $L_{45} \equiv L/10^{45}$ erg s^{-1}. Thus, the IR source must be compact if the turnover is caused by synchrotron self-absorption. If the IR-optical continuum (and possibly the IR-X continuum) consists of synchrotron radiation with a spectral index $\alpha \equiv -d\log f_\nu/d\log \nu \sim 1$, then the relativistic electrons must be distributed in energy according to $N_\gamma \propto \gamma^{-3}$, implying that relativistic electrons are injected continuously at all

$$\gamma \gtrsim \gamma_t \sim 60\, B_3^{-1/2} \left(\frac{\nu_t}{5 \times 10^{12}\ \mathrm{Hz}}\right)^{1/2} \quad . \tag{3}$$

Here γ_t is the minimum Lorentz factor of a synchrotron-emitting electron which radiates most of its energy above the turnover frequency. In the absence of continuous injection the spectral index would be 0.5. Under the assumed conditions, electrons with $\gamma < \gamma_t$ must lose energy at a rate comparable to the IR luminosity, but are unable to do so through synchrotron radiation. De Kool, Begelman snd Sikora (1989) and Ghisellini, Guilbert and Svensson (1988) have recently demonstrated in detail how synchrotron self-absorption causes the electron distribution to build a thermal peak until some other mechanism is able to drain away the accumulating energy. Where does this energy go?

Several possible sinks for the energy can be eliminated on observational or theoretical grounds. The most obvious loss mechanism is Compton scattering. However, synchrotron self-Compton scattering of the IR continuum will generally lead to a spectrum which is too flat, $\alpha < 1$ (Ghisellini 1987). If the electrons cool by Comptonizing the UV bump, then we would expect $L_{2keV} \gtrsim L_{IR}$, which is not observed (Kriss 1988). These arguments can be used to place a lower limit on the magnetic field strength of a few hundred Gauss. We cannot rule out the possibility that the electrons cool by Comptonizing hard X-rays. To avoid flattening the IR continuum spectrum, the X-ray luminosity would have to exceed both L_{IR} and the energy flux injected in electrons with $\gamma \lesssim \gamma_t$. The excess energy could also be lost through Coulomb interactions with thermal plasma. For this mechanism to be effective, the density in thermal electrons must be $\gtrsim 10^{10}$ cm^{-3}. To avoid unacceptable spectral distortions the emitting region would have to be a thin shell or a corona on the surface of a disk (de Kool and Begelman 1988). It is unlikely that the physical removal of electrons by accretion or in a wind can seriously affect their energetics. The characteristic cooling time scale of an electron with $\gamma \sim 10$ is $\lesssim 100$ seconds, which is far shorter than the light travel time across the emission region. If the radiating particles are actually electron-positron pairs, then annihilation will occur too slowly to affect the energetics unless the electron scattering optical depth is unacceptably high.

We can use existing observations to place limits on the amount of thermal plasma that may be mixed with the relativistic electrons. The most stringent constraint seems to be that imposed by induced Compton scattering, which distorts the far-IR spectrum if the electron scattering optical depth in thermal plasma satisfies

$$\tau_{es} > \frac{m_e c^2}{kT_b} \sim 0.05\ B_3^{1/2} \left(\frac{\nu_t}{5 \times 10^{12}\ \mathrm{Hz}}\right)^{-1/2} \qquad (4)$$

(Sunyaev 1971; Blandford and Rees, private communication), where $T_b \sim \gamma_t m_e c^2/3k$ is the brightness temperature at the far-IR turnover. The effect of induced Compton scattering is to "chop the top" off the self-absorbed synchrotron spectrum, increasing the turnover frequency and making the turnover more gentle. This does not seem to be consistent with recent observations suggesting that the turnover is sudden and steep. Of course, a contribution by dust could mask the effects of induced Compton scattering. Two less stringent constraints on the thermal plasma are: 1) the Compton y-parameter must satisfy

$$y \equiv \max\ [\tau_{es}, \tau_{es}^2]\ \frac{4kT}{m_e c^2} < 1 \qquad (5)$$

to avoid flattening the $\alpha \sim 1$ power law; and 2) free-free absorption must not cut off the spectrum at a frequency above the observed turnover. Guilbert and Rees (1988) and Ferland and Rees (1988) have studied additional constraints associated with X-ray photoelectric absorption in $\sim 10^5$ K thermal gas.

It is instructive to compare these observationally inspired constraints with the amount of thermal plasma which is likely to be present on theoretical grounds. At the very least, the thermal background will consist of those cooled relativistic electrons which have not been reaccelerated or removed by some other means. The optical depth in thermal electrons satisfies

$$\tau_{es} > \left(\frac{\sigma_T L_{IR}}{m_e c^3 R}\right) \left(\frac{t_{removal}}{R/c}\right) \left[\frac{1}{\pi\gamma_{min} \ln(\gamma_{max}/\gamma_{min})}\right]_{inj}, \qquad (6)$$

where σ_t is the Thomson cross section and γ_{max}, γ_{min} are the maximum and minimum injected electron energies, respectively. The first factor in brackets is similar to the "compactness parameter" familiar from studies of pair production, and has a value $\sim 30\ L_{45}(R/10^{15}\mathrm{cm})^{-1}$. The second factor compares the removal time of the thermal electrons with the light travel time across the emission region, while the third factor reflects the fact that the total number of injected electrons can be reduced if the injection process cuts off below some minimum γ. Note that γ_{min} must be at least as large as γ_t. To avoid spectral distortions by induced Compton scattering, it is apparent that either most electrons must be reaccelerated before being physically removed from the system, or few electrons are injected with $\gamma \lesssim 10$.

5. HYDROMAGNETIC ACCELERATION AND THE BLAZAR CONTINUUM

I noted earlier that a hydromagnetically driven wind from the central region of an accretion flow is a promising mechanism for the initial stage of jet production. It has long been suspected that emission from the inner region of a relativistic jet accounts for the wildly variable, polarized radiation from blazars (Blandford and Rees 1978, Blandford and Königl 1979). The standard model presumes that the jet emission is some combination of synchrotron radiation and synchrotron self-Compton radiation. In this picture there is little direct connection between the blazar radiation component and the radiation from the underlying accretion flow, which presumably is also present but is swamped by that from the jet. Marek Sikora and I (1987) proposed that there may be a deeper connection between the radiation from the accretion flow and that from the jet. If the outflow is accelerated to relativistic speeds close to the black hole, it must interact strongly with the intense radiation field of the accretion flow on its way out.

This interaction is illustrated in Fig. 3a. In a frame of reference which is stationary with respect to the black hole, the ambient radiation may be roughly isotropic and weakly polarized (Fig. 3b). Viewed from the relativistic jet frame, however, the ambient radiation will be concentrated into a narrow cone about the apex of bulk particle motion (Fig. 3c). Therefore, after scattering by cold electrons in the jet, it will be highly polarized, and in the observer frame, strongly beamed as well (Fig. 3b). We performed detailed calculations to illustrate the angular distributions of polarization and intensity of radiation scattered by jets having a variety of velocity and density profiles. Because of relativistic beaming, the observed quantities are sensitive to small asymmetries and fluctuations in jet properties; this could give rise to the high degree of variability that is observed. We have also found that the mean direction of polarization can serve as a diagnostic of jet structure, if the mean jet axis is known (e.g., from observations of a VLBI jet). "Pencil beam" jets, i.e., jets with velocity peaking on the axis of symmetry, always produce perpendicular polarization. On the other hand, "hollow cone" jets, i.e., jets with velocity peaking off-axis, will generally be observed to yield polarization parallel to the projected jet axis. A number of observational studies suggest that parallel polarization is favored on average, suggesting that the hollow cone description is more apt. Such a configuration would arise naturally if jets originate as centrifugally-driven hydromagnetic winds.

The radiation drag associated with the scattering process may extract a significant fraction of the jet's kinetic energy during the acceleration process. Li, Chiueh and I have recently begun to investigate the dynamical effects of radiation drag on the hydromagnetic acceleration of relativistic winds. Our initial results suggest that in the presence of drag a larger amount of rotational energy can be extracted from the central engine than in the drag-free case. This extra energy ends up as scattered radiation, and is therefore potentially observable. Thus, the magnetic properties of the central engine may affect its observable features in yet another surprising way.

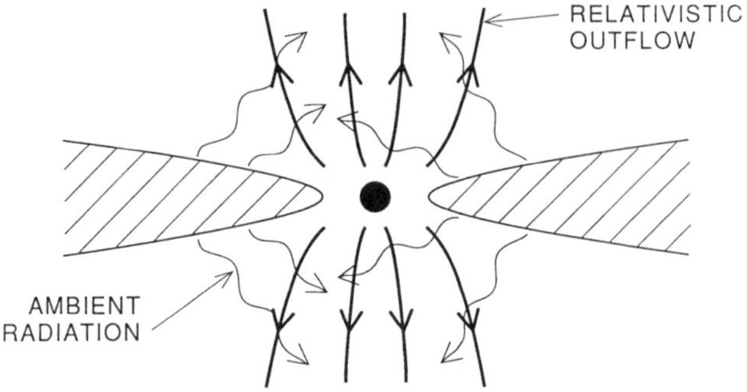

Figure 3(a). Interaction of relativistic outflow with accretion disk radiation.

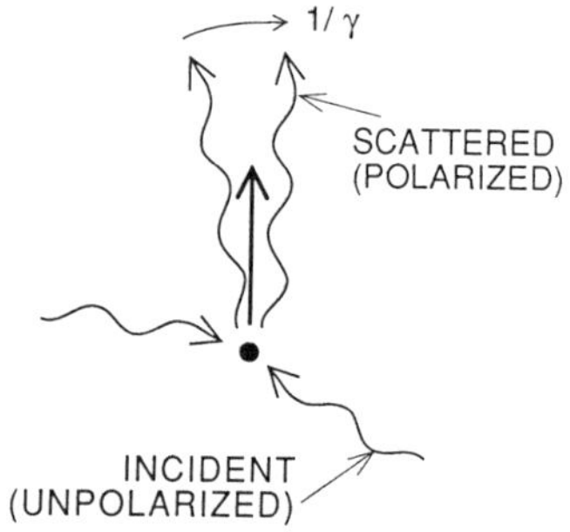

Figure 3(b). Electron scattering viewed from observer frame.

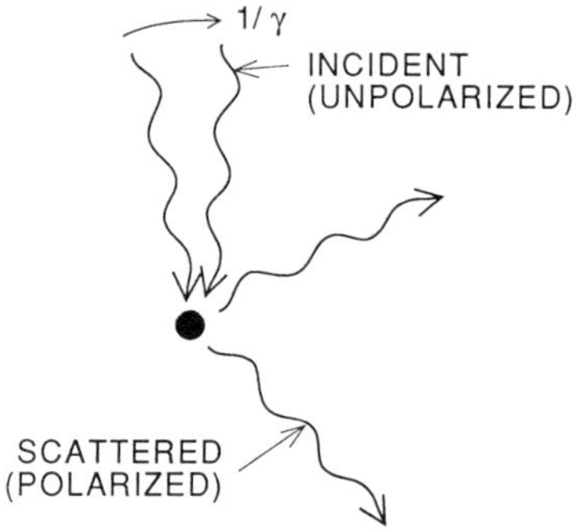

Figure 3(c). Electron scattering viewed from frame of relativistic electron in outflow.

ACKNOWLEDGMENTS

The views presented in this article have been particularly influenced by discussions and/or collaboration with R. Blandford, T. Chiueh, M. de Kool, M. Rees, and M. Sikora. AGN research at the University of Colorado is supported in part by NSF grant AST83-51997, NASA Astrophysical Theory Center grant NAGW-766, and grants from the Alfred P. Sloan Foundation and Rockwell International Corporation.

REFERENCES

Begelman, M. C. 1985, in Astrophysics of Active Galaxies and Quasi-Stellar Objects, ed. J. S. Miller (Mill Valley, CA: University Science Books), p. 411.

Begelman, M. C., Blandford, R. D., and Rees, M. J. 1984, Rev. Mod. Phys., **56**, 255.

Begelman, M. C., and Chiueh, T. 1988, Ap. J., **332**, 872.
Begelman, M. C., and Sikora, M. 1987, Ap. J., **322**, 650.
Blandford, R. D. 1983, A. J., **88**, 245.
Blandford, R. D., and Königl, A. 1979, Ap. J., **232**, 34.
Blandford, R. D., and Payne, D. G. 1982, M.N.R.A.S., **199**, 883.
Blandford, R. D., and Rees, M. J. 1978, in Pittsburgh Conference on BL Lac Objects, ed. A. M. Wolfe (Pittsburgh: University of Pittsburgh), p. 328.
Blandford, R. D., and Znajek, R. L. 1977, M.N.R.A.S., **179**, 433.
de Kool, M., and Begelman, M. C. 1988, submitted to Ap. J.
de Kool, M., Begelman, M. C., and Sikora, M. 1989, Ap. J., in press.
Dressler, A., and Richstone, D. O. 1988, Ap. J., **324**, 701.
Ferland, G., and Rees, M. J. 1988, Ap. J., **332**, 141.
Ghisellini, G. 1987, Ph.D. Thesis, SISSA, Trieste, Italy.
Ghisellini, G., Guilbert, P. W., and Svensson, R. 1988, Ap. J. (Letters), **334**, L5.
Guilbert, P. W., and Rees, M. J. 1988, M.N.R.A.S., **233**, 475.
Kormendy, J. 1988, Ap. J., **325**, 128.
Kriss, G. 1988, Ap. J., **324**, 809.
MacDonald, D., and Thorne, K. S. 1982, M.N.R.A.S., **198**, 345.
Malkan, M. A. 1983, Ap. J., **268**, 582.
Malkan, M. A., and Sargent, W. L. W. 1982, Ap. J., **254**, 22.
Phinney, E. S. 1981, in Plasma Astrophysics, ed. T.Guyenne, ESA SP-161, p. 337.
Phinney, E. S. 1983, Ph.D. Thesis, University of Cambridge.
Pringle, J. E. 1981, Ann. Rev. Astr. Ap., **19**, 137.
Rees, M. J. 1984, Ann. Rev. Astr. Ap., **22**, 471.
Rees, M. J., Begelman, M. C., Blandford, R. D., and Phinney, E.S. 1982, Nature, **295**, 17.
Shapiro, S. L., Lightman, A. P., and Eardley, D. M. 1976, Ap. J., **204**, 187.
Sunyaev, R. A. 1971, Soviet Astron. AJ, **15**, 190.

TIME VARIABILITY OF THE X-RAY EMISSION FROM SEYFERT GALAXIES

F. FIORE, E. MASSARO
Istituto Astronomico, Universita' di Roma "La Sapienza"
Via G. Lancisi, 29
00161 Roma
Italy

ABSTRACT. We present the preliminary results on the analysis of some X-ray light curves of Seyfert galaxies. Two different types of variability are distinguished: one characterized by coherent components and the other by chaotic processes. Phase space analysis indicates also that the latter is related to deterministic chaos.

1. Introduction

Variability is a powerful source of information on the physical process occurring in accreting compact objects, such those are likely to produce the X-ray emission from Active Galactic Nuclei (AGNs). The particular and unique capability of the EXOSAT satellite to perform long and nearly uninterrupted observations of the same target allows one, at least for a few bright Seyfert galaxies, to study in some detail the flux variations with several kinds of algorithms and to attempt a first rough characterization of the nature of these variations.

In this contribution we present some recent results about the time analysis of two long ($\geq 10^5$ s) observations, performed with the Medium Energy (ME) experiment on board the EXOSAT satellite (Turner, Smith and Zimmerman 1981) of the two Seyfert galaxies NGC 4151 and NGC 4051. As we discuss in the following, these two galaxies could be considered the prototypes of two classes of variability: the first characterized by one (or more) coherent components and the latter with light curves of chaotic appearence and without any typical time scale. In particular, an interesting finding of our analysis is that the behaviour of NGC 4051 seems governed mainly by deterministic chaos rather than a pure stochastic process (i.e. random events of random amplitude).

2. Observations and Results

We discuss first the case of NGC 4151. In Fig. 1 we show the 2-10 keV light curve of the source counts recorded on 1985 Jan. 2 together with that of the simultaneouly acquired background. The light curve in Fig. 1a shows the "typical" behaviour of this source: it is characterized by a fairly regular trend with a rate of change of

G. Belvedere (ed.), Accretion Disks and Magnetic Fields in Astrophysics, 267–272.

1-4 % hr^{-1} to which rapid fluctuations of small amplitude are superimposed. The background light curve (Fig. 1b) is remarkably regular and the gap in coicidence of the swap of the two ME array reflects the different levels affecting the two monitors. In order to study the spectral properties of the fluctuations we have subtracted the trend (estimated by means of a 2nd order polynomial fit) from the source light curve and have calculated the power spectral density of the resulting time series. This power spectrum is shown in Fig. 2; it was divided by half of the total variance in the observation because with this normalization it is possible to estimate quickly the statistical significance of each point from the χ^2 distribution with two d.o.f. (Leahy et al. 1983). We see that one peak at the frequency f = $1.75 \cdot 10^{-4}$ Hz (corresponding to the period of 5700 s) is particularly prominent: the probability of a chance occurrence of such a feature, with respect to the poisson noise, which dominates the high frequencies part of the power spectrum, is about 10^{-5}. It is therefore unlikely that this peak were a statistical fluctuation. On the basis of an accurate analysis (Fiore et al. 1988) it is possible to exclude that this feature is of instrumental origin or that it could be due to imperfect corrections for the different levels of the background count rates on the two monitors. Moreover, a similar effect was found in another observations of the same source confirming that coherent variations of small amplitude are occasionally present in the X-ray emission from NGC 4151.

In two other cases of active galaxies the spectral analysis of the X-ray light curves show the existence of characteristic time scale for the variations: Branduardi-Raymont and Mittaz (1988) report a time scale of 12000 s in the light curve of a long ($\sim 10^5$ s) EXOSAT observation of the Seyfert galaxy NGC 6814, and Barone et al. (1988) show with an autoregressive analysis that the conspicuous changes in the count rate of NGC 7314 (Turner 1987) can be explained by a decaying periodical component with a frequency varying from $9.7 \cdot 10^{-5}$ Hz to $8.7 \cdot 10^{-4}$ Hz in the course of the observation.

A rather different behaviour has been observed in a few other Seyfert galaxies, for which no clear indication of privileged frequencies was found. In the three cases of NGC 4051 (Lawrence et al. 1987), NGC 5506 (McHardy and Czerny 1987) and MCG 6-30-15 (Pounds and Turner 1987) the power spectra have low frequency excesses moderately well fit by power laws with slopes close to -1. The variations of the X-ray emission seem therefore to be due to some stochastic process as those originting the "flicker noise" (Dutta and Horn 1981). In Fig. 3 is shown the light curve of NGC 4051 obtained with the ME counters (1.5-7 keV) on 1985 Dec. 3: the quite "chaotic" behaviour of the rapid fluctuations (some of them not completely resolved) is evident. In Fig. 4 is given the corresponding power spectrum: the slope between 10^{-5} and 10^{-3} Hz is - 1.2 $\pm$ 0.1. Recent results from GINGA (Matsuoka et al. 1988) suggest that this range could extend to about 10^{-2} Hz.

We have tried to characterize the flux fluctuations of this galaxy by means of the so called "phase space recontruction" starting from the observed time series (already applied, for instance, to signals of different nature by Nicolis and Nicolis, 1986, and to the light curve of the X-ray binary Her X-1 by Voges et al., 1987). With this method one is able to state whether or not a signal is stochastic and to obtain information about possible deterministic processes underlying the chaotic behaviour. The result depends on the duration of the time series and on the signal to noise ratio; in our case the latter figure is about 10 and the time series is about 400 point long. Atmanspacher et al. (1988), however, have shown that in similar conditions reliable results can be obtained.

We compute the correlation integral

$$C(r) = \frac{1}{n^2} \sum_{i=j} \theta(r - |X_i - X_j|) \tag{1}$$

where θ is the Heaviside step function and X_i (or X_j) is a vector defined in a D dimensional phase space by extracting n elements from the time series with an arbitrary time lag Δt. $C(r)$ turns out to be proportional to the number of points in the phase space with a distance less than r. It was shown (Grassberger and Procaccia, 1983) that for small values of r , $C(r) \sim r^{\nu}), \nu$ measuring the dimension of the region in the phase space within the points are clustered. The asympthotic value of the exponent

$$\nu = lim_{r \to 0} \frac{log_2 C(r)}{log_2 r} \tag{2}$$

can be then used to discriminate between a pure stochastic (uncorrelated) process and deterministic chaos. In the first case one finds $\nu = D$, while in the latter $\nu \leq D$. In Fig. 5 are plotted the values of ν .vs. $log_2 r$ for Δt equal to 600 s (corresponding to one bin in the light curves of Fig. 3) and two different choices of the dimension (D=2 and 20). We see that for the lower value of D the curves of the source and the background reach the same asympthotic value $\nu = 2$. For D=20, the value of ν for the background series increases up to about D as expected, while that for the source signal remains well below this figure and converges to about 6. The rapid increase for $r < 8.5$ is due to the degrading of the S/N for low r that makes the signal's behaviour similar to that of the background. This result can be considered a first indication of the existence of deterministic chaos in AGNs: the estimated dimension, however, is somewhat high. An extension of this analysis to other sources of the same kind could provide useful information on this respect.

3. Conclusions

It is not simple to explain the observed different behaviours within the widely considered models. Choerent phenomena could be related, for instance, to the orbital motion of matter accreting onto a central massive object (Sunyaev, 1973). In this hypothesis we can estimate its mass that, in the case of NGC 4151, turns out to be $M \leq 1.3 \cdot 10^7 M_\odot$ (for a Schwarzschild metric). A figure significatively less than that obtained by Wandel and Mushotzky (1986) with the classic causality argument for the time scale derived from large amplitude variations ($M \leq 5 \cdot 10^8 M_\odot$). Chaotic variability could be not simply correlated with the size of the source, and then with its mass. This behaviour does not seem to be related to the source luminosity: the mean luminosity in the 2-10 keV band of the three sources that show self similar variations, ranges from about $3 \cdot 10^{41}$ erg s^{-1} (NGC 4051) to $1 - 1.5 \cdot 10^{43}$ erg s^{-1} for MCG 6-30-15 and NGC 5506. For comparison, the luminosities of NGC 4151

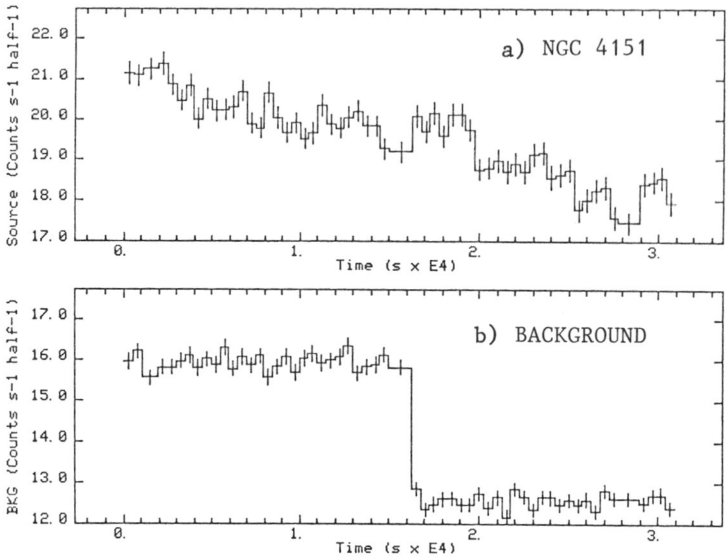

Fig. 1
The X-ray light curves of NGC 4151 and the simultaneous background.

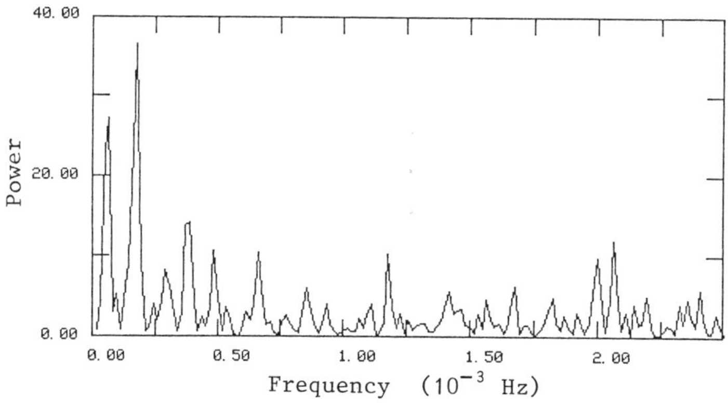

Fig. 2
Power spectrum of the source time series of Fig. 1 after detrending.

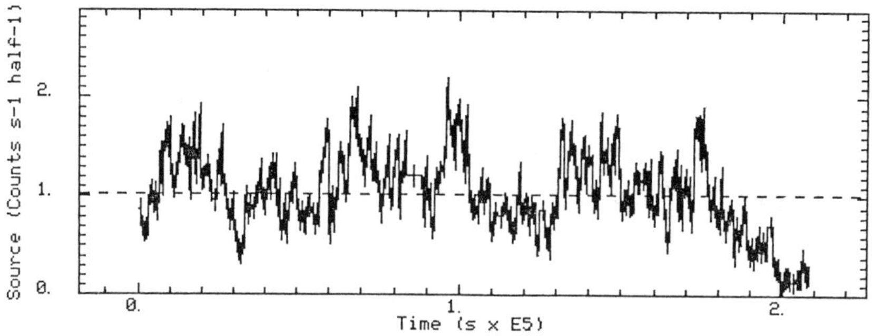

Fig. 3. The X-ray light curve of NGC 4051.

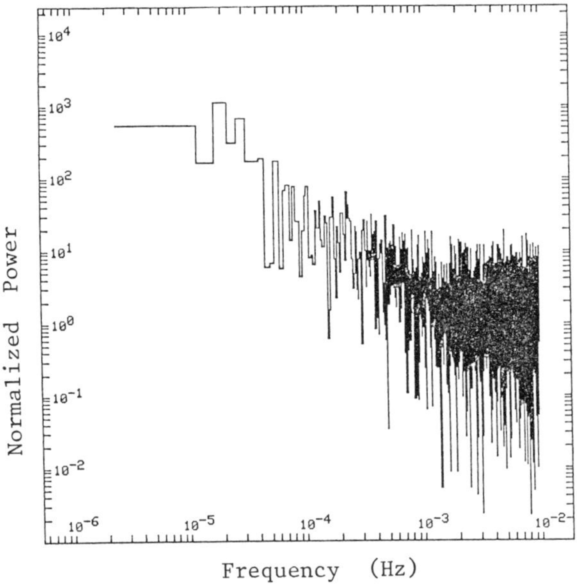

Fig. 4
The Power Spectrum of the time series of NGC 4051 shown in Fig. 3.

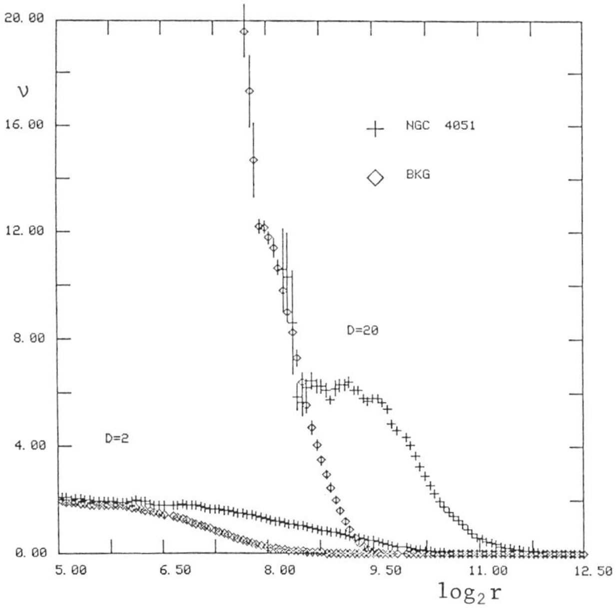

Fig. 5
The slope of the correlation integral vs $\log_2 r$ for two values of D.

and NGC 6814 are $5 \cdot 10^{42}$ erg s^{-1} and $1.5 \cdot 10^{42}$ erg s^{-1} respectively. A further extension of the "chaos analysis" of the light curves of these sources is in progress. We think that in this way should be possible to gain more insight on the complex phenomena occurring in the innermost region of AGNs.

Reference

Atmanspacher, H., Scheingraber, H. and Voges, W., 1988, *Phys. Rev. A*, **37**, 1341.
Barone, P., Massaro, E., Polichetti, A., 1988, *Astron. Astrophys.* in press.
Branduardi-Raymont, G. and Mittaz, J. P. D., 1988, *Mon. Not. R. astr. Soc.*, in press.
Dutta, P. and Horn, P. M., 1981, *Rev. Mod. Phys.* **53**, 497.
Fiore, F., Massaro, E., Perola, G. C. and Piro, L., 1988, subm. to *Astrophys. J.*.
Grassberger, P., and Procaccia, I., 1983, *Phys. Rev. Letters*, **51**, 346.
Lawrence, A., Watson, M. G., Pounds, K. A. and Elvis, M., 1987, *Nature*, **325**, 694.
Leahy, D. A. et al., 1983, *Astrophys. J.*, **266**, 160.
Matsuoka, M. et al., 1988 preprint.
McHardy, I. and Czerny, B., 1987, *Nature*, **325**, 696.
Nicolis, C., and Nicolis, G., 1986, *Proc. Natl Acad. Sci. USA*, **83**, 536.
Pounds, K. and Turner, T. J., 1987, in *Variability of Galactic and Extragalactic X-ray sources* ed. A. Treves (Milano: Associazione Avanzamento Astronomia) p.1.
Sunyaev, R. A., 1973, *Sov. Astron.* -AJ, **16**, 941.
Turner, T. J., 1987, *Mon. Not. R. astr. Soc.*, **226**, 9p.
Turner, M. J., Smith, A., and Zimmerman, H., 1981, *Space Sci. Rev.*, **30**, 479.
Voges, W., Atmanspacher, H., and Scheingraber, H., 1987, *Astrophys. J.*, **320**, 794.
Wandel, A. and Mushotzky, R., 1986, *Astrophys. J. Letters*, **306**, L61.

SLIM ACCRETION DISCS

J.P. LASOTA
Département d'Astrophysique Relativiste et Cosmologie
Observatoire de Paris, Section de Meudon
92195 Meudon Principal Cedex, France

ABSTRACT. In a recent paper, Abramowicz, Czerny, Lasota and Szuszkiewicz (1988) have found a new branch of equilibrium solutions for stationary accretion discs around black holes. These solutions correspond to moderately super-Eddington accretion rates. The existence of the new branch is a consequence of cooling due to general relativistic Roche lobe overflow and horizontal advection of heat. On an accretion rate versus surface density plane the new branch forms, together with the two "standard" branches (corresponding to to the Shakura-Sunyaev accretion discs models) a characteristically S-shaped curve. The relation between this "S-curve" and the variability of accretion discs is under study and seems to allow, at least in principle, for a limit cycle behaviour.

1. INTRODUCTION

Accretion discs are observed in cataclysmic variables and in several types of X-ray close binaries. (Wade & Ward 1985, Mason 1986). They are often invoked in the context of active galactic nuclei (AGN) and quasars (Collin-Souffrin 1987) although for those objects the existence of accretion discs is less certain. The standard model describes a geometrically thin, keplerian accretion disc (Frank, King & Raine 1985). There is evidence that such discs are present around accreting white dwarfs in cataclysmic variables (Horne and Marsh 1986). Detailed models, however are still not very satisfactory: they fail to reproduce correctly the spectra (Wade 1988), variability (Verbunt 1986) and other observed properties (Horne & Marsh 1986) of accretion discs around white dwarfs.

The case of Low Mass X-ray Binaries (LMBX) (accreting neutron stars and black holes) and AGNs (accreting supermassive black holes) is more complicated than that of CVs: because of the increased efficiency, the luminosities released by accretion are much higher and effects of radiation pressure, X-ray illumination etc. have to be taken into account. It is therefore not surprising, that models of disc accretion on to neutron stars and black holes are even less successful in reproducing observations than disc models of CVs. A fundamental difficulty concerns the structure of discs in stationary equilibria: for sufficiently high accretion rates the disc thickness becomes comparable to its radial dimension and since it is neither geometrically thin nor keplerian the standard model

G. Belvedere (ed.), Accretion Disks and Magnetic Fields in Astrophysics, 273–278.

does not apply.

Here I will describe an attempt (Abramowicz, Czerny, Szuszkiewicz & Lasota 1988, hereafter ACLS) to include some non-standard effects into the modelling of accretion discs around black holes. Our model, like all others proposed until now, suffers from the limited knowledge of the physical mechanism providing the viscosity which drives accretion in discs (we use the "α prescription" of Shakura & Sunyaev 1973, hereafter S&S), but its predictions may help in understanding the physics of accretion on to compact objects.

2. EQUILIBRIA AND STABILITY

The inner region (close to the black hole) of the S&S standard disc becomes thermally (and viscously) unstable when radiation pressure dominates over gas pressure. This happens typically for accretion rates $\dot{M} \geq 10^{-2}\dot{M}_{Edd}$, where $\dot{M}_{Edd} = L_E/c^2 = 1.7\times10^{17} M/M_{\odot} gs^{-1}$ is the Eddington accretion rate. For accretion rates close to $10^{-2}\dot{M}_{Edd}$ the disc is still geometrically thin since the ratio of its thickness H to its radius r (Frank, King & Raine 1985):

$$\frac{H}{r} \approx \frac{\dot{M}}{\dot{M}_{Edd}} \cdot \frac{R_G}{r} \tag{1}$$

(where $R_G = 2GM/c^2$), is very small compared to one.

The thermal instability is due to an insufficient dependence of the rate of radiative cooling Q^- on temperature, or equivalently on H. We have $Q^- \sim H$, while the rate of viscous heating $Q^+ \sim H^2$; thus overheating produces expansion that leads to a thermal runaway. The condition for the disc to be thermally stable is $(\partial lnQ^+/\partial lnH)_\Sigma < (\partial lnQ^-/\partial lnH)_\Sigma$ (where Σ is the surface density). Abramowicz (1981) remarked that advective cooling (heat transport by the bulk motion of the matter) at the inner edge of an accretion disc would have a sufficiently steep dependence on H ($Q^-_{adv} \sim H^3$) to quench the thermal runaway. The extent of the stabilized region of the disc depends on accretion rate; only for sufficiently high rates will the whole region subject to thermal instability be stabilized by advective cooling. This can be understood by the following simple argument. Consider a fluid element of the disc, located at a distance Δr from the inner edge of the disc in the thermally unstable region. The element expands vertically in a thermal time $t_{th} \approx 2\pi/\Omega\alpha$, where $\Omega(r)$ is the disc angular speed, and α is the viscosity parameter ($\alpha \leq 1$). Simultaneously the fluid element is drifting radially towards the black holes in a viscous time $t_{vis} \approx t_{th} \cdot (H/R)^{-2}$. If the element is able to arrive to the inner edge of the disc (where, beacuse of general relativistic "Roche lobe" overflow it can be considered as being swallowed by the black hole, see e.g. Abramowicz 1981) before expanding over a distance ~ H, the thermal runaway is suppressed and no instability occurs. One requires therefore that $t_{vis} \approx t_{th}$, or in other words $H \approx R$. Thus from (1) one gets a condition $\dot{M} \geq \dot{M}_{Edd}$. The stabilization by advection is therefore possible at high accretion rates where the standard, geometrically thin model is clearly no longer applicable. It was however realised by Abramowicz and Lasota (1985, unpublished, see also Abramowicz, Lasota & Xu 1986) that inclusion of advective cooling and general relativistic effects at the inner edge of the disc allows for a construction of a new *family* of stationary equilibria ("slim discs") for moderately super-Eddington accretion rates. ACLS found those equilibrium solutions numerically by using non-standard equations proposed by Paczyński & Bisnovatyi-Kogan (1981), later developped by Muchotrzeb & Paczyński (1982).

3. BASIC ASSUMPTIONS AND EQUATIONS

As mentioned in the introduction ACLS use the S&S prescription for the relevant component of the viscous stress tensor:

$$\tau_{\phi r} = -\alpha P \tag{2}$$

where P is the total pressure. Note that (2) is equivalent to the other version of the S&S prescription: $\nu = \alpha c_s H$ (where c_s is the speed of sound) only for keplerian motions.

The thermal runaway discussed above takes place in the standard model only if $\tau_{\phi r}$ is proportional to the *total* pressure; if one assumes that $\tau_{\phi r}$ is proportional only to the gas pressure no runaway occurs. It is an open question which form of the viscosity law should be used and "physical" arguments can be put forward in favour of both. Here I take the attitude that studying various models and comparing them with observations can teach us something useful about viscosity in accretion discs. From this point of view it is interesting to study the consequences of equation (2).

General relativistic effects are simulated by the use of the so-called pseudo-Newtonian potential (Paczyński & Wiita 1980):

$$\Psi = -\frac{GM}{r - R_G}. \tag{3}$$

This is sufficient to represent the relevant properties of accretion discs structure around non-rotating black holes.

The structure of slim discs is described by a set of equations in which, as in the standard case, the vertical and horizontal structures are decoupled. The radial momentum (Euler) equation takes the form:

$$\frac{1}{\rho}\frac{dP}{dr} - (\Omega^2 - \Omega_K^2) + v_r \frac{dv_r}{dr} = 0 \tag{4}$$

where ρ is the density, Ω_K the keplerian orbital speed (corresponding to the potential (3)) and v_r is the radial speed. In the standard case this equation reduces to $\Omega = \Omega_K$.

The energy equation contains a non-standard advective term:

$$\dot{M}(l - l_o) \cdot (-\frac{d\Omega}{dr}) + B_3 \dot{M} T \frac{dS}{dr} = 4\pi r F^- \tag{5}$$

where l is the specific angular momentum, l_o its value at $r = R_G$, T the temperature, S the entropy and F^- the vertical radiative flux. B_3 is one of the three correcting factors B_1, B_2, B_3 which appear in the vertically integrated equations (Paczyński & Bisnovatyi-Kogan 1980). All other equations are as in S&S, the boundary conditions however take properly into account the presence of the black hole at the inner boundary.

4. RESULTS

The solutions are presented in Fig. 1 in which accretion rate measured in units of $\dot{M}_{Edd}/16$ (corresponding to the efficiency of accretion $\eta = 1/16$ of a non-rotating pseudo-Newtonian black hole) is plotted (for a given radius r) as a function of surface density Σ. The models were calculated for $M = 10M_\odot$ and $\alpha = 0.001$.

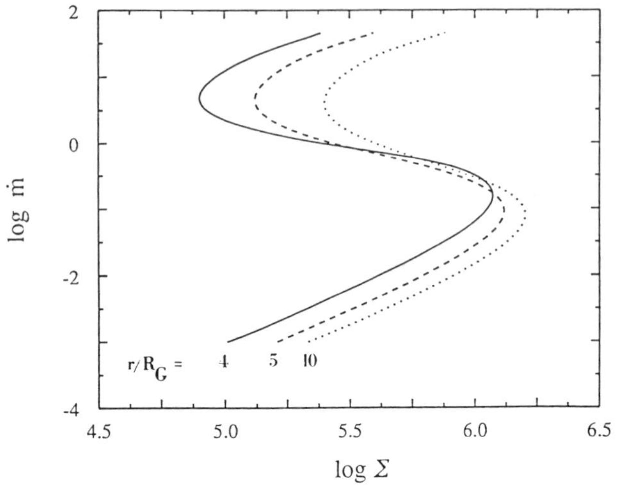

Fig. 1. *The $\dot{m}(\Sigma)$ relation for slim discs at three fixed radii, r/R_G=4, 5 and 6. Σ is in gcm^{-2}.*

The $\dot{M}(\Sigma)$ curves have a characteristic S shape. The lower branch (positive slope) corresponds to the "gas pressure, electron scattering dominated" regime of S&S. The middle branch (negative slope) is close to the "radiative pressure, electron scattering dominated" solution of S&S. The change of slope reflects the change in the stability properties: negative slope corresponds to unstable solutions (Abramowicz, Lasota & Xu 1986). The upper branch (positive slope) represents the new, non-standard slim disc solutions. The positive slope is the result of the advective cooling and relativistic "Roche lobe" overflow. Note that the upper turnover appears at $\dot{M} \approx 0.3\dot{M}_{Edd}$ in excellent agreement with the simple-minded argument of §.1. The slope of the upper branch of the S-curve can be estimated from an equally simple argument. Assuming that radiation pressure dominates and that rotation is keplerian one can replace the entropy gradient term in the energy equation (5) by $8P/r\rho$, using the first law of thermodynamics and putting $dS/dr \approx S/r$. All equations are now linear algebraic. The solution for the surface density gives $\Sigma = f(r)\alpha^{-1}\dot{M}$ where $f(r)$ is a function of r, known with accuracy up to a constant factor, the value of which is rather poorly estimated in the present approximation. However the slope of the $\dot{M}(\Sigma)$ curve on the upper branch is exactly recovered.

From mass conservation and equation (4) one gets

$$\frac{dlnv_r}{dlnr} = \frac{c_S^2(1 + dlnH/dlnr) + v_\phi^2(1 - \Omega_K^2/\Omega^2)}{v_r^2 - c_S^2} \tag{6}$$

where c_S is a sound speed defined by $c_S^2 = (dP/dr)(d\rho/dr)^{-1}$. The numerator and the denominator of (6) vanish simultaneously at the sonic radius r_S. For low viscosity cases the sonic radius is located below the cusp of the critical equipotential surface (the "Roche lobe") as seen in Fig. 2.

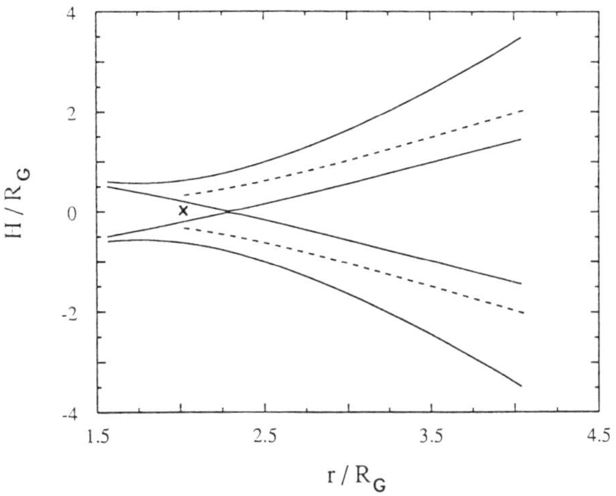

Fig. 2. *The equipontential surfaces (solid lines) and the surface of the disc (dashed lines). The cross indicates the position of the sonic point.*

Figure 2. shows the principal ingredient of the advective cooling in slim discs: transonic accretion through the critical point of the equipotential surfaces. It is a general relativistic analogue of the well known Roche lobe overflow in binary systems. Accretion via Roche lobe overflow occurs only for low viscosities; at high viscosities the physical picture is different (see ACLS and references therein).

The existence of the S-shaped equilibrium curves for radiation pressure dominated accretion discs allows, in principle, for limit cycle behaviour similar to that found in the case of accretion discs in dwarf novae (see e.g. Verbunt 1986). These time-dependent processes and their relevance to observed properties of objects accreting at high rates are currently under investigation.

References

Abramowicz, M.A. (1981) *Nature,* **294**, 235.

Abramowicz, M.A., Lasota, J.P. & Xu, C. (1986) 'Variability and Instabilities of Accretion Discs' in G. Swarup & V.K. Kapahi *Quasars*, I.A.U. Symposium No. 119, p.371.

Abramowicz, M.A., Czerny, B., Lasota, J.P. & Szuszkiewicz, E. (1988) *Astrophys. J.,* **332**, 646.

Collin-Souffrin, S. (1987) *Astr. Astrophys.,* **179**, 60.

Frank, J., King, A.R. & Raine, D.J. (1985) *Accretion Power in Astrophysics*, Cambridge University Press, Cambridge.

Horne,K. & Marsh, T.R. (1986) 'Indirect Imaging of Accretion Disks in Binaries', in K.O. Mason, M.G.Watson & N.E. White *The Physics of Accretion onto Compact Objects*, Springer Verlag, Berlin, p.1.

Mason, K.O. (1986) 'Accretion Discs in Low Mass X-ray Binaries', in K.O. Mason, M.G.Watson & N.E. White *The Physics of Accretion onto Compact Objects*, Springer Verlag, Berlin, p.29.

Muchotrzeb, B. & Paczyński, B. (1982) *Acta Astron.,* **32**, 1.

Paczyński, B. & Bisnovatyi-Kogan, B. (1981) *Acta Astron.*, **31**, 283.
Paczyński, B. & Wiita, P.J. (1980) *Astr. Astrophys.*, **88**, 23.
Shakura, N.I. & Sunyaev, R.A. (1973) *Astr. Astrophys.*, **24**, 337.
Verbunt, F. (1986) 'Theory and Observations of Time-Dependent Disks' in K.O. Mason, M.G.Watson & N.E. White *The Physics of Accretion onto Compact Objects*, Springer Verlag, Berlin, p.59.
Wade, R.A. & Ward, M.J. (1985) 'Cataclysmic variables: Observational Overview' in J.E. Pringle & R.A. Wade *Interacting Binary Stars*, Cambridge University Press, Cambridge, p.129.
Wade, R.A., (1988), *Astrophys. J.*, , . in press.

SUMMARY OF THE CONFERENCE: STARS

L Mestel
Astronomy Centre, University of Sussex
Falmer, Brighton BN1 9QH
England

Our conference has the title "Accretion Disks and Magnetic Fields". In my review on the early stages of star formation, I emphasised the accumulated strong evidence for the presence of dynamically important flux from the galactic magnetic field in the molecular clouds which are clearly observed to be a locale of contemporary star formation. I noted also the several papers reporting on flattened structures in some clouds. Ralph Pudritz has predicted that the discovery of disk-like regions in star-forming regions will be a growth industry: he anticipates a wealth of new observations at higher resolution.

A priori one would expect a cool cloud with a sub-critical mass/flux ratio, prevented from collapse by the essentially anistropic forces exerted by a large-scale magnetic field, to be highly oblate about the overall direction of the field. The fact that the observed flattening appears to be fairly modest links up with the observation of line widths that are plausibly interpreted as "Alfvénic turbulence". One would like to see more work clarifying the structure and the maintenance of these motions. The expected strong dissipation of super-Alfvénic motions ensures that the slow but locally accelerated flux leakage through ambipolar diffusion is accompanied by a decline in turbulent support, leading to a plausible picture of fragmentation. Magnetic braking is likely to be very efficient for slowly contracting regions, whereas super-critical bodies in free-fall will effectively conserve their angular momentum. The density at which a super-critical body has spun up sufficiently to yield strong centrifugal forces may thus be critically dependent on its pre-history. New observations may soon give clear indication of rotational support for disk-like bodies; perhaps there will also emerge evidence for two scales for such structures, indicating sharply differing efficiencies of angular momentum transport in earlier phases.

On the observational side, I cannot recall our discussing

G. Belvedere (ed.), Accretion Disks and Magnetic Fields in Astrophysics, 279–281.

direct evidence for magnetic fields in star forming disks. In the literature there are reports — e.g. by groups at Durham and Edinburgh — of optical polarization measurements consistent with an aligning magnetic field in a disk or torus. A tentative interpretation is in terms of a parameter domain where the kinetic energy of rotation is dominant, with the shear generating and maintaining against buckling instabilities a mainly toroidal field. The basic poloidal component could be again a relic of the original galactic flux, re-frozen into the gas at the epoch when thermal ionization becomes important; or it could be generated and maintained by dynamo action in a turbulent disk.

The magnetic field thus appears to be essential for our understanding of the support of cool molecular clouds; it can also resolve much of the "angular momentum problem", and it probably affects the ultimate mass spectrum. Inevitably the question arises: is a magnetic field essential for star formation? Martin Rees has recently emphasized that the first generation of stars may have formed before the presumed galactic dynamo has had time to build up a strong field, so one may need an alternative process for the redistribution of angular momentum. I was particularly intrigued by Henk Spruit's disk model with spiral shocks, in which each element of inflowing gas feels an azimuthal pressure gradient, its angular momentum being ultimately carried away along the shocks. Would he suggest his model as a possible non-magnetic mode for the later stages of star formation, when either relic or dynamo-built fields are likely to be comparatively weak?

Henk describes his model in terms of an equivalent accretion-disk α-parameter. I was grateful to Andrew King for his clarificatory talk on the basics of disk structure, not least for his characterization of α as the "ignorance parameter". We have been reminded of still unanswered questions on the stability of Keplerian disks. Nevertheless, α retains its phenomenological usefulness, and Emmi Meyer-Hofmeister showed how in the dwarf nova context, detailed comparison with observation, such as the lag in the rise of the ultra-violet flux behind that of the optical, can put constraints on its value.

We have heard of some detailed theoretical developments on disk theory. Friedrich Meyer discussed behaviour analogous to solar phenomena in accretion disk atmospheres, inspired by the need to explain occultation features in high inclination cataclysmic variables and low-mass X-ray binaries. Bram Achterberg has pointed out both the utility and the limitations of the modelling of mhd behaviour in disks in terms of isolated flux tubes, developed and exploited by Henk Spruit and Adriaan van Ballegooijen. Andrew King, Jean-Pierre Lasota and colleague relate the presence or absence of disks in magnetic cataclysmic variables to the strength of the magnetic field emerging from the white dwarf component, which in turn is related to other work on the appearance or non-appearance of internal flux above the surface of a rotating white dwarf. Arnold Benz and colleague propose masering via coherent cyclotron emission for the radio emission from some dwarf novae.

Arieh Königl, Y. Uchida (both *in absentia*) and Richard Lovelace have developed detailed models of mhd disks and jets in proto-stellar systems, with magnetic transfer of angular momentum playing a crucial role. It is emphasized that there must be regions in the disk where perfect conductivity breaks down, so that trans-field drift can occur. There is a partial analogy with the pulsar magnetosphere problem, where the adoption of the "pulsar equation" everywhere outside a perfectly conducting star leads to difficulties, implying that the Goldreich-Julian ansatz $\underset{\sim}{E}.\underset{\sim}{B} = 0$ must likewise break down somewhere.

It is striking how basic problems in mhd recur. Instabilities of strongly distorted magnetic fields may lead to reconnection of field lines and an accelerated, dynamical destruction of flux. Nigel Weiss gave an account of work by David Hughes and Fausto Cattaneo on the non-linear development of a magnetic Rayleigh-Taylor instability, motivated initially from solar physics but possibly a paradigm for other structures with trapped strong internal fiels. Adriaan van Ballegooijen and John Raymond calculate a rather short time within which the magnetic flux initially trapped within an accretion disk will be expelled by the action of a strong turbulent resistivity. I was left wondering whether this result is valid only if the initial field is too weak to interfere significantly with the motions that presumably arise spontaneously and are responsible for the turbulent viscosity and resistivity in a weakly magnetic disk: maybe weak fields are expelled and strong fields survive. Adriaan's and John's results put a severe constraint on possible dynamo action in a disk. In his review of dynamo theory Nigel warned us against too glib an extrapolation of current solar dynamo models both to more rapid rotators and to bodies of different geometry. One should also beware of thinking that primeval and dynamo-built flux within a given body are always mutually exclusive. Roger Tayler has recently argued that the observations of pre-main sequence stars, showing a spread in the level of magnetic activity compared with the tighter correlation with rotation found in main sequence stars, may imply that the magnetic flux in these stars contains a decaying primeval component. And several people have suggested that the enormous flux of mechanical energy in the T Tauri phenomenon is powered by the destruction of a reservoir of the primeval flux.

There is a lot more that could be said, but as we are all rather tired, let me leave it at that. We should all be returning home with a lot more problems for ourselves, graduate students and post-docs.

CONFERENCE SUMMARY : ACTIVE GALACTIC NUCLEI

R.N. Henriksen
Astronomy Group
Queen's University at Kingston
Ontario, Canada, K7L3N6

ABSTRACT. We have had intense and far-ranging discussions in these last few days, broken only occasionally by our sunny interludes on the beach! I regret that I could not join you there today as I was busy preparing these few remarks. I hope that they will prove worthy of the sacrifice! In any event I trust that errors and oversights will be attributed more to distraction than to malice or to incompetence.

We should begin by defining 'AGN' to include both the central electromagnetic spectrum and the extended particulate-wave-magnetohydrodynamic structure inferred to have been emitted from it. Moreover I shall discuss these two aspects separately, beginning with the nuclear region.

1. Nuclear Activity

This certainly takes place on a scale $\ell < 1$ pc (as the VLBI map shows so dramatically for M87) and may well originate on scales from a few, to a thousand AU. The evidence presented here by Fiore for the black hole mass in the relativity innocuous Seyfert NGC 4151 suggests $\sim$ 0.2 AU as a definite lower limit for the scale of the activity. It is amusing that this range of scales is much the same as that arising in the discussion of proto stellar discs.

But what important questions have we addressed? I suggest the following candidates:

1. 1. What is the evidence for a low entropy central engine?

For this 'engine' we may suppose following our discussions either;

(i) accretion disc bounded, electromagnetically driven outflows, in which binding energy relative to infinity is tapped, let us call them 'gravars',

or

(ii) Electromagnetically dominated outflows originating in the 'magnetospheres' of black holes, neutron stars, et cet., in which the spin energy of the central object is the ultimate source. Let us call them 'spinars'.

Turning however to the evidence we have

a) The 'jet phenomenon' itself reveals low entropy, by its preference for restricted angles, by its apparently high specific energy, and by the presence of non-thermal electrons. The preferred range of angle about an axis is generally attributed to an angular momentum vector $\underline{L}$, either that of the central object or that of the disc (if they differ). It is worth noting that if $\underline{L} = \underline{L}_{disc}$

G. Belvedere (ed.), Accretion Disks and Magnetic Fields in Astrophysics, 283–287.

then it might in reality be a mean over some quite chaotic structure, that is $\underline{L} = <\underline{L}_{matter}>$. A mean magnetic field might arise in the same way according to dynamo theory.

b) The remarkable, hollow, filamentary structure exhibited by the M87 jet on scales from ~ 1 pc to several kpc as reported to us by Hardee, suggests a highly organized origin, or at least an organizing internal structure that is already simply scaling (having forgotten the details of its origin) at 1 pc. In this connexion we should remember the high apparent equipartition magnetic fields in the filaments, although their presence is not inevitable if they represent oblique shocks (see e.g. Henriksen + Westbury, 1988). The confinement problem is not serious in such a case because the expansion speed will only be a small fraction of the translation speed. This latter is already known to be $\geq 0.1c$ at 1 pc from the VLBI data.

c) The apparent existence of Keplerian discs (as judged from fits to the double peaked line profiles reported by Lasota and Meyer) reveals at least a highly concentrated central mass. It should be noted however that a more general nuclear circulation pattern may be induced by the jets, according to the entrainment solutions presented here and elsewhere (Henriksen, 1987) by myself. Real Keplerian discs are not dominated by pressure or other chaotic effects and are therefore low-entropy flows. However we do not know how close to the nuclear action they really are.

d) The high energy synchrotron particles (revealed particularly in the infra-red spectrum shown to us by Begelman) require 'accelerators' or entropy reducers. This usually requires such organization as particle beams or 'cascades' of waves and/or shocks from large scales.

e) The super-luminal VLBI motion speaks of either phase organization or high specific energy (relativistic bulk motion) near the line of sight. In 'shot-gun' models of the nuclear ejecta (attractive now because of super luminality in the centres of wide classical double sources), the latter organization may simply be selected by this line of sight.

1. 2. What is the evidence for a high entropy central engine?

a) The electromagnetic spectrum reviewed for us by Begelman reveals complex and varied behaviour. In particular the UV-soft X-ray 'bump' appears to be thermal. This might be the signature of a thermal 'cauldron' at the centre of a thick disc, or it might occur farther out where relativistic jets decelerate. Begelman has pointed out however, that the hot ions may not remain decoupled from the electrons in a thick disc, which would lead to its destruction.

b) The radiation 'explosion' expected with $T_{eff} \geq 10^5$ K and high opacity emphasized for us by King, suggests dynamical disorder. The calculations of radiation transfer by Madej, Loken and Henriksen (1987)) that I reported here do give radiative forces sufficient to disrupt a disc for $T_{eff} \sim 10^5$ K.

c) The broad line regions (BLR) and possibly the narrow line regions (NLR) may also be related to jet dissipation. Entrainment of galactic material is equivalent to the entrainment of negative momentum in Ferrari's calculations. This can lead to the transition to a 'breeze' solution from a 'wind' solution at a critical point and so trap and dissipate the jet. This can give one answer to why there are no jets in some cases.

d) Finally, one must take note of the 'red noise' X-ray variability with $P \propto f^{-1}$ reported here for NGC 4051 by Fiore. This suggests deterministic chaos. I would like to point out that I have argued at this meeting that the stochastic diffusion of magnetic field from a po int source into a surrounding hierarchically turbulent medium gives for the 2-point magnetic correlation

$$U^{i,j} \equiv < B^i B'^j > = \text{const} + \text{const}/\ell$$

under very general conditions. If U is proportional to a power per unit band width, and if there is a scale invariant signal speed (such as the Alfven speed) then the observations might be understood directly in terms of a stochastic region. This is a speculation!

1. 3. Now what have we had to say regarding theories of 'clean' (low entropy) and 'dirty' (high entropy) processes?

a) 'Clean" models of the spinar variety have been mostly developed by the Pulsar theorists, as I reviewed earlier in the week, but also by Blandford and Znajek(see my contribution for references) . In terms of the key parameter $\sigma = 0(e\Delta\psi/mc^2), \equiv \frac{\Phi^2}{4\pi r_c^2 c}\frac{1}{f}$ for monopolar winds, the asymptotic charged particle energy is $\gamma \sim \sigma$ at infinity (see my review for definitions). However there is no globally consistent solution for the pulsar magnetosphere that permits particles to escape to infinity. Mestel has argued that particles must return after radiating their angular momentum just outside the light cylinder. Various authors, including myself and Don Rayburn, have argued that $\gamma \sim \sigma^{2/3}$ is a robust estimate of the energy of particles accelerated near the light cylinder. Cylindrical helicon wave beams can even be produced which are confined within the light cylinder by self-screening. The minimum mass flux in such models is essentially f $\simeq \rho_e$/e mc where ρ_e is the Goldreich-Julian charge density, which leads to very hig h γ's, more suitable for the non-thermal electrons in the beam than for the bulk motion.

However, if more than the minimum number of particles to carry the Goldreich-Julian charge density are available, then they obtain lower specific energies as $\gamma \propto f^{-2/3}$. Here is where the 'gravar' models enter the picture as discussed at this meeting by Lovelace and Camenzind (non-relativistically these models are like those of Blandford + Payne, Pudritz + Norman et cet.). These use f = $\varepsilon\dot{M}$, where $\varepsilon < 1$ and $\dot{M}$ is the disc-accretion rate, related to the bolometric luminosity. Thus bulk flows with γ characteristic of the super-luminal phenomenon may be produced. It is likely that the 'spinar' and 'gravar' limits may operate simultaneously. Thus a small number of very high energy particles may be embedded in the bulk flow, emanating from the central regions. It is interesting that this picture is similar to the Norman-Pudritz jet model for protostellar jets, although their model is 'gravar-gravar' rather than 'spinar-gravar'.

Moreover Lovelace and Camenzind have reported here real progress towards solving the global problem with disc boundary conditions, by using the relativistic Grad (Shluter) - Shafranov equation. This is really the transverse dynamical equation in a field line orthogonal (FLOT) reference system. It is most important that they continue this work to show that they can deal with finite discs, the associated return current problem, and realistic boundaries.

b) 'Thermal' or 'dirty' models may be permitted to invoke T $\simeq 10^9$ K. A 'nozzle' (which may be gravity as in a wind - but then subsequent collimation is required) is then employed as an ' entropy reducer'. I have reported on such models for protostellar jets, but they follow from the original models for AGN by Blandford and Rees, and also Begelman and Rees. The recent discovery of gas at the base of SS433 with T $\sim 10^8$ K is intriguing in this connexion. However such models fail to produce relativistic jets of 'ions', unlike the clean models. In SS433 and M87 we know this to be necessary. This is a strong argument for the clean models. However one wonders whether a mechanism of the sort described here by Benz for the solar wind might be relevant. This is that in a flow with a collisional-collisionless transition, the particles out on the tail of the Maxwellian can break free and rush ahead in a high energy 'beam'. Such beams are observed in the solar wind. A dense gas at temperatures of 10^8 - 10^9 K could produce small quantities of moderately relativistic ions in this way. There is no VLBI evidence for large quantities of ions. This is a natural explanation for why the jets often seem (in non-turbulent examples) to be 'light' jets. However, we must remember that enough energy must be available to power the 'lobes' not just the jet emissivity.

However it may be that we require a 'clean jet' to be punching through a thermal 'furnace' or 'cauldron'. The cauldron might be radiation driven from a central source, or conceivably it might be a 'star burst' region. This would then be tapping thermo-nuclear energy. Even if the region were due to tidal stellar destruction (and compression, e.g. Carter and Luminet) thermonuclear energy might still be available for the cauldron.

c) Belvedere raised the important question of scaling in our discussions. It is clear that the clean models do scale if the magnetospheres exist, but one still has to depend on the environment for $\dot{M}$, which adds a dimension to the scaling. In general one might expect this to depend on ρ, T in the environment and on M of the central mass. We might also guess that B should be important. However if ℓ is the scale of the central mass, it is arguable that $M = M(\ell)$ and $B = B(\ell)$. Although the field scaling may be a universal, it is clear that $M(\ell)$ is very different for black holes than it is for white dwarfs and neutron stars. Thus we can expect universality to be broken by $M(\ell)$, and probably by $\rho(\ell)$ and $T(\ell)$, which would seem to be much more contingent. Thus we do not need to expect the same phenomenon for every gravitating object!

2. The extended structure of AGN

There were various questions and results on this topic, even though it did not represent the main thrust of our meeting.

2. 1. Jet Stability

a) The simulations presented here by Hardee and Norman do suggest that the linear stability analysis is useful in delineating the internal jet geometry. In the pre-turbulent state, the dominant mode analysis seems to predict internal structures in the non-linear simulation of sl ab jets. This can be important in predicting the kind of spatial coherence available for models of VLBI superluminality, although of course real jets are unlikely to be slab jets.

b) We also saw the Hardee-Norman simulations of the non-linear growth of the slab-jet instabilities. It seems to me that the real conclusion is that turbulence results (much as it would for 2 parallel shear layers). It is not correct to assume that the jet is destroyed. It probably does that by interacting strongly with the ambient medium until a 'lobe' is formed.

2. 2. Jet Termination

Norman showed some interesting results concerning the effect of an ambient shock on a jet traversing it. This clearly demonstrates the ability of a pressure discontinuity to disrupt the jet. The ultimate termination is probably again by turbulent dissipation until the lobes form. Such mechanisms may relieve the pressure on the 'working surfaces' somewhat.

2. 3. Acceleration of the Synchrotron Particles.

a) In jets with high frequency emission it is clear that reacceleration must be taking place 'in situ', just as it must be in the nuclear emitting regions. The turbulent termination of jets is useful in this connexion, because it provides a natural way of producing shock ensembles (which Achterberg has told us accelerate particles efficiently) or wave cascades (as in the work of Eilek + Henriksen). The shock wave mechanism is likely to dominate when present, as it is more efficient, being a first order Fermi process as opposed to a second order resonant scattering. Such mechanisms might also be relevant in the central 'cauldron' although they are density dependent. Begelman has suggested proton beams for the photon production (by way of pion production and decay), but the beams must still be produced. This could be a 'spinar' beam.

b) It is conceivable that a helical version of Spruit's spiral disc shocks might also be instrumental in accelerating particles. These have been suggested recently as a way of producing the filamentary structure in M87 (Henriksen + Westbury, 1988; Hardee, this meeting).

2. 4. Is the magnetic field dynamically important?

a) The Chan-Henriksen model (1980) already showed that magnetic energy at the level of $\geq$ 0.1% of the jet kinetic energy would subject the jet to a violent pinching instability. Nevertheless even at this level the jet may be gradually collimated by the field. This was used by Achterberg et al. in an application to SCO X-1, which otherwise remains a mystery. Thus the field is never likely to be dynamically important to the jet motion in the extended structure. It may have significant collimating effects for laminar jets, but as these are generally high velocity (such as M87) the confinement problem is not critical. Thus there is as yet no compelling argument for magnetic collimation.

b) The conclusion of part (a) raises the question of how magnetically dominated clean jets could make the transition to the observed jets. In this regard expansion and $B \propto \ell^{-1}$ probably provide the link between the VLBI and the VLA conditions, but this scaling does not change the relative energy fluxes. For that, some dramatic dissipation of the field or augmentation of the KE flux at the expense of the field is required. As mass entrainment will also result in the entrainment of magnetic flux in general (Henriksen, 1987), it seems that the magnetic fields must 'leak' out. This is also the only way to understand the clean VLBI structure of M87 which is nevertheless not magnetically dominated (by analogy with the scaled VLA structure). We conclude that leakage mechanisms should be put into the 'clean' jet models.

Well, the evening wears on. I think we have done enough. Let me say finally how much I personally have enjoyed this meeting and offer a vote of thanks to the organizers on behalf of us all. I think we have all become conscious of the weight of history resting on the towns and countryside of Sicily. Perhaps we can not justifiably run naked through the streets shouting 'Eureka' as a result of our deliberations, but I feel sure that at least we have not disgraced the spirit or the legacy of Archimedes in our proceedings. It has been a good meeting in a good location with good papers and with some thoughtfull exchanges. I can think of nothing that the Local Organizing Committee might have done better. We all should thank them for that.

References

Blandford, R.D., and Znajek, R.L. 1977, M.N.R.A.S. **179**,433.
Chan, K.L., and Henriksen, R.N. 1980, Ap.J. **241**,534.
Henriksen, R.N., and Westbury, C. 1988, Ap.J. **327**,50.
Henriksen, R.N. 1987, Ap.J. **314**,33.

SUBJECT INDEX